CRYSTAL-QUASICRYSTAL TRANSITIONS

North-Holland
Delta Series

NORTH-HOLLAND
AMSTERDAM • LONDON • NEW YORK • TOKYO

Crystal-Quasicrystal Transitions

Edited by

M.J. Yacamán

Instituto de Física
Universidad Nacional Autónoma de México
México

M. Torres

Instituto de Electrónica de Comunicaciones
Consejo Superior de Investigaciones Científicas
Madrid, Spain

1993

NORTH-HOLLAND
AMSTERDAM • LONDON • NEW YORK • TOKYO

North-Holland
ELSEVIER SCIENCE PUBLISHERS B.V.
Sara Burgerhartstraat 25
P.O. Box 211, 1000 AE Amsterdam
The Netherlands

ISBN: 0-444-89827-1

Printed in The Netherlands.

PREFACE

The integers lie at the basis of our accounting systems – so many head of cattle, so many jars of oil. The Egyptians were familiar with fractions and the Babylonians could calculate square roots numerically with indefinitely great accuracy. However, in Classical Greece, the idealisation of the integers led to the awful discovery that the diagonal of a unit square is irrational and cannot be expressed as the ratio of two integers. This was the first scandal of incommensurability and of it Proclus Diadochus (412-485), the commentator on Euclid, wrote:

> "It is well known that the man who first made public the theory of irrationals perished in a shipwreck in order that the inexpressible and unimaginable should ever remain veiled. And so the guilty man, who fortuitously touched on and revealed this aspect of living things, was taken to the place where he began and there is for ever beaten by the waves".

The whole present-day division between Western and Eastern Europe might even be traced to the incommensurability of the periods of the Sun and the Moon and the necessity of reconciling solar and lunar aspects of religion. (A rational approximation, that 19 solar periods are almost equal to 235 lunations is known as the Metonic cycle.) At the Council of Nicaea (325) the Eastern and Western Churches split apart on this issue (inter alia) and adopted different calendars with different dates for Easter.

From the discovery by the Abbé Haüy that crystals are made up by the repetition of unit cells – "molécules intégrantes" – crystallography has acquired a dogma based on integers. The immense success of the methods of crystal structure analysis has obscured the obvious fact that not everything is crystalline and the present period is one in which, following the aims of J.D. Bernal, a more generalised view of the structure of matter is developing.

The celebrated discovery by Linus Pauling about 1951 of the α-helix, a helical protein chain where the number of residues per turn of the helix is not an integer, was by way of being a second scandal of incommensurability. W.L. Bragg confessed to having missed this solution to the problem of protein structure by considering only those rational helices which were compatible with the symmetries of the 230 space groups.

The discovery by D. Shechtman in 1984 of five-fold and icosahedral diffraction symmetry in certain rapidly-quenched alloys, now known as quasi-crystals, make a third scandal of incommensurability, since five-fold symmetry finds no place in the International Tables for Crystallography. The usage in the name "quasi-crystals" is old and Shakespeare (in Love's Labours Lost) refers to someone as "a quasi-person" (as if a person, not quite a real person). It turns out that five-fold symmetry in infinite patterns an incommensurability are closely related. Quasi-crystals have two possible steps in the same direction (where orthodox crystals have one) and these steps are in the golden ratio ($(1 + \sqrt{5})/2 = 1.618...$).

There is a considerable pre-history to Shechtman's discovery. About 1945, Hendricks and Teller considered the diffraction to be expected from a random stack of layers of two thicknesses. Unfortunately, they missed the special case where the one-dimensional series is a section through a two-dimensional lattice. Still earlier, Hume-Rothery had studied what he called "electron alloys" where the ratio of valence electrons to atoms was 13 to 8 or 8 to 5 or 3 to 2 (all rational approximants to the golden ratio). Would an alloy, with the golden ratio of electrons to atoms, necessarily be non-periodic?

A two-dimensional pattern, which represented a step in the generalisation of the concept of the crystal lattice, appeared about 1974, from two different directions. Roger Penrose sought a jigsaw puzzle, with many tiles of the minimum number of types, where the rules for joining the pieces would force the production of a non-periodic tiling. The now well-known Penrose tiling was the result and later R. Ammann produced a three-dimensional analogue. At the same time I developed the hierarchic packing of pentagons (in two-dimensions) and icosahedra/dodecahedra (in three-dimensions) to fill space according to rules which would be outside the framework of the classical Bravais lattices. From this it emerged that two types of tile sufficed and that the vertices of these tiles could be given integer indices with respect to 5 (in 2-D) or 6 (in 3-D) axes. I called the result a quasi-lattice (after calling it first a "mock-lattice").

It soon appeared that the Penrose tiling could be generated in several ways, in particular, as the projection of a 5-dim. hypercube lattice on to 2-dim. (or a 6-dim. lattice on to 3-dim.) and that rational (crystallographic) tilings could also be obtained by projection in rational directions. Accordingly, changing the direction of projection offered a putative description of a transformation from an incommensurate to a commensurate structure.

This does not necessarily mean that the intermediate stages represent a real physical path from one structure to another. F.C. Frank pointed out that the ordinary custom of using four indices h,k,i,l, for a hexagonal crystal represents it as a projection (in a rational direction) of a four-dimensional lattice. A cubic cell can also be obtained by projecting the same lattice in another direction. S. Hyde and S. Anderson have also produced a different one-parameter path between a face-centred cubic cell and a body-centred cubic cell, based on the Bonnet transformation of a periodic minimal surface, which may also represent martensitic transformations.

Incommensurable modulation of a crystal makes what were identical unit cells different and is thus a step in the direction of Schrödinger's "aperiodic crystal" which appeared at first sight to be a contradiction in terms. It is by solving such contradictions that science makes progress.

Madrid, December 1992 A.L. Mackay

TABLE OF CONTENTS

Crystal-Quasicrystal Transitions
M.J. Yacamán and M. Torres (Editors)
1993 Elsevier Science Publishers B.V.

Continuous transformation of Al-Mn-Si and Al-Cr-Si decagonal quasicrystals to a new approximant

H. Zhang[a,b], X.Z. Li[a,c] and K.H. Kuo[a]

[a]Beijing Laboratory of Electron Microscopy, Chinese Academy of Science, P.O. Box 2724, 100080 Beijing, China

[b]Departmet of Materials Engineering, Dalian University of Science and Technology, 116024 Dalian, China

[c]Department of Materials Physics, University of Science and Technology, Beijing, 100083 Beijing, China

Abstract

The decagonal quasicrystal in rapidly solidified Al-Mn-Si and Al-Cr-Si alloys was found to transform to a new B-centered orthorhombic phase with a=1.24, b=1.24, c=3.79 nm. These lattice parameters belong to a Penrose-tiling approximant predicted earlier using 3/2 and 8/5, respectively, to substitute for the irrational τ in two orthogonal directions in a decagonal quasicrystal [1]. The continuous transformation of the decagonal quasicrystal to this new phase was followed experimentally by selected-area electron-diffraction and theoretically by simulations based on the phason strain analysis. This transformation occurs simultaneously along some of the ten twofold directions of the decagonal phase as multiple twin domains. The existence of intermediate structures during this continuous transformation of a decagonal quasicrystal to tenfold twins of a crystalline structure excludes the multiple-twin interpretation of the quasicrystal proposed earlier by Linus Pauling [2].

1. INTRODUCTION

In our previous papers various Fibonacci [3] and Penrose-tiling [1] approximants (for these terms see [4]) of the decagonal quasicrystal (QC) have been discussed from the phason strain point of view [5,6]. Decagonal QC is a two-dimensional (2D) one which is periodic along its tenfold axis but aperiodic in the plane normal to it. If one of the irrational or aperiodic directions in this plane becomes rational or periodic, the 2D QC will change to a 1D one [3,7]. If two aperiodic directions in this plane become periodic, the Penrose-tiling of this plane will change to a 2D periodic lattice and the QC to a crystalline phase called Penrose-tiling approximant or simply approximant [1,4].

Electron diffraction evidences of the continuous transformation of the 2D decagonal QC to various 1D QCs have been compared with the simulated results [3]. The aperiodic distribution of diffraction spots gradually becomes periodic in one of the twofold directions and the periodicity in this direction reduces in successive steps following reversely the Fibonacci series: 1, 1, 2, 3, 5, 8, 13,..., F_n,..., where $F_{n+2}= F_{n+1}+F_n$. During this series of transformation, the irrational number τ characteristic of the Fibonacci sequence and also of the decagonal QC is replaced successively by F_{n+1}/F_n with smaller and smaller n. A similar study of the continuous transformation of the 2D decagonal QC to a relevant crystalline phase, by approximating τ in two directions with ratios of two successive Fibonacci numbers, is presented in this paper.

The presence of an intermediate state displaying local periodic translation order as a step of the continuous transformation of the icosahedral QC to a crystalline phase has already been observed earlier [8,9]. The specific case of the continuous transformation of an icosahedral QC to a cubic crystalline phase has been studied both theoretically [10,11] and experimentally [12,13]. Such a transformation of an octagonal QC to the β-Mn structure has also been addressed recently [14,15]. A similar study on the decagonal QC certainly will enrich our knowledge of this kind of transformation and possibly also deepen our understanding of the nature of QCs.

Both icosahedral and decagonal QCs have been found in rapidly solidified $Al_{13}Cr_4Si_4$ [16] and $Al_{74}Mn_{21}Si_5$ alloys. One of the crystalline phases coexisting with the AlCrSi decagonal QC was found to be the cubic $Al_{13}Cr_4Si_4$ which, as analyzed before [1], is a Penrose-tiling approximant. In the present study, another hitherto unknown approximant with an orthorhombic or a corresponding monoclinic lattice has been identified in this alloy. The same crystalline phase has also been found coexisting with the decagonal QC in the $Al_{74}Mn_{21}Si_5$ alloy. Furthermore, this phase very often occurs as micro twins displaying five- or ten- fold symmetry not unlike the monoclinic $Al_{13}Fe_4$ [17]. The continuous transformation of the decagonal QC to this new crystalline phase in these two alloys has been followed by a careful electron diffraction study and this new crystalline phase is analyzed as a Penrose-tiling approximant based on the phason theory of QCs.

The nominal compositions of the alloys used in the present study are $Al_{74}Mn_{21}Si_5$ and $Al_{13}Cr_4Si_4$, respectively. The former was kindly supplied by Dr. J.-M. Dubois [18] and the latter has been used in a previous study [16]. The experimental details of the alloy preparation and heat treatment can be found in the original papers.

2. PENORSE-TILING APPROXIMANT

After rapid solidification, the main constituent in these two alloys is the icosahedral QC, though some decagonal QC also exists as a minor constituent. A new crystalline phase was found to occur at the periphery of the decagonal nodules. The electron diffraction patterns (EDPs) of this phase are closely related to those of the decagonal QC. Figure 1 (a) shows the twofold D pattern of the decagonal QC, in which the periodicity along the tenfold direction, as reported

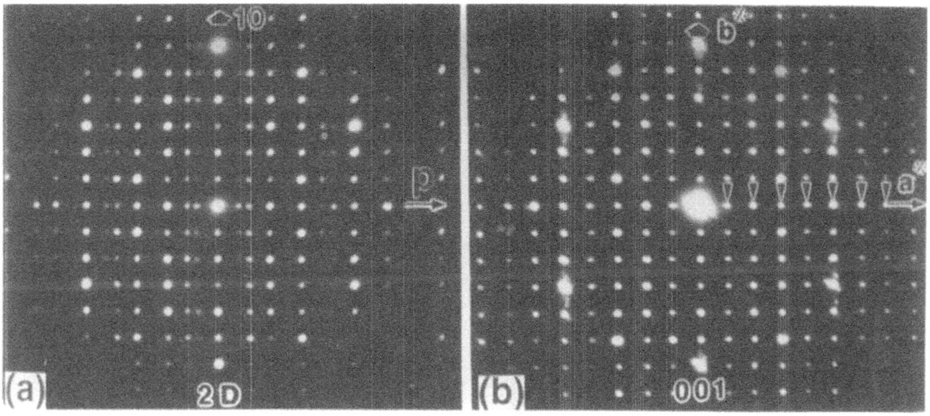

Figure 1. (a) Twofold 2D EDP of the decagonal QC, spots being periodic along the tenfold axis and aperiodic in the direction P normal to it. (b) Corresponding EDP of the crystalline phase with a cross-grid of spots. The hexagon of strong spots and the periodicity in the vertical direction are similar in both EDPs.

earlier [16], is 1.24 nm, whereas the direction P orthogonal to it is quasiperiodic. On the other hand, the corresponding EDP of the crystalline phase shows a rectangular array of spots (Fig. 1 (b)). In the $\mathbf{b}^*$ direction corresponding to the tenfold axis of the decagonal QC the lattice parameter b is also 1.24 nm, whereas in the $\mathbf{a}^*$ direction orthogonal to it the spots become also periodic (indicated by arrowheads).

Figure 2(a) is the EDP taking along the b axis and the angle between the two shortest reciprocal vectors is about 72 or 108°, similar to the well-known [010] EDP of the monoclinic $Al_{13}Fe_4$ with $\beta = 107.71°$ [19]. For a crystalline lattice with $\beta = 72$ or 108°, it can either be described as monoclinic or B-centred orthorhombic and the cooresponding lattice parameters thus calculated are shown in the following:

Monoclinic a_m=1.24 nm, b_m=1.24 nm, c_m=1.99 nm, $\beta \approx 108°$
Orthorhombic a_o=1.24 nm, b_o=1.24 nm, c_o=3.79 nm.

This ambiguity in general can not be solved adequately by selected-area electron diffraction. The crystal size is too small to allow a detailed convergent-beam electron diffraction study to ascertain its lattice symmetry. For the sake of convenience in the approximant description, the "orthorhombic" lattice will be used in the following discussion.

As discussed in details before [1], this "B-centered orthorhombic" lattice can also be considered as a rational Penrose-tiling approximant derived from the decagonal QC with an irrational quasilattice (see the corresponding reciprocal lattice in Figs.

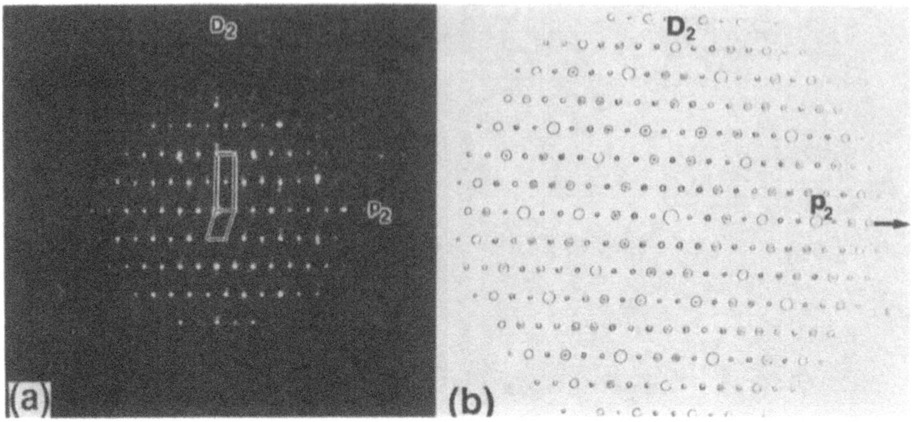

Figure 2. (a) The [010] EDP of the new crystalline phase with the monoclinic (β≈108°) and the B-center orthorhombic unit cells outlined. (b) Simulated [010] EDP with periodic arrangements of diffraction spots along the two orthogonal D and P directions.

3 and 4) in which the two orthogonal aperiodic directions D and P gradually become almost periodic. Or in other words, using ratios of two successive Fibonacci numbers 3/2 and 8/5, respectively, substituting for the irrational golden mean τ in the QC. Figure 2(b) is the simulated [010] EDP derived from the tenfold EDP of the decagonal QC by such an approximation and it matches well with the experimental EDP shown in Fig. 2(a). The calculated lattice parameters a = 1.23 and c = 3.79 nm given in Ref. 1 for an approximant of the decagonal QC agree quite well with the experimental data given above. In other words, this AlMnSi or AlCrSi intermetallic compound belongs to the family of giant Al-*M* (*M*=transitional metals) crystals with orthorhombic and akin lattices derived from the decagonal QC, and their orientation relationship is as the following:

$[010]_{Ortho}$ // 10_{Deca}
$[100]_{Ortho}$ // $2P_{Deca}$
$[001]_{Ortho}$ // $2P_{Deca}$

In this respect it is of interest to note the presence of another member of this family, namely, the newly found π-Al_4Mn with a=0.77, b=1.24 and c=2.36 nm, whose a and c are about τ times smaller than a=1.24 and c=3.79 nm given above for the new phase found in the present study (their b parameters are equal). As discussed before [1,20], this π-Al_4Mn is also an approximant of the decagonal QC with τ in D and P directions substituted by 2/1 and 5/3, respectively, compared with 3/2 and 8/5 in the present case (see Table 1 in [1]). In the case of π-Al_4Mn, there are 2 and 5 spots from the center to D_2 and P_2, respectively, whereas there are 3 and 8 in the present case, as shown in Figs. 2(b), 3(d) and 4(d). Moreover, this new crystalline phase can also be compared with the orthorhombic AlMnCu phase with a=1.48, b=1.26, and c=1.24 nm (isostructural with Al_3Mn) reported recently [1,20]. These two giant crystals have two lattice parameters almost equal, being all close to 1.24 nm, and the third one in a ratio of 3.79:1.48 $\approx \tau^2$. The lattice parameter of 1.48 nm is close to the value of 1.45 nm obtained by the substituting 3/2 for τ in the P direction.

In the Al-Mn-Si system, Audier and Guyot [21] and Elser and Henley [22] earlier have pointed out the close structureal relation between the icosahedral QC and the *bcc* α-AlMnSi phase and derived the structural model of the former from that of the latter. Henley [23] even made the proposal that the latter is an approximant of the former. In this respect, the presence of another approximant in the same system, this time orthorhombic and coexisting with the decagonal QC, should be of interest.

The similarity of the distribution of strong spots, such as the hexagon of strong spots in Fig. 1 and the several sets of ten strong spots in Figs. 3 and 4, also supports this structureal relationship. But the strongest evidence comes from the direct observation of the continuous transformation of the decagonal QC to this orthorhombic crystalline phase and its simulation based on phason strain theory. These will be presented below.

3. CONTINUOUS TRANSFORMATION

In some specimens different stages of the transformation of the decagonal QC to this orthorhombic crystalline phase have been observed. Figure 3(a) is the tenfold EDP of the decagonal QC displaying tenfold, aperiodic arrangement of diffraction spots. In Fig.3(b) the ten spots corresponding to those marked with arrowheads in Fig. 4(b), originally lying on an inner circle around the central beam as shown in Figs. 3(a) and 4(a), are now located on an ellipse with its long axis outlined with an arrow. This becomes more so in Fig. 3 (c) in which the spots in the two orthogonal directions P and D marked with an arrow and an arrowhead, respectively, become more or less periodic. Moreover, the weak spots on the horizontal rows with a zigzag appearance, indicated by black-white bars, are approaching each other. In Fig. 3 (d) the spots lie almost on a 2D cross-grid, though the parallel lines are somewhat wavy or zigzag. This implies that this transformation is almost but not yet coming to the end. It should be pointed out that the ten strong spots on the periphery of these EDPs inherited from the decagonal QC can still be clearly seen.

As it has been pointed out earlier [1], the shifts and coalescence of diffraction spot can best be followed by simulations based on phason strain theory [5,6]. The spot position at the end of reciprocal vector in the physical space $\mathbf{G}^{\parallel}$ now changes to

$$\mathbf{G}^{\parallel'} = \mathbf{G}^{\parallel} + \mathbf{M} \bullet \mathbf{G}^{\perp} \qquad (1)$$

where

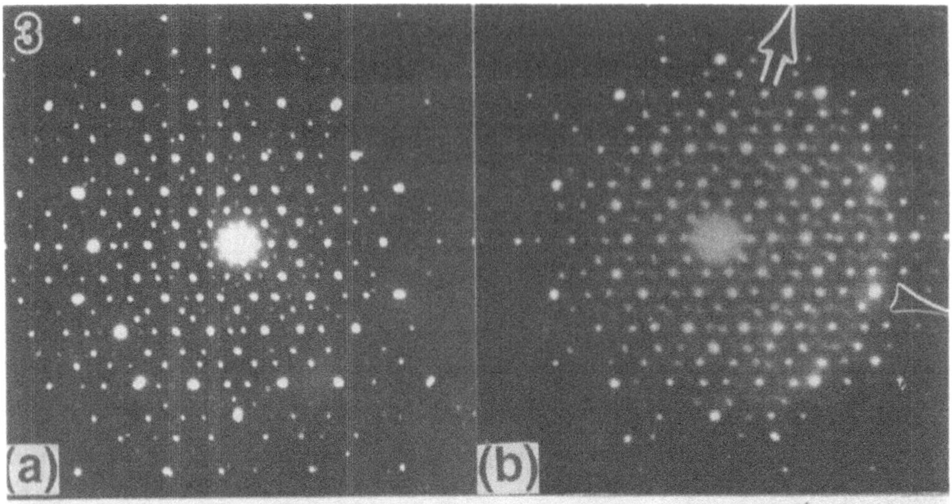

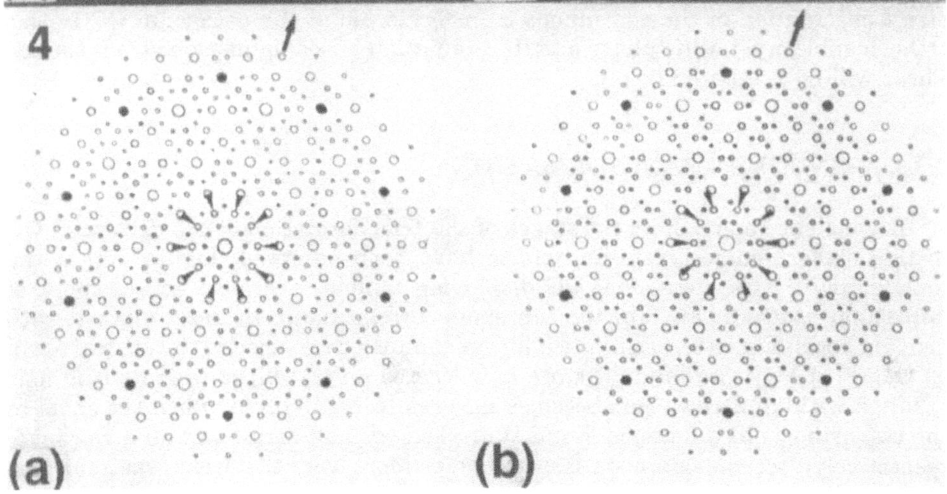

Figures 3. (above) and 4. (below) Experimental EDPs shows the continuous shift of spots from the tenfold orientation of the decagonal QC (a), through the intermediate stages (b) and (c), to almost the [010] orientation of the B-centered orthorhombic phase (d). The inner ten weak spots change from a circle arrangement to an ellipse (the long and short axes are marked with P and D, respectively), whereas the ten strong spots on the periphery are almost the same during this transformation. (to be continued)

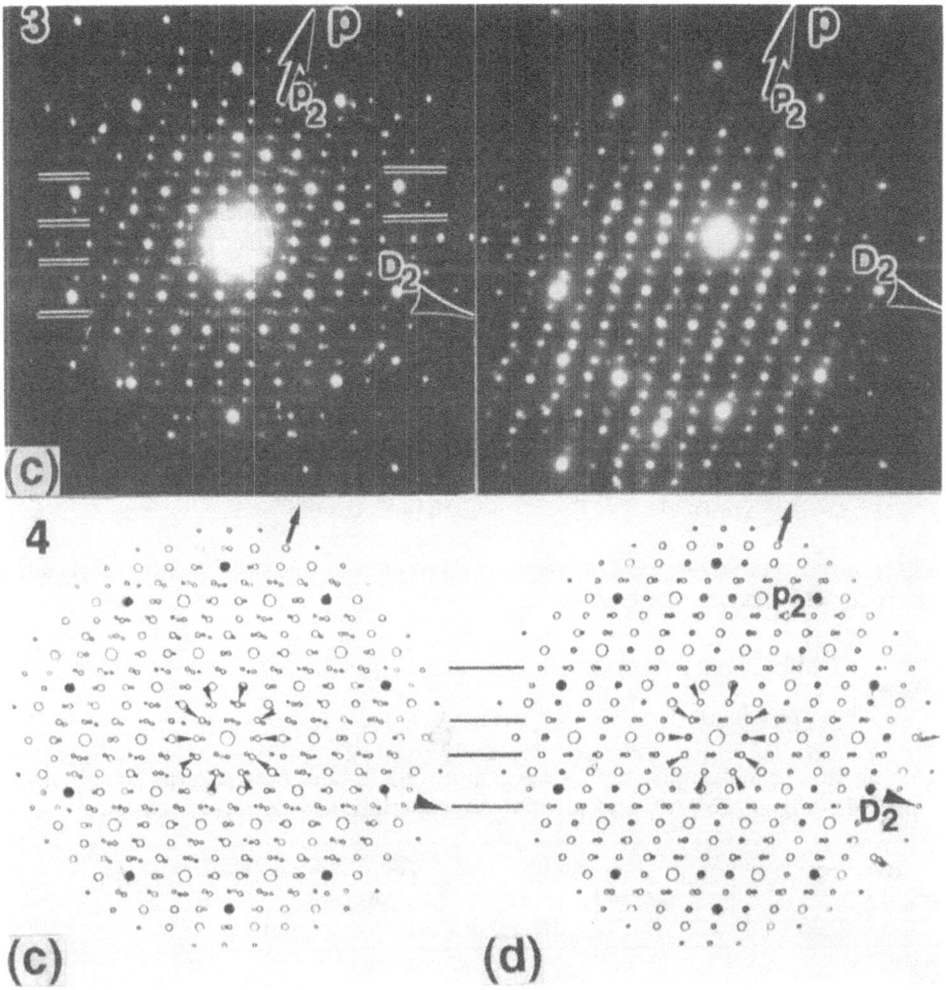

Figures 3.(above) and 4.(below) (continued) Figure 4. (below) shows simulated EDPs with increasing of phaseon strain to be compared with the corresponding ones in Figure 3.

$$\mathbf{M} = \begin{bmatrix} m_{11} & m_{12} \\ m_{21} & m_{22} \end{bmatrix} \tag{2}$$

is the phason strain tensor and $\mathbf{G}^{\perp}$ the reciprocal vector in the complementary space [5]. It has been given before [1] that for the transformation of the decagonal QC to a Penrose-tiling approximant with the unit cell and orientation as shown in Fig. 2, $m_{11} = 0.0557$, $m_{22} = -0.236$, and $m_{12} = m_{21} = 0$. Now the crystalline phase has an orientation as shown in Figs. 3(d) and 4(d), about -72° from Fig. 2, the phason strain tensor becomes (see Appendix 1)

$$\mathbf{M}_{\Theta} = \mathbf{T}^{\parallel}{}_{\Theta} \bullet \mathbf{M} \bullet (\mathbf{T}^{\perp}{}_{2\Theta})^{-1} \tag{3}$$

For $\Theta = -72°$,

$$\mathbf{T}^{\parallel}{}_{\Theta} = \begin{bmatrix} \cos72° & \sin72° \\ \sin72° & \cos72° \end{bmatrix}, \quad (\mathbf{T}^{\perp}{}_{2\Theta})^{-1} = \begin{bmatrix} \cos144° & -\sin144° \\ \sin144° & \cos144° \end{bmatrix} \tag{4}$$

The phason tensor for the transformation of the decgonal QC to the crystalline approximant in this case is (Θ=-72°)

$$\mathbf{M}_{\Theta} = \begin{bmatrix} -0.146 & 0 \\ 0.171 & 0.09 \end{bmatrix} \tag{5}$$

During this transformation of a decagonal QC to this new crystalline phase, as shown in Figs. 4 (b), (c), and (d), the corresponding **M'** are, respectively,

$$\mathbf{M}_b = \begin{bmatrix} -0.04 & 0 \\ 0.06 & 0 \end{bmatrix}, \quad \mathbf{M}_c = \begin{bmatrix} -0.08 & 0 \\ 0.08 & 0.06 \end{bmatrix}, \quad \mathbf{M}_d = \begin{bmatrix} -0.146 & 0 \\ 0.11 & 0.09 \end{bmatrix}$$

Obviously, the m_{11} and m_{22} in $\mathbf{M}_d$ are the same as those in $\mathbf{M}_{\Theta}$, whereas the m_{21} = 0.11 in $\mathbf{M}_d$ is somewhat less than 0.171 in $\mathbf{M}_{\Theta}$ in (5), implying that this transformation is still not complete yet.

As shown in Fig. 4, the simulated EDPs compare favorably with the corresponding experimental EDPs in Fig. 3. The ten spots marked with arrowheads change gradually from their positions on a circle first to an ellipse with its long axis in the arrowed direction and finally to an almost 2D cross-grid pattern. The spots on the marked horizontal wavy lines show the process of grouping and merging together of several spots. It is clear that the ten strong spots on the periphery of these simulated EDPs, drawn as filled circles in Fig. 4, and other sets of ten strong spots generally change only slightly their positions during this transformation. This proves the preservation of the icosahedral units of the decagonal QC in the newly formed crystalline phase.

4. ROTATIONAL DOMAINS

Since the "orthorhombic" crystal has two axes parallel to the two orthogonal twofold D and P directions, respectively, of the decagonal QC and there are ten sets of such direction pairs, the orthorhombic crystal should have ten possible orientations. Since the twin domains in general are very small, we have to adopt the micro-beam diffraction technique. Using a small aperture of 30 μm in diameter, the irradiated size of the focused beam is about 100 nm. Figure 5 shows the micro EDPS of four neighboring areas. In all of them the ten inner spots, or rather discs, no longer lie on a circle, but the long axes of the ellipses lie in different directions, with an angle of mostly 36° between two of them. In other words, it seems that the decagonal QC is transforming into orthorhombic crystals of ten different orientations with equal possibilities. Figure 5 shows the intermediate stage of this transformation and in the long run tenfold rotational twins will result from this transformation.

The occurrence of such 36° rotational twins should be general in the formation of Penrose-tiling approximants from the decagonal QC. This is indeed the case. The monoclinic $Al_{13}Fe_4$ with β (107.71°) close to 108° can be considered to have a base-centered orthorhombic lattice with a=0.7745 and c=2.377 nm [1]. The axis ratio 2.377:0.7745 = 3.07 is similar to that of c:a=3.79:1.24=3.06 of the present case. This explains why tenfold rotational twins can easily be observed together with the decagonal QC in rapidly solidified Al-Fe alloy [17]. In this respect the newly found 36° rotational twins of Al_3Pd should also be mentioned [24]. This simple orthorhombic phase (a=2.34, b=1.67, and c=1.23 nm) is also a Penrose-tiling approximant with its a and c axes parallel, respectively, to the two orthogonal twofold D and P directions. From these findings it can perhaps be concluded that all rational approximants of the decagonal QC may grow along the ten sets of D-P direction pairs and form tenfold twins.

In the past Pauling [2] and others have advocated a multiple twin (five-, ten-, or twenty-fold) interpretation of the QC to account for the EDPs with fivefold or tenfold rotational symmetry. The evidences presented above of the continuous transformatin of a substance with tenfold symmetry (the decagonal QC) to 36° rotational twins, especially the presence of intermediate states which are neither quasicrystalline nor crystalline, may perhaps rule out the multiple twin hypothesis of the decagonal QC.

A similar study on the transformation of the AlCuCo decagonal QC to a new orthorhombic approximant (a=1.97 nm, b=0.40 nm and c=2.33nm) has been published recently [25].

5. CONCLUSIONS

1. A new B-centered orthorhombic phase with a=1.24, b=1.24, and c=3.79 nm or its corresponding monoclinic phase with a=1.24, b=1.24, c=1.99 nm, and $\beta \approx 108°$ has been found in $Al_{13}Cr_4Si_4$ and $Al_{74}Mn_{21}Si_5$ alloys. From their EDPs compared with those of the decagonal QC, the following orientation relationship has been found:

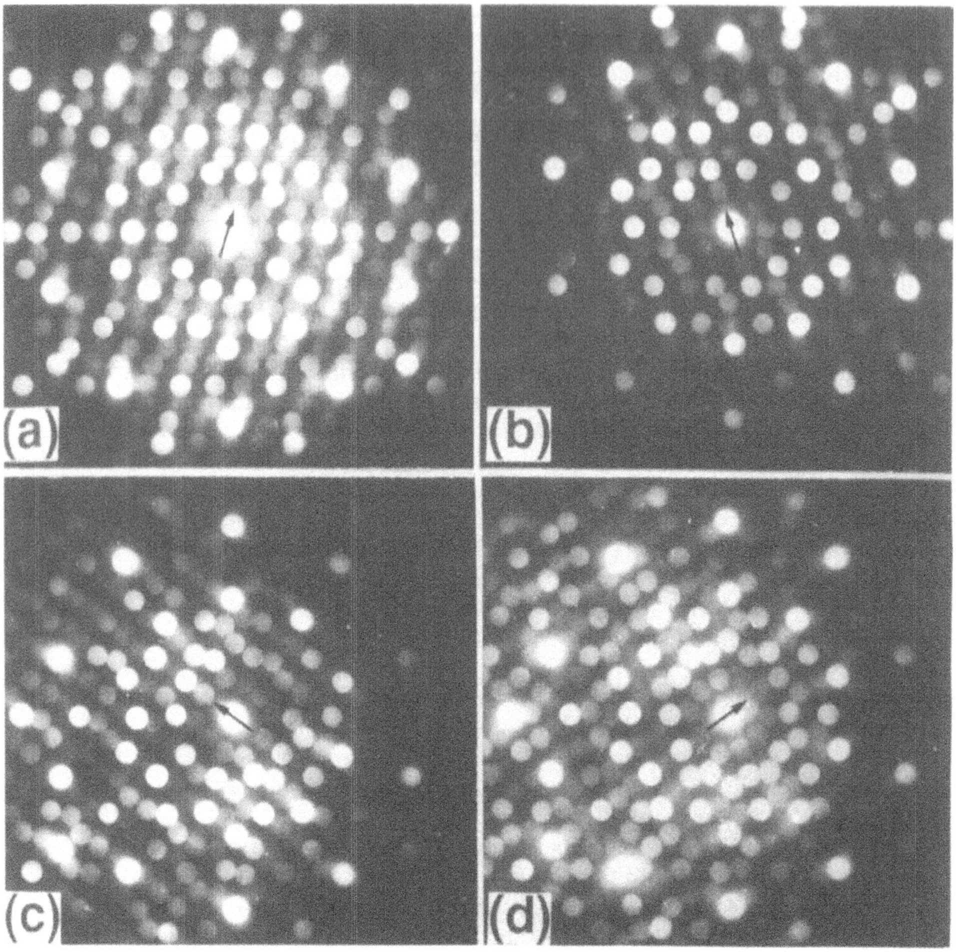

Figure 5. Micro-beam EDPs of the intermediate stage of transformation where the ten inner spots on an ellipse are marked. The long axis of the ellipse differs by 36° or a multiple of it, implying that 36° rotation twins may result from this transformation.

	$[010]_O$ // 10 axis	
Crystal	$[100]_O$ // 2 axis (D)	Quasicrystal
	$[001]_O$ // 2 axis (P).	

2. If this crystalline phase is described as having a B-centered orthorhombic lattice, its lattice parameters can be derived from the decagonal QC by substituting rational ratios 3/2 and 8/5, respectively, for the irrational τ in two orthogonal twofold directions of the decagonal QC. In other words, this new crystalline phase is also a Penrose-tiling approximant of the decagonal QC.

3. The continuous transformation of the decagonal QC to this new crystalline phase has been followed experimentally by selected-area electron-diffraction and theoretically by simulations with increased phason strains in the decagonal QC. The match between the experimentally and simulated EDPs is fairly good.

4. This crystalline phase can grow on ten equivalent twofold directions in the decagonal QC and consequently tenfold rotational twins with a rotation angle of $n \times 36^{\circ}$ around the *b* axis can form from the decagonal QC. An intermediate stage between the decagonal QC and these multiple twins can frequently been observed. This proves definitely that QC and multiple twins are two different structures and the multiple twin interpretation of the QC advocated by Linus Pauling is not well founded.

Acknowledgements

This project was funded by the Chinese Academy of Sciences and also by the Chinese National Natural Science Foundation. The authors are grateful to Dr. J.-M. Dubois for the supply of the Al-Mn-Si sample used in the present work.

6. REFERENCES

1 H. Zhang and K.H. Kuo, Phys. Rev. B42 (1990) 8907.
2 L. Pauling, Phys. Rev. Lett., 58 (1987) 365.
3 H. Zhang and K.H. Kuo, Phys. Rev. B41 (1990) 3482.
4 G. Entin-Wohlman, M. Klemen and A. Pavlovitch, J. Phys. (Paris) 49 (1988) 587.
5 T.C. Lubensky, J.E.S. Socolor, P.J. Steinhardt, P.A. Bancel and P.A. Heiney, Phys. Rev. Lett., 57 (1986) 1440.
6 P.M. Horn, W. Menzfeldt, D.P. DiVincenzo, J. Toner and R. Gambino, Phys. Rev. Lett., 57 (1986) 1444.
7 L.X. He, X.Z. Li, Z. Zhang and K.H. Kuo, Phys. Rev. Lett., 61 (1988) 1116.
8 Z. Zhang and K.H. Kuo, J. Microscopy, 146 (1987) 313.
9 D.S. Zhou, H.Q. Ye and K.H. Kuo, Phil. Mag. Lett., 56 (1987) 209.
10 P. Kramer, Acta Crystallog., 43 (1987) 486.
11 J. Fayos, I. Jimenez, G. Pastor, E. Gancedo and M. Torres, Z. Kristallogr., 185 (1988) 283.
12 Z.H. Mai, S.Z. Tao, L.Z. Zeng and B.S. Zhang, Phys. Rev. B38 (1987) 1293.
13 F.H. Li, G.Z. Pan, S.Z. Tao, M.J. Hui, Z.H. Mai, X.S. Chen and L.Y. Cai, Phil. Mag. B59 (1989) 535.
14 Z.H. Mai, L. Xu, N. Wang, K.H. Kuo, Z.C. Jin and G. Cheng, Phys. Rev. B40 (1989) 12183.

15 N. Wang and K.H. Kuo, Philos. Mag. Lett. 61 (1990) 63.
16 H. Zhang, D.H. Wang and K.H. Kuo, J. Mater. Sci. 24 (1989) 2981.
17 X.D. Zou, K.K. Fung and K.H. Kuo, Phys. Rev. 35 (1987) 4526.
18 J.-M. Dubois, Ch. Janot and M De Boissieu, Quasicrystalline Materials, Eds. Ch. Janot and J.M. Dubois, World Scientific, Singapore, 1988, p. 97.
19 P.J. Black, Acta Crystallogr., 8 (1955) 175.
20 X.Z. Li and K.H. Kuo, Philos. Mag. B65 (1992) 525; B66 (1992) 117.
21 P. Guyot, M. Audier and R. Lequette, J. Phys. (Paris), C3 (1986) 389.
22 V. Elser and C.L. Henley, Phys. Rev. Lett., 55 (1985) 2883.
23 C. Henley, J. Non-cryst. Solids, 75 (1985) 91 .
24 L. Ma, R. Wang and K.H. Kuo, J. Less-Common Metals, 163 (1990) 367.
25. X.Z. Liao, K.H. Kuo, H. Zhang and K. Urban, Philos. Mag. B66 (1992) 549.

Appendix 1

It can be seen that Figs. 3 and 4 are rotated $\Theta=-72^\circ$ from Fig. 2. Now we derive the relation between the phason strain $\mathbf{M}_\Theta$ after a rotation of Θ angle in the reciprocal space and the original phason strain $\mathbf{M}$.

The reciprocal vector $\mathbf{G}_\Theta$ after the rotation is related to the original $\mathbf{G}$ by

$$\mathbf{G}_\Theta = \mathbf{T} \bullet \mathbf{G} \tag{A}$$

where $\mathbf{T} = \begin{bmatrix} \cos\Theta & -\sin\Theta \\ \sin\Theta & \cos\Theta \end{bmatrix}$ is the rotation matrix of coordination transformation. After introducing a linear phason strain the reciprocal vector $\mathbf{G}^{\parallel}_\Theta$ becomes $\mathbf{G}^{\parallel'}_\Theta$

$$\mathbf{G}^{\parallel'}_\Theta = \mathbf{G}^{\parallel}_\Theta + \mathbf{M}_\Theta \bullet \mathbf{G}^{\perp}_{2\Theta} \tag{B}$$

where $\mathbf{M}_\Theta$ is the phason strain tensor after the rotation. A rotation of Θ in the parallel or real space has a corresponding rotation of 2Θ in the perpendicular or pseudo space. Using the relation in Eq. (A) we obtain

$$\mathbf{T}^{\parallel}_\Theta \bullet \mathbf{G}^{\parallel'} = \mathbf{T}^{\parallel} \bullet \mathbf{G}^{\parallel} + \mathbf{M}_\Theta \bullet \mathbf{T}^{\perp}_{2\Theta} \bullet \mathbf{G}^{\perp}$$

or

$$\mathbf{G}^{\parallel'} = \mathbf{G}^{\parallel} + (\mathbf{T}^{\parallel}_\Theta)^{-1} \bullet \mathbf{M}_\Theta \bullet \mathbf{T}^{\perp}_{2\Theta} \bullet \mathbf{G}^{\perp}$$

Comparing with Eg. (1) we obtain

$$\mathbf{M} = (\mathbf{T}^{\parallel}_\Theta)^{-1} \bullet \mathbf{M}_\Theta \bullet \mathbf{T}^{\perp}_{2\Theta}$$

and

$$\mathbf{M}_\Theta = \mathbf{T}^{\parallel}_\Theta \bullet \mathbf{M} \bullet (\mathbf{T}^{\perp}_{2\Theta})^{-1} \tag{3}$$

Crystal-Quasicrystal Transitions
M.J. Yacamán and M. Torres (Editors)

Relation between icosahedral quasicrystal and its (1/1) cubic approximant*

F. H. Li

Institute of Physics, Chinese Academy of Sciences, Beijing 100080, China

Abstract

The structure formulation for perfect and phason defected quasicrystals is described in the cut description. It is shown that the transformation from icosahedral (I) quasilattice into body-centered-cubic (b.c.c.) lattice can be performed by introducing a special linear phason strain and that the action of a linear phason strain is equivalent to the rotation of physical space. Intermediate states between I quasicrystal and b.c.c. crystal observed in Al-Cu-Li and Al-Cu-Mg systems with electron diffraction and high resolution electron microscopy are described. Their structures possess both the character of quasicrystal and that of b.c.c. crystal. The derivation of six-dimensional crystal structure model corresponding to the I quasicrystal from the structure of its b.c.c. crystalline approximant is introduced. It is demonstrated that electron diffraction and high resolution electron microscope image simulation for perfect and phason defected quasicrystals based on high-dimensional crystal is straightforward, and it is shown that the local phason strain can be identified by image simulation.

1. INTRODUCTION

Soon after the discovery of quasicrystal[1] the disorder in quasicrystals, which leads to shift, split and broadening of diffraction peaks[2-4] was drawn attention. It was explained as a result of the frozen phason strain[5-7]. In some cases the shift of diffraction peaks is so large that the peaks are arranged equidistantly along a certain direction[8,9], or the peaks are located at equispaced lines. Such phenomena were interpreted as the presence of local periodical translation[8] or the presence of intermediate states between the quasicrystal and a crystal which is close to the quasicrystal in composition and local atomic arrangement[9]. Recently, such crystals are named crystalline approximants. It was found by X-ray and electron diffraction study on Al-Cu-Li alloy that the icosahedral(I) symmetry $m\bar{3}\bar{5}$ may degenerate to cubic symmetry m3[10,11].

* This project is partially suported by National Natural Science Fundation of China.

Accordingly, the I phase T2-Al_6CuLi_3 transforms into the body-centered-cubic(b.c.c.) phase R-Al_5CuLi_3 [11]. In addition, a series of intermediate states between T2 phase and R phase was recorded by electron diffraction [11]. They may occupy a wide range between T2 phase and R phase. The transformation from I quasilattice into b.c.c. lattice and the existence of intermediate states can be interpreted by introducing a special linear phason strain [10-13]. The phason strain was also used to interpret the transformation of the octagonal phases[14], decagonal phases [15,16] and other icosahedral phases [17-19] into their crystalline approximants. Evidently, to reveal the relation between quasicrystals and their crystalline approximants is important for understanding quasicrystals themselves.

The present paper aims at illustrating the relation between quasicrystals and their approximants. In the following, for convenience, the relation between I quasicrystal and its b.c.c. or almost b.c.c. approximant is described as an example. In section 2 the phason defected quasicrystal structure formulation is derived in the cut description. In section 3 the transformation from I quasilattice into b.c.c. lattice is interpreted by introducing a special linear phason strain. The section 4 is about electron diffraction and high resolution electron microscopy investigation on intermediate states produced during the phase transformation from I phases into their cubic approximants. In section 5 it is shown that the 6D structure which is related to both the I phase and its b.c.c. or almost b.c.c. approximant can be derived from the structure of the approximant. Section 6 is about the identification of phason strain in quasicrystals on the basis of high-dimensional crystal structure model.

2. QUASICRYSTAL STRUCTURE FORMULATION IN CUT DESCRIPTION

The cut method was firstly proposed by de Wolf [20] in describing the incommensurate modulated structure. It gives the same result as the projection method in describing quasicrystals [21-25] so that the term 'cut and projection' is popular in literature concerning quasicrystals. However, the cut method shows some advantages in describing the quasicrystal structure [13,26,27], especially in describing the quasicrystal structure distorted by the phason strain and the structural relationship between the quasicrystalline phase and its crystalline approximants[13].

2.1. Real and reciprocal quasilattice formulation

Let $L_0(\mathbf{r}_{\parallel}, \mathbf{r}_{\perp})$ represent the n-dimensional or high-dimensional(HD) lattice function which consists of a series of δ functions with their centers at the position of lattice nodes

$$L_0(\mathbf{r}_{\|},\mathbf{r}_{\perp}) = \sum_{\mathbf{R}_{\|}}\sum_{\mathbf{R}_{\perp}} \delta(\mathbf{r}_{\|} - \mathbf{R}_{\|})\delta(\mathbf{r}_{\perp} - \mathbf{R}_{\perp}). \qquad (2-1)$$

Here $\mathbf{r}_{\|}(x_{\|}, y_{\|}, z_{\|})$ and $\mathbf{r}_{\perp}(x_{1\perp}, x_{2\perp}, ..., x_{(n-3)\perp})$ denote coordinate vectors, $\mathbf{R}_{\|}$ and $\mathbf{R}_{\perp}$ denote lattice vectors in physical space $E_{\|}$ and pseudo space $E_{\perp}$ respectively. Axes $x_{\|}, y_{\|}, z_{\|}$ and $x_{1\perp}, x_{2\perp}, ..., x_{(n-3)\perp}$ are orthogonal to one another, but they are generally not parallel to the basic vectors of the HD unit cell. When all lattice nodes have a definite size and shape described by the window function

$$S(\mathbf{r}_{\perp}) = \begin{cases} 1, & \text{inside a certain region in } E_{\perp}, \\ 0, & \text{elsewhere,} \end{cases} \qquad (2-2)$$

the lattice function becomes the convolution of $L_0(\mathbf{r}_{\|}, \mathbf{r}_{\perp})$ with $S(\mathbf{r}_{\perp})$:

$$L(\mathbf{r}_{\|},\mathbf{r}_{\perp}) = S(\mathbf{r}_{\perp}) * L_0(\mathbf{r}_{\|},\mathbf{r}_{\perp}), \qquad (2-3)$$

where $*$ denotes the operation of convolution. The corresponding reciprocal lattice function is written as

$$l(\mathbf{g}_{\|},\mathbf{g}_{\perp}) = s(\mathbf{g}_{\perp}) l_0(\mathbf{g}_{\|},\mathbf{g}_{\perp}), \qquad (2-4)$$

where $s(\mathbf{g}_{\perp})$ and $l_0(\mathbf{g}_{\|}, \mathbf{g}_{\perp})$ are Fourier transforms(FT's) of $S(\mathbf{r}_{\perp})$ and $L_0(\mathbf{r}_{\|}, \mathbf{r}_{\perp})$ respectively. Obviously, the real and reciprocal lattices are quite different in the shape of lattice nodes and in the boundary condition. The real lattice is unlimited. Its lattice nodes have definite size and shape along $E_{\perp}$, but they are sharp along $E_{\|}$. The reciprocal lattice is limited and its lattice nodes are sharp along both subspaces. In the real space the section of HD lattice with $E_{\|}$ gives the 3D quasilattice. When the HD lattice is a six-dimensional(6D) hypercubic lattice and the lattice nodes have the shape of a triacontahedron, the quasilattice is a standard 3D Penrose tiling(3DPT)[28]. Because the FT of a section of any function equals the projection of the FT of this function along the direction perpendicular to the sectional plane, it is easy to write down the FT of the quasilattice function as follows:

$$V(\mathbf{g}_{\|}) = \int l(\mathbf{g}_{\|},\mathbf{g}_{\perp}) dv_{\mathbf{g}_{\perp}} = \sum_{\mathbf{G}_{\|}}\sum_{\mathbf{G}_{\perp}} s(\mathbf{G}_{\perp})\delta(\mathbf{g}_{\|} - \mathbf{G}_{\|}), \qquad (2-5)$$

where $\mathbf{G}_{\|}$ and $\mathbf{G}_{\perp}$ are the components of HD reciprocal-lattice vectors in $E_{\|}$ and $E_{\perp}$ respectively. The inverse FT of $V(\mathbf{G}_{\|})$ gives the formula for quasilattices identical to the density wave expression:

$$u(\mathbf{r}_{\|}) = \sum_{\mathbf{G}_{\|}} \sum_{\mathbf{G}_{\perp}} s(\mathbf{G}_{\perp}) \exp(2\pi i \mathbf{r}_{\|} \mathbf{G}_{\|}). \qquad (2-6)$$

Equations (2-5) and (2-6) are the same as those obtained by other methods. The principle and procedure for deriving these equations by the cut method are shown schematically in figure 1.

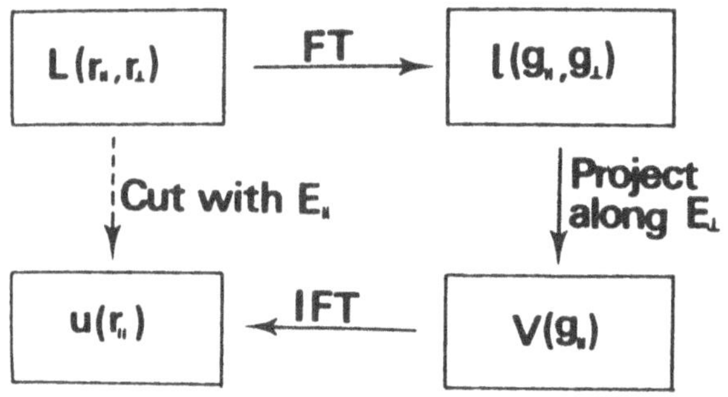

Figure 1. Schematic diagram showing the principle of deriving the quasiperiodic lattice function in the cut description.

2.2. Quasilattice with linear phason defect

Now a special displacement strain which has component only in the pseudo space and hence is called the phason strain [5] is introduced in the HD space. When the phason is linear it is expressed as follows [5]:

$$w(\mathbf{r}_{\perp}) = \mathbf{M}\mathbf{r}_{\|}, \qquad (2-7)$$

where $\mathbf{M}$ is a second-rank tensor, or

$$\mathbf{M} = \mathbf{e}_{\perp} M \mathbf{e}_{\|}, \qquad (2-8)$$

$\mathbf{e}_{\|}$ and $\mathbf{e}_{\perp}$ denote unit vectors in $E_{\|}$ and $E_{\perp}$ respectively, M is a matrix of order $3 \times (n-3)$ and will be called the phason matrix.

Equations (2-7) and (2-8) imply that to introduce a linear phason strain into the HD lattice makes all lattice nodes which have a definite shape and size in $E_\perp$ and are sharp in $E_\parallel$ shift along $E_\perp$. The shift amount depends on both the phason matrix and the component of the lattice vector in $E_\parallel$. Hence, the HD lattice and also the 3D quasilattice obtained by cutting the HD lattice with $E_\parallel$ will be distorted. In principle, the formulation of such a distorted quasilattice, called phason defected quasilattice, should be derived on the basis of the distorted HD lattice. However, we are not interested in the 6D lattice itself, but only interested in its section with the physical space. In the following it will be illustrated that to introduce a linear phason strain into quasilattices is equivalent to rotate the physical space relatively to the high dimensional space so that the phason defected quasilattice can also be obtained by cutting the undistorted HD lattice with a new 3D hyperplane.

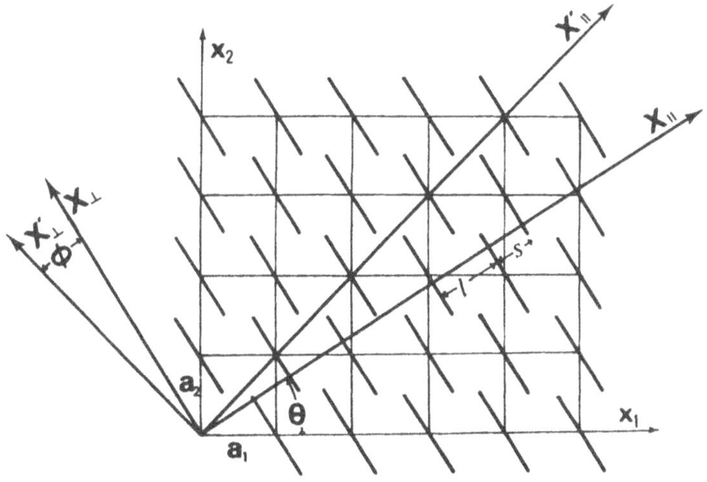

Figure 2. Schematic diagram showing that the transformation from a 1D quasilattice to a periodic lattice is equivalent to the rotation of physical space.

Figure 2 shows a two-dimensional (2D) square lattice with basis vectors $\mathbf{a}_1$ and $\mathbf{a}_2$ of the same length a. All lattice nodes elongate along axis $x_\perp$ (pseudo space) to form lattice bars. The one-dimensional (1D) quasilattice is obtained by cutting the 2D lattice with axis $x_\parallel$ (physical space) which is perpendicular to axis $x_\perp$ and irrationally orientated to the 2D lattice. It consists of two elemental segments of length $l = a \cos\theta$ and $s = a \sin\theta$. When arccot $\theta = \tau$ and the length of lattice bars equals $(l+s)$, the obtained 1D quasilattice is identical to the Fibonacii sequence. If a linear phason strain $w = mx_{\|}$ is introduced, all lattice bars will be displaced along the direction parallel to $x_{\perp}$ by a distance $m\mathrm{X}_{\|}$; $\mathrm{X}_{\|}$ is the component of the corresponding lattice vector in the physical space. A distorted 1D quasilattice is obtained by cutting the displaced lattice bars with axis $x_{\|}$. It is none other than the 1D quasilattice obtained, firstly, by cutting the undistorted lattice bars with an axis $x'_{\|}$ which form an angle $\phi = arctan\ m$ with axis $x_{\|}$, and ,secondly, by multiplying the obtained sequence by a factor $1/(1+m^2)^{1/2}$. The formulation of such a phason defected quasilattice is written straightforward by setting up a new Cartesian coordinate system with axes $x'_{\|}$ and $x'_{\perp}$:

$$\begin{pmatrix} x'_{\|} \\ x'_{\perp} \end{pmatrix} = \frac{1}{1+m^2}\begin{pmatrix} 1 & -m \\ m & 1 \end{pmatrix}\begin{pmatrix} x_{\|} \\ x_{\perp} \end{pmatrix} = T_m \begin{pmatrix} x_{\|} \\ x_{\perp} \end{pmatrix}. \qquad (2-9)$$

T_m is the rotation matrix describing the rotation transformation from the coordinate system $(x_{\|}, x_{\perp})$ to $(x'_{\|}, x'_{\perp})$.

$$\begin{pmatrix} x_{\|} \\ x_{\perp} \end{pmatrix} = T_m^{-1}\begin{pmatrix} x'_{\|} \\ x'_{\perp} \end{pmatrix} = \begin{pmatrix} 1 & m \\ -m & 1 \end{pmatrix}\begin{pmatrix} x'_{\|} \\ x'_{\perp} \end{pmatrix}. \qquad (2-10)$$

Accordingly, in the reciprocal space

$$(g'_{\|}, g'_{\perp}) = (g_{\|}, g_{\perp})\begin{pmatrix} 1 & m \\ -m & 1 \end{pmatrix}. \qquad (2-11)$$

For deriving the FT of the 1D phason defected quasilattice which is the intersection of the 2D square lattice with axis $x'_{\|}$ defined by equation (2-9), it should be noticed that the FT of the window function remains unchanged in the new coordinate system. This is because the lattice nodes should always remain sharp in the physical space so that the lattice bars must always be considered as perpendicular to the physical space. Hence, the FT of a phason defected 1D quasilattice can be simply rewritten as

$$V'(g'_{\|}) = \int s(g'_{\perp}) \times \sum_{G'_{\|}}\sum_{G'_{\perp}} \delta(g'_{\|} - G'_{\|})\delta(g'_{\perp} - G'_{\perp})dg'_{\perp}, \qquad (2-12)$$

here $G'_{\parallel}, G^{p}_{\perp}rime$ are coordinates of reciprocal lattice nodes.

It should be emphasized that the new coordinate system is only a tool for formulating the phason defected quasilattice. In general, equation (2-9) gives only the relationship between two coordinate systems but not give the coordinate relationship between the idealized quasilattice nodes obtained by cutting the 2D square lattice with axis $x_{\parallel}$ and the phason defected quasilattice nodes obtained by cutting the 2D square lattice with axis $x'_{\parallel}$. The only exception is the case when axis $x'_{\parallel}$ intercepts lattice bars at their center so that the 1D quasilattice turns into a periodic lattice. However, equation (2-11) always gives the right coordinate relationship between the reciprocal-lattice nodes of the idealized and the phason defected quasilattice. This is because all lattice nodes are sharp in the 2D reciprocal space and the reciprocal-quasilattice nodes are projections of the sharp 2D lattice nodes. Thus the reciprocal-quasilattice vectors of the phason defected quasilattice are expressed as

$$G'_{\parallel} = G_{\parallel} - mG_{\perp}. \qquad (2-13)$$

To illustrate clearly the change of Fourier spectra of the phason defected quasilattice, a coordinate transformation is made so that

$$V'(g_{\parallel}) = \int s(g_{\perp}) \sum_{G_{\parallel}} \sum_{G_{\perp}} \delta[g_{\parallel} - (G_{\parallel} - mG_{\perp})] \times \delta[g_{\perp} - (G_{\perp} + mG_{\parallel})] dg_{\perp}, \qquad (2-14)$$

or

$$V'(g_{\parallel}) = \sum_{G_{\parallel}} \sum_{G_{\perp}} s(G_{\perp}) \delta[g_{\parallel} - (G_{\parallel} - mG_{\perp})], \qquad (2-15)$$

here the function $s(G_{\perp} + mG_{\parallel})$ is replaced by $s(G_{\perp})$ owing to the same reason mentioned above that in the real space lattice bars are always perpendicular to the physical space. The density wave expression of the phason defected 1D quasilattice is

$$u'(x_{\parallel}) = \sum_{G_{\parallel}} \sum_{G_{\perp}} s(G_{\perp}) \exp[2\pi i x_{\parallel}(G_{\parallel} - mG_{\perp})]. \qquad (2-16)$$

In the case of a 3D quasilattice with a phason defect the new HD orthogonal coordinate system $(\mathbf{r}'_{\parallel}, \mathbf{r}'_{\perp})$ is related to the old one by

$$\begin{pmatrix} \mathbf{r}'_{\parallel} \\ \mathbf{r}'_{\perp} \end{pmatrix} = \frac{1}{k} \begin{pmatrix} I & -\mathbf{M} \\ \mathbf{M} & I \end{pmatrix} \begin{pmatrix} \mathbf{r}_{\parallel} \\ \mathbf{r}_{\perp} \end{pmatrix} = T_m \begin{pmatrix} \mathbf{r}_{\parallel} \\ \mathbf{r}_{\perp} \end{pmatrix}, \qquad (2-9')$$

and

$$\begin{pmatrix} \mathbf{r}_{\parallel} \\ \mathbf{r}_{\perp} \end{pmatrix} = T_m^{-1} \begin{pmatrix} \mathbf{r}'_{\parallel} \\ \mathbf{r}'_{\perp} \end{pmatrix} = \begin{pmatrix} I & \tilde{\mathbf{M}} \\ -\tilde{\mathbf{M}} & I \end{pmatrix} \begin{pmatrix} \mathbf{r}'_{\parallel} \\ \mathbf{r}'_{\perp} \end{pmatrix}, \qquad (2-10')$$

where I is the unit matrix of order $3 \times (n-3)$, $\tilde{\mathbf{M}}$ the transpose of $\mathbf{M}$ and k the multiplying factor depending on the elements of $\mathbf{M}$. In the reciprocal space

$$(\mathbf{g}'_{\parallel}, \mathbf{g}'_{\perp}) = (\mathbf{g}_{\parallel}, \mathbf{g}_{\perp}) \begin{pmatrix} I & \tilde{\mathbf{M}} \\ -\tilde{\mathbf{M}} & I \end{pmatrix} \qquad (2-11')$$

and

$$\mathbf{G}'_{\parallel} = \mathbf{G}_{\parallel} - \mathbf{G}_{\perp}\tilde{\mathbf{M}}. \qquad (2-12')$$

The FT of the 3D quasilattice with a linear phason defect is

$$V'(\mathbf{g}_{\parallel}) = \sum_{\mathbf{G}_{\parallel}} \sum_{\mathbf{G}_{\perp}} s(\mathbf{G}_{\perp}) \delta[\mathbf{g}_{\parallel} - (\mathbf{G}_{\parallel} - \mathbf{G}_{\perp}\tilde{\mathbf{M}})]. \qquad (2-13')$$

The corresponding density wave expression is

$$u'(\mathbf{r}_{\parallel}) = \sum_{\mathbf{G}_{\parallel}} \sum_{\mathbf{G}_{\perp}} s(\mathbf{G}_{\perp}) \exp[2\pi i \mathbf{r}_{\parallel}(\mathbf{G}_{\parallel} - \mathbf{G}_{\perp}\tilde{\mathbf{M}})]. \qquad (2-14')$$

The phase change caused by the phason strain is

$$\Delta\mathbf{G} = -\mathbf{G}_{\perp}\tilde{\mathbf{M}}. \qquad (2-17)$$

According to (2-5) the diffraction intensity of quasicrystals is modulated by $|s(\mathbf{G}_{\perp})|$ so that strong diffraction peaks generally correspond to small $\mathbf{G}_{\perp}$ and *vice versa.* Equation (2-17) indicates that the shift of diffraction peaks is proportional to the absolute value of $\mathbf{G}_{\perp}$. Hence, a weak diffraction peak is bound to have a large shift while a strong peak has a small shift as was pointed out by Lubensky et al[5]. This has been confirmed by experimental results.

2.3. Structure factor of quasicrystal

The atomic structure of quasicrystals can be obtained by cutting a HD periodic structure when hyperatoms are decorated in the HD unit cell. Function $\phi(\mathbf{r}_{\|}, \mathbf{r}_{\perp})$ which describes the HD crystal structure can be written as the convolution of the lattice function $L(\mathbf{r}_{\|}, \mathbf{r}_{\perp})$ with the function $\phi_0(\mathbf{r}_{\|}, \mathbf{r}_{\perp})$ which describes the atomic structure inside the HD unit cell:

$$\phi(\mathbf{r}_{\|}, \mathbf{r}_{\perp}) = S(\mathbf{r}_{\perp}) * L_0(\mathbf{r}_{\|}, \mathbf{r}_{\perp}) * \phi_0(\mathbf{r}_{\|}, \mathbf{r}_{\perp}). \qquad (2-18)$$

As far as the structure analysis by diffraction methods is concerned, $\phi_0(\mathbf{r}_{\|}, \mathbf{r}_{\perp})$ and $\phi(\mathbf{r}_{\|}, \mathbf{r}_{\perp})$ are electron density functions for X-rays and are potential distribution functions for electrons. The function $S(\mathbf{r}_{\perp})$ describes the shape of lattice nodes and the shape of hyperatoms in the pseudo space. The intersection of any hyperatom with $E_{\|}$ gives the real atom. The expression for the 3D quasicrystal structure $\rho(\mathbf{r}_{\|})$, which is the intersection of $\phi(\mathbf{r}_{\|}, \mathbf{r}_{\perp})$ with $E_{\|}$, can be obtained by the procedure shown in figure 1, namely, firstly by doing the FT of $\phi(\mathbf{r}_{\|}, \mathbf{r}_{\perp})$, secondly by projecting the product along $E_{\perp}$ and finally by performing the inverse FT.

$$\rho(\mathbf{r}_{\|}) = \sum_{\mathbf{G}_{\|}} \sum_{\mathbf{G}_{\perp}} s(\mathbf{G}_{\perp}) F_0(\mathbf{G}_{\|}, \mathbf{G}_{\perp}) \exp(2\pi i \mathbf{r}_{\|} \mathbf{G}_{\|}), \qquad (2-19)$$

where $F_0(\mathbf{G}_{\|}, \mathbf{G}_{\perp}) \equiv F_0(\mathbf{G})$ is the FT of function $\phi_0(\mathbf{r}_{\|}, \mathbf{r}_{\perp})$ and is the structure factor of HD crystal. The quasicrystal structure factor is

$$F(\mathbf{G}) = s(\mathbf{G}_{\perp}) F_0(\mathbf{G}). \qquad (2-20)$$

Hence, there is a one-to-one correspondence between the structure factor of quasicrystal and the structure factor of the HD crystal. The structure factor of the quasicrystal equals the structure factor of the HD crystal modulated by the function $s(\mathbf{G}_{\perp})$.

When atoms are of different shapes in the HD space, the HD crystal structure is described as

$$\phi(\mathbf{r}_{\|}, \mathbf{r}_{\perp}) = \sum \phi_j(\mathbf{r}_{\|}) * S_j(\mathbf{r}_{\perp}), \qquad (2-21)$$

where $S_j(\mathbf{r}_\perp)$ is the window function describing the shape and size of the j kind of atoms. Let $s_j(\mathbf{g}_\perp)$ denote the FT of $S_j(R_\perp)$ and $f_j(G_\parallel)$ the atomic scattering factor of the j kind of atoms. Thus the structure factor of quasicrystals is written as

$$F(\mathbf{G}) = \sum_j s_j(\mathbf{G}_\perp) f_j(\mathbf{G}_\parallel) \exp(2\pi i \mathbf{r}\mathbf{G}). \qquad (2-22)$$

2.4. Structure factor of quasicrystal with linear phason strain

When the quasicrystal structure is distorted by a linear phason strain, the structure factor can be derived by the similar procedure as for deriving that of a perfect quasicrystal. The difference is only in the projection direction of the FT of $\phi(\mathbf{r}_\parallel, \mathbf{r}_\perp)$. The FT of $\phi(\mathbf{r}_\parallel, \mathbf{r}_\perp)$ should be projected along the newly defined pseudo space $E'_\perp$, but not $E_\perp$. The expression of the structure factor of phason defected quasicrystal is

$$F'(\mathbf{G}) = s(\mathbf{G}_\perp) \sum_j f_j(\mathbf{G}_\parallel - \mathbf{G}_\perp \tilde{\mathbf{M}}) \exp(2\pi i \mathbf{r}\mathbf{G}), \qquad (2-20')$$

or

$$F'(\mathbf{G}) = \sum_j s_j(\mathbf{G}_\perp) f_j(\mathbf{G}_\parallel - \mathbf{G}_\perp \tilde{\mathbf{M}}) \exp(2\pi i \mathbf{r}\mathbf{G}), \qquad (2-22')$$

3. RELATION BETWEEN I QUASICRYSTAL AND b.c.c. LATTICE

3.1. Transformation from I quasilattice to b.c.c. lattice

The I quasilattice can be obtained by cutting a six-dimensional (6D) hypercubic lattice with the irrationally oriented $E_\parallel$. The 6D hypercubic lattice is described by a 6D orthogonal coordinate system $(x_1, x_2, x_3, x_4, x_5, x_6)$ with six basis vectors $\mathbf{a}_1, \mathbf{a}_2, \mathbf{a}_3, \mathbf{a}_4, \mathbf{a}_5$ and $\mathbf{a}_6$, which form a 6D unit cell of edge length a. The components of the six basis vectors in $E_\parallel$ are six basis vectors $\mathbf{e}_1, \mathbf{e}_2, \mathbf{e}_3, \mathbf{e}_4, \mathbf{e}_5$ and $\mathbf{e}_6$ in I coordinate system [29,30]. Another set of six basis vectors in the 6D space can be selected such that three vectors, for instance, $\mathbf{a}_{\parallel x}, \mathbf{a}_{\parallel y}$ and $\mathbf{a}_{\parallel z}$ are in $E_\parallel$, while the other three, $\mathbf{a}_{\perp x}, \mathbf{a}_{\perp y}$ and $\mathbf{a}_{\perp z}$ are in $E_\perp$. The transformation from

$\mathbf{a}_{\|x}, \mathbf{a}_{\|y}, \mathbf{a}_{\|z}, \mathbf{a}_{\perp x}, \mathbf{a}_{\perp y}, \mathbf{a}_{\perp z}$ to $\mathbf{a}_1, \mathbf{a}_2, \mathbf{a}_3, \mathbf{a}_4, \mathbf{a}_5, \mathbf{a}_6$ is performed by a matrix T:

$$(\mathbf{a}_{\|x}, \mathbf{a}_{\|y}, \mathbf{a}_{\|z}, \mathbf{a}_{\perp x}, \mathbf{a}_{\perp y}, \mathbf{a}_{\perp z}) = (\mathbf{a}_1, \mathbf{a}_2, \mathbf{a}_3, \mathbf{a}_4, \mathbf{a}_5, \mathbf{a}_6)T. \qquad (3-1)$$

A convenient form of the matrix T is

$$T = \frac{1}{\sqrt{2+2\tau^2}} \begin{pmatrix} \tau & 0 & 1 & -1 & 0 & \tau \\ \tau & 0 & -1 & -1 & 0 & -\tau \\ 0 & 1 & -\tau & 0 & \tau & 1 \\ -1 & \tau & 0 & -\tau & -1 & 0 \\ 0 & 1 & \tau & 0 & \tau & -1 \\ 1 & \tau & 0 & \tau & -1 & 0 \end{pmatrix}, \qquad (3-2)$$

where $\tau = (1+\sqrt{5})/2$. According to equation (3-2), vectors $\mathbf{a}_{\|x}, \mathbf{a}_{\|y}, \mathbf{a}_{\|z}$ and $\mathbf{a}_{\perp x}, \mathbf{a}_{\perp y}$, $\mathbf{a}_{\perp z}$ are of the same length as vectors $\mathbf{a}_j (j = 1, 2, ..., 6)$. They are parallel to three twofold axes of the basis icosahedron in $\mathrm{E}_\|$ and those in $\mathrm{E}_\perp$, respectively. The $\mathrm{E}_\|$ defined by equations (3-1) and (3-2) is irrationally orientated and named $(\tau/1)$ physical space.

In the following it will be proved that a special linear phason strain with phason matrix

$$M = \begin{pmatrix} m_1 & 0 & 0 \\ 0 & m_2 & 0 \\ 0 & 0 & m_3 \end{pmatrix} \qquad (3-3)$$

makes the 6D lattice distort such that the I quasilattice can transform into b.c.c. lattice. The 6D space can also be divided into another pair of subspaces $\mathrm{E}'_\|$ and $\mathrm{E}'_\perp$ with basis vectors $\mathbf{a}'_{\|x}$, $\mathbf{a}'_{\|y}$, $\mathbf{a}'_{\|z}$, and $\mathbf{a}'_{\perp x}$, $\mathbf{a}'_{\perp y}$, $\mathbf{a}'_{\perp z}$ respectively. The transformation from $\mathrm{E}_\|$ and $\mathrm{E}_\perp$ into $\mathrm{E}'_\|$ and the corresponding new pseudo space $\mathrm{E}'_\perp$ is performed by the matrix

$$T_m^{-1} = \begin{pmatrix} I & M \\ -M & I \end{pmatrix} \qquad (3-4)$$

and

$$(\mathbf{a}'_{\|x}, \mathbf{a}'_{\|y}, \mathbf{a}'_{\|z}, \mathbf{a}'_{\perp x}, \mathbf{a}'_{\perp x}, \mathbf{a}'_{\perp x}) = (\mathbf{a}_{\|x}, \mathbf{a}_{\|y}, \mathbf{a}_{\|z}, \mathbf{a}_{\perp x}, \mathbf{a}_{\perp y}, \mathbf{a}_{\perp z})T_m^{-1}. \qquad (3-5)$$

The intersection of the 6D hypercubic lattice with $E'_{\|}$ gives the 3D quasilattice distorted by the linear phason strain. The transformation from $(\mathbf{a}_1, \mathbf{a}_2, \mathbf{a}_3, \mathbf{a}_4, \mathbf{a}_5, \mathbf{a}_6)$ to $(\mathbf{a}'_{\|x}, \mathbf{a}'_{\|y}, \mathbf{a}'_{\|z}, \mathbf{a}'_{\perp x}, \mathbf{a}'_{\perp y}, \mathbf{a}'_{\perp z})$ is performed by the product matrix $T_m^{-1}T$:

$$(\mathbf{a}'_{\|x}, \mathbf{a}'_{\|y}, \mathbf{a}'_{\|z}, \mathbf{a}'_{\perp x}, \mathbf{a}'_{\perp y}, \mathbf{a}'_{\perp z}) = (\mathbf{a}_1, \mathbf{a}_2, \mathbf{a}_3, \mathbf{a}_4, \mathbf{a}_5, \mathbf{a}_6)TT_m^{-1}. \quad (3-6)$$

When $m_1 = m_2 = m_3 = m$,

$$TT_m^{-1} = \frac{1}{\sqrt{2+2\tau^2}} \begin{pmatrix} t_1 & 0 & t_2 & -t_2 & 0 & t_1 \\ t_1 & 0 & -t_2 & -t_2 & 0 & -t_1 \\ 0 & t_2 & -t_1 & 0 & t_1 & t_2 \\ -t_2 & t_1 & 0 & -t_1 & -t_2 & 0 \\ 0 & t_2 & t_1 & 0 & t_1 & -t_2 \\ t_2 & t_1 & 0 & t_1 & -t_2 & 0 \end{pmatrix}, \quad (3-7)$$

where $t_1 = \tau + m$ and $t_2 = 1 - m\tau$. When m=0, *i.e.* when the phason strain vanishes, the product matrix degenerates to matrix T so that the intersection of the 6D hypercubic lattice with the subspace $E'_{\|}$ gives a perfect or idealized I quasilattice. When t_1/t_2 becomes a ratio of two integers, the section would be a 3D periodic lattice. When $t_1/t_2 = 1$ or $m = -(\tau - 1)/(\tau + 1) = -0.236$,

$$TT_m^{-1} = \frac{\sqrt{2+2\tau^2}}{2\tau^2} \begin{pmatrix} 1 & 0 & 1 & -1 & 0 & 1 \\ 1 & 0 & -1 & -1 & 0 & -1 \\ 0 & 1 & -1 & 0 & 1 & 1 \\ -1 & 1 & 0 & -1 & -1 & 0 \\ 0 & 1 & 1 & 0 & 1 & -1 \\ 1 & 1 & 0 & 1 & -1 & 0 \end{pmatrix}, \quad (3-8)$$

Equation (3-8) defines a rationally orientated physical space((1/1) physical space). In this case the corresponding index relationship in the reciprocal space is

$$(h'_{\|}, k'_{\|}, l'_{\|}, h'_{\perp}, k'_{\perp}, l'_{\perp}) = (h_1, h_2, h_3, h_4, h_5, h_6)TT_m^{-1}. \quad (3-9)$$

Let $h \equiv h'_{\|}, k \equiv k'_{\|}, l \equiv l'_{\|}$, then

$$(h,k,l) = \frac{\sqrt{2+2\tau^2}}{2\tau^2}(h_1,...,h_6)\begin{pmatrix} 1 & 0 & 1 \\ 1 & 0 & -1 \\ 0 & 1 & -1 \\ -1 & 1 & 0 \\ 0 & 1 & 1 \\ 1 & 1 & 0 \end{pmatrix}. \qquad (3-10)$$

Equation (3-10) shows that the three reciprocal basis vectors for the physical space are of the same length. Therefore, they form a cubic reciprocal unit cell.

In order to have integer indices h, k, l, the reciprocal-lattice parameter a_c^* should be

$$a_c^* = \frac{\sqrt{2+2\tau^2}}{2\tau^2}a^*, \qquad (3-11)$$

where $a^* = 1/a$. Then the corresponding lattice parameter in real space is

$$a_c = \frac{2\tau^2}{\sqrt{2+2\tau^2}}a = \frac{2\tau^2}{\sqrt{1+\tau^2}}a_r, \qquad (3-12)$$

where a_r is the edge length of rhombohedra which form the 3DPT. Let $\mathbf{a}_c, \mathbf{b}_c$ and $\mathbf{c}_c$ represents the three basis vectors of the cubic lattice in the real space, then the relationship of $\mathbf{a}_c, \mathbf{b}_c, \mathbf{c}_c$ with the six basis vectors $\mathbf{a}_1, ..., \mathbf{a}_6$ is obtained from (3-6), (3-8) and (3-12) as follows:

$$\begin{aligned} \mathbf{a}_c &= \tau^2(\mathbf{a}_1 + \mathbf{a}_2 - \mathbf{a}_4 + \mathbf{a}_6)/\sqrt{2+2\tau^2}, \\ \mathbf{b}_c &= \tau^2(\mathbf{a}_3 + \mathbf{a}_4 + \mathbf{a}_5 + \mathbf{a}_6)/\sqrt{2+2\tau^2}, \\ \mathbf{c}_c &= \tau^2(\mathbf{a}_1 - \mathbf{a}_2 - \mathbf{a}_3 + \mathbf{a}_5)/\sqrt{2+2\tau^2}. \end{aligned} \qquad (3-13)$$

The diagonal of the cubic unit cell is expressed as

$$|\mathbf{a}_c + \mathbf{b}_c + \mathbf{c}_c| = \tau^2|2\mathbf{a}_1 - 2\mathbf{a}_5 + 2\mathbf{a}_6|/\sqrt{2+2\tau^2}. \qquad (3-14)$$

Equation (3-14) implies that there is a lattice node at the center of the body diagonal. Hence, the cubic lattice is of the b.c.c. type.

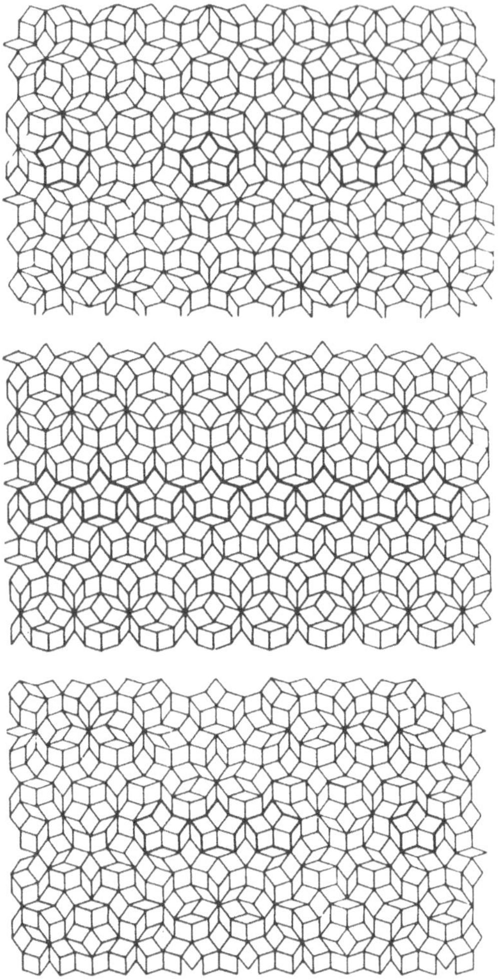

Figure 3. Tiling sections perpendicular to the fivefold axis for $m=0$ (top), $m = -0.236$ (middle) and $m = -0.12$ (bottom). They correspond to I quasilattice, b.c.c. lattice and intermediate states between I and b.c.c. lattices, respectively.

3.2. Intermediate states between I and b.c.c. lattices

It is interesting to see the feature of a phason defected quasilattice when the diagonal elements of the phason matrix equal to a value between 0 and -0.236. Figure 3 shows the change of tiling sections perpendicular to the fivefold axis with the change of the phason strain. The top picture corresponds to the perfect I quasilattice with $m_1 = m_2 = m_3 = 0$. It possesses the fivefold symmetry. The distance ratio of adjacent fivefold symmetrical figures with bold profiles aligning along the horizontal equals the golden mean τ. The middle picture corresponds to b.c.c. lattice with $m_1 = m_2 = m_3 = -0.236$, where the bold fivefold symmetrical figures arranged periodically along the horizontal. The aperiodicity along the vertical is due to the reason that when the I quasilattice transforms into a b.c.c. lattice the fivefold axis becomes to the direction [1 0 τ] rather than any zone axis. The bottom picture corresponds to $m_1 = m_2 = m_3 = -0.12$. It shows the local periodicity and the local fivefold symmetry. This means that to cut the 6D hypercubic lattice with the $(\tau/1)$ 3D hyperplane gives the I quasilattice, to cut it with the (1/1) 3D hyperplane defined by equations (3-6) and (3-7) gives the b.c.c. lattice, and to cut it with a 3D hyperplane located between the $(\tau/1)$ and (1/1) hyperplanes gives a distorted I quasilattice which in fact is an intermediate state between I quasilattice and b.c.c. lattice. The intermediate state is neither a quasiperiodic nor a periodic lattice. It possesses both the local quasiperiodicity and the local periodicity.

4. ELECTRON MICROSCOPY STUDY OF INTERMEDIATE STATES BETWEEN I QUASICRYSTALS AND (1/1) CUBIC APPROXIMANTS

The intermediate states between I quasicrystals and b.c.c. or almost b.c.c. crystals were found in some alloy systems such as Al-Cu-Mg [9,17], Al-Cu-Li [11] and Al-Mn-Si [18]. In these alloy systems the I phase usually coexists with a b.c.c. or an almost b.c.c. phase. The composition of the I phase is similar to that of the cubic phase. In addition, the structure of these cubic phases mainly consists of atomic clusters with I symmetry. Namely, the local structure of the I quasicrystal is similar to that of the corresponding cubic crystal((1/1) cubic approximant).

4.1. Al-Cu-Mg [9,17]

The electron diffraction observation indicates that the sample of Al-Cu-Mg prepared by rapidly solidifying a homogeneous melted alloy of

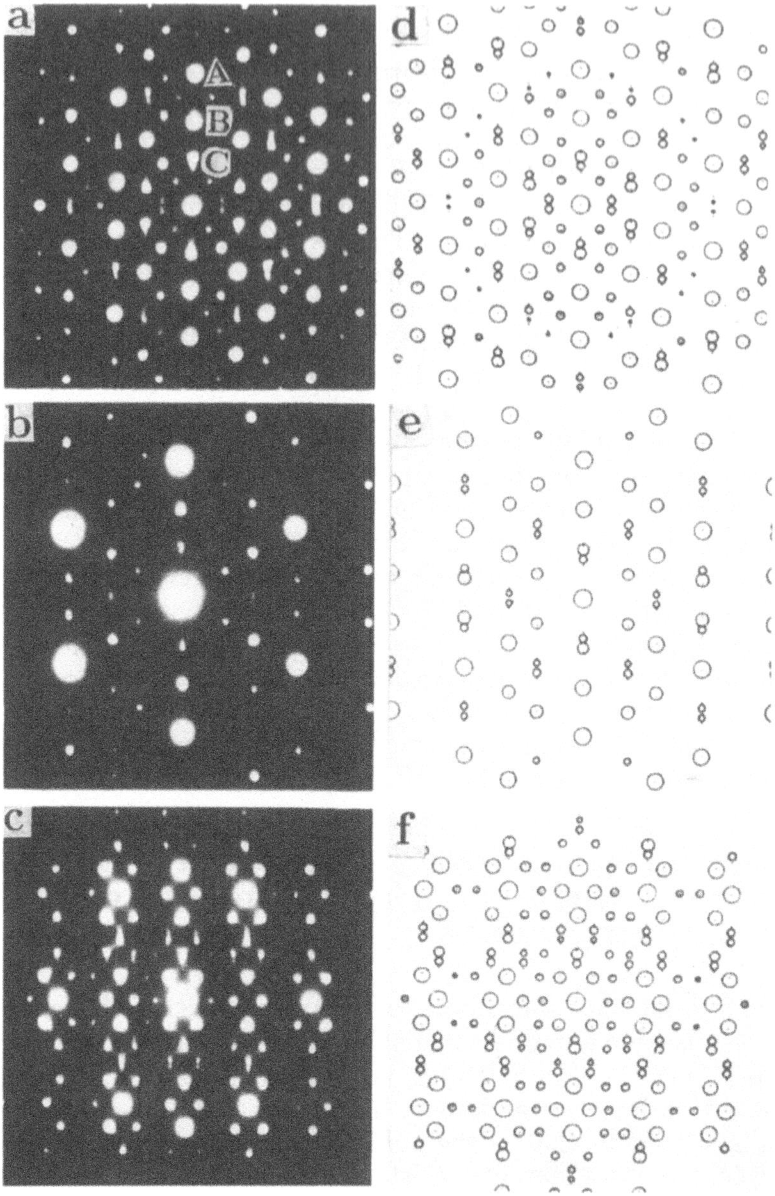

Figure 4. EDP's of an intermediate state between I phase and b.c.c. phases in Al-Cu-Mg alloy taken along (a) degenerated fivefold axis, (b) threefold and (c) twofold axes respectively. (e), (d) and (f) are distorted reciprocal quasilattice planes corresponding to (a), (b) and (c), respectively, they are calculated by introducing a special linear phason strain.

composition 47.6 wt% Al - 19.9 wt% Cu - 32.5 wt% Mg contains distorted I quasicrystalline grains.

Figures 4(a), 4(b) and 4(c) are electron diffraction patterns(EDP's) of a distorted quasicrystalline grain taken along the fivefold, threefold and twofold axes respectively. In figure 4(a) the fivefold symmetry is broken and degenerates to 2mm symmetry approximately with the vertical and horizontal center lines as lines of symmetry if some diffused reflections are ignored. It can be seen that all spots almost align periodically along the vertical direction owing to the shift of some diffraction spots, especially the shift of weak spots. Some adjacent spots displace toward each other and unite to form elongated spots. The distance between two adjacent spots along the vertical direction almost equal to the spacing of (200) plane in the crystalline phase of $Mg_{32}(Al, Cu)_{49}$ (R phase), while the distances between two adjacent vertical reciprocal lattice lines still obey the golden ratio. In figures 4(b) and 4(c), an approximate periodicity along the vertical direction and an elongation of diffraction spots similar to that shown in figure 4(a) can be found. The spacings between vertically aligned spot rows also obey the golden ratio. Hence, these diffraction patterns show the structural feature of both a crystal and a quasicrystal. They can be simulated by introducing a special linear phason strain with the phason matrix expressed as

$$M_1 = \begin{pmatrix} 0 & 0 & 0 \\ 0 & 0 & 0 \\ 0 & 0 & m \end{pmatrix}, \qquad (4-1)$$

where $m = -(\tau - a)/(\tau a + 1)$. The value of a can be determined from the ratio of distances between diffraction spots by the expression $a = d_{BC}/d_{AB}$, where d_{AB} denotes the distance between diffraction spots A and B shown in figure 4(a), and d_{BC} the distance between diffraction spots B and C. Obviously, when $a = \tau$, m becomes zero and the phason strain vanishes. Figures 4(d), 4(e), and 4(f) are calculated reciprocal-quasilattice planes with a=1.25, which correspond to figures 4(a), 4(b) and 4(c), respectively. The small dots denote the reciprocal-quasilattice nodes for the perfect I quasicrystal. The open circles represent the shifted reciprocal-quasilattice nodes under the action of linear phason strain. The radii of the circles are proportional to the weight of reciprocal-quasilattice nodes. Evidently, figures 4(d), 4(e) and 4(f) are in good agreement with figures 4(a), 4(b) and 4(c), respectively. The misfit seen from the shape of some diffraction spots can be interpreted that inside the observed area there are micro domains with phason strains slightly different one another. The size difference for some weak spots between the observed and simulated EDP's is acceptable, because here the weight of reciprocal lattice nodes are calculated only on the basis of a quasilattice, while the decoration of atoms is not considered.

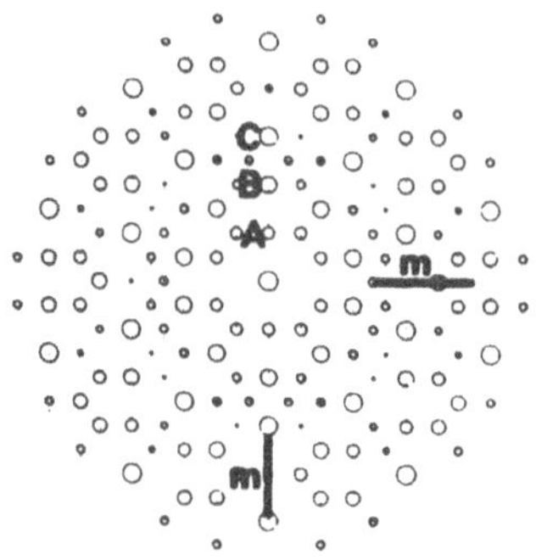

Figure 5. Calculated distorted reciprocal-quasilattice planes having 1D periodicity along the vertical.

Figure 5 shows a simulated EDP projected along the fivefold axis with a=1, namely with m=-0.236. It can be seen that all circles align along the vertical direction with a strict periodicity and there are two mirror planes lying along the vertical and horizontal, respectively. This means that this simulated EDP corresponds to a quasilattice with 1D periodicity, or in other words, corresponds to a 2D quasilattice or a one-dimensionally crystallized quasilattice. However, this kind of 2D quasicrystal is different from the decagonal phase which has periodicity along the tenfold axis, while in figure 5 the periodicity is not along the direction of the fivefold axis but along one of the twofold axes, and in fact there is no fivefold symmetry in this state. In this sense figure 4 indicates an intermediate state between the I quasicrystal and a 2D quasicrystal which itself is an intermediate state between I phase and R phase.

Figure 6 shows a high resolution electron microscope image of the same sample projected along the degenerated fivefold axis. The inserted EDP is similar to that shown in figure 4(c). Namely, all diffraction spots are almost aligned equidistantly along the vertical direction. Accordingly, the image reveal contrast bands, for instance those marked with dark arrows, running horizontally with the spacing equal to the spacing of

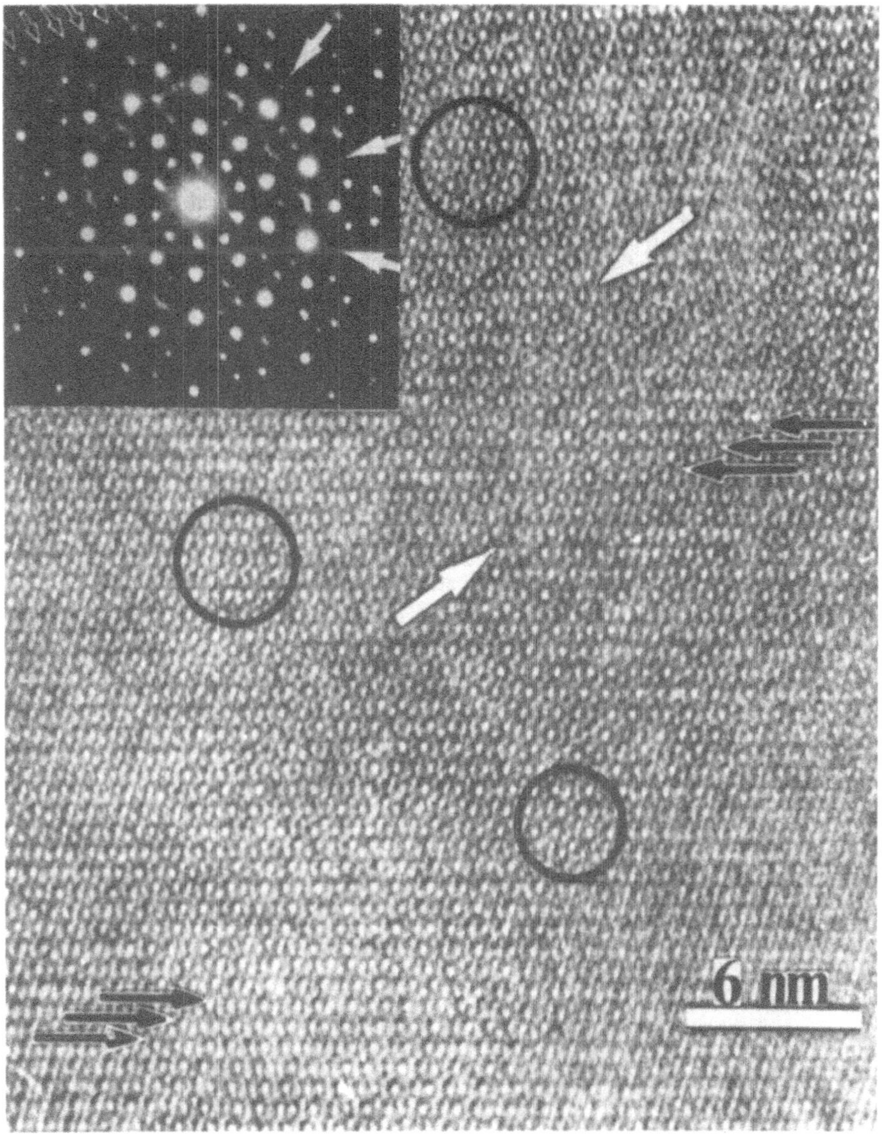

Figure 6. High resolution electron microscope image of phason defected quasicrystal in Al-Cu-Mg alloy. It shows both periodic and quasiperiodic structure features. Inset is the corresponding EDP.

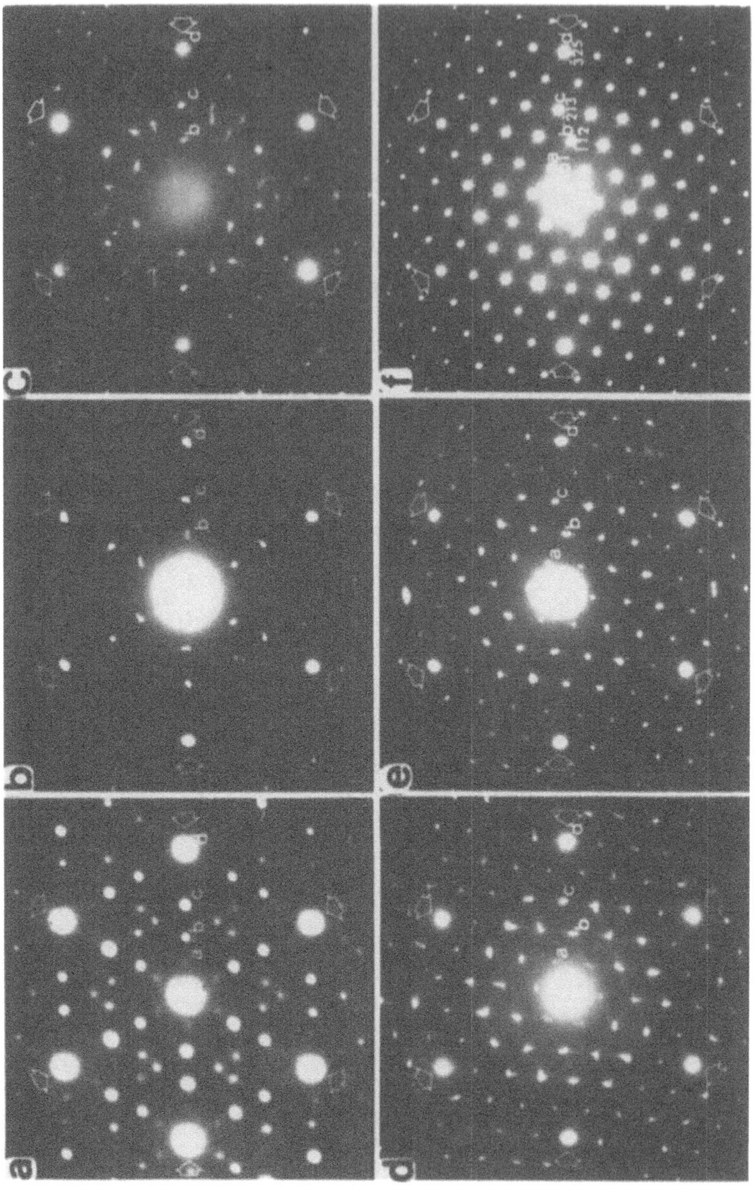

Figure 7. Series of EDP's of Al-Cu-Li alloy taken along the threefold axis showing the continuous transformation from I phase to b.c.c. phase.

(200) planes of R phase. Another feature of the inserted EDP is that all spots are located in parallel equispaced lines marked with small black arrows at the top left. The correspondence in the image is the existence of some equidistant white dot strings as pointed by the white arrows. In some regions of the image, for instance those marked by circles, the local fivefold symmetry can be seen. Hence, this image shows both the periodic and quasiperiodic structure features.

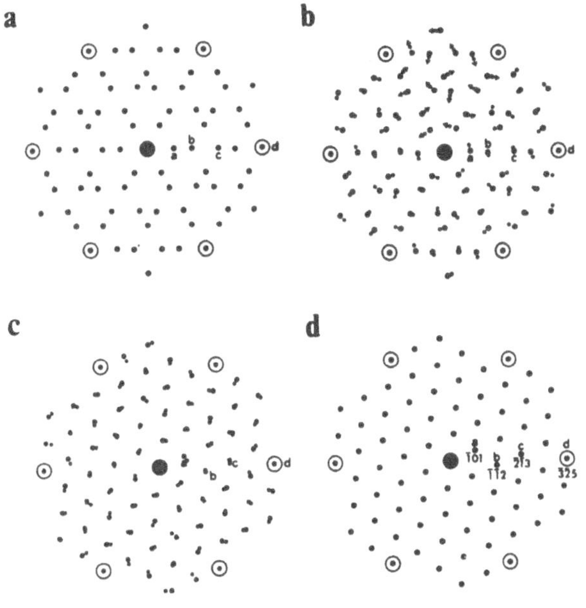

Figure 8. Calculated reciprocal-quasilattice planes for interpreting figure 7(see text).

4.2. Al-Cu-Li [11]

Samples of I phase (T2 – Al_6CuLi_3) were obtained by ingot casting, followed by 1 *hr* heating and quenching into water. Most areas of the sample showed the I symmetry, but in a very small area the EDP's seriously deviate from the I symmetry. Series of EDP's corresponding to different stages of the transition from T2 phase to b.c.c. phase ($R - Al_5CuLi_3$)) were obtained by moving the sample distances of around 10 μm.

Figure 7 shows a series of EDP's taken along the threefold axis. The deviation from I symmetry increases gradually from figure 7(a) to 7(f), and figure 7(f) is coincident with the EDP of [111] zone in R phase. Figure 8 shows the calculated reciprocal-quasilattice planes which are distorted by introducing a linear phason strain with the phason matrix expressed as

$$M_2 = \begin{pmatrix} m & 0 & 0 \\ 0 & m & 0 \\ 0 & 0 & m \end{pmatrix}. \qquad (4-2)$$

Figure 8(a) which fits to figure 7(a) is calculated with m=-0.15. Figure 8(b) gives two superimposed reciprocal-quasilattice planes with m=-0.02 and -0.15. The arrows indicate the shift direction of reciprocal-quasilattice nodes on further increase of the absolute value of parameter m. On comparing figure 8(b) with figures 7(b) and 7(c) a good fit can be seen. It is easy to realize that the comet shape of the diffraction spots is associated with the inhomogeneity of the phason strain inside the grain. Figure 8(c) which fits figures 7(d) and 7(e) is the superposition of two reciprocal-quasilattice planes calculated with m=-0.15 and -0.22. It is interesting that with the increase of the absolute value of m the arrangement of reciprocal-lattice nodes becomes close to the reciprocal-lattice plane corresponding to [111] zero-order zone of R phase. When m=-0.236, the former is coincident with the latter(figure7 (f) and figure 8(d)).

Figure 9 shows a series of EDP's taken with the same sample as figure 8 but along the twofold axis. Figure 10 gives the calculated reciprocal-quasilattice planes corresponding to figure 9. Figure 10(a) which is calculated for m=-0.01, fits figure 9(a). Figure 10(b), with m=-0.08, is similar to figures 9(b) and 9(c). In figure 10(c), with m=-0.20, the shifted reciprocal-quasilattice nodes form approximately three superimposed periodic reciprocal-lattice planes corresponding to the [123] zero-order zone and two first-order zones of R phase. This figure can be used to some extent for interpreting figure 9(d). Figure 10(d) shows a slightly inclined [123] zone of R phase and fits figure 9(e). This is in agreement with the orientational relationship between T2 phase and R phase [31,32]. Lattice nodes B, D, F, G and I belong to the zero-order zone, and nodes C and E belong to the first-order zone. It can be seen that the number of

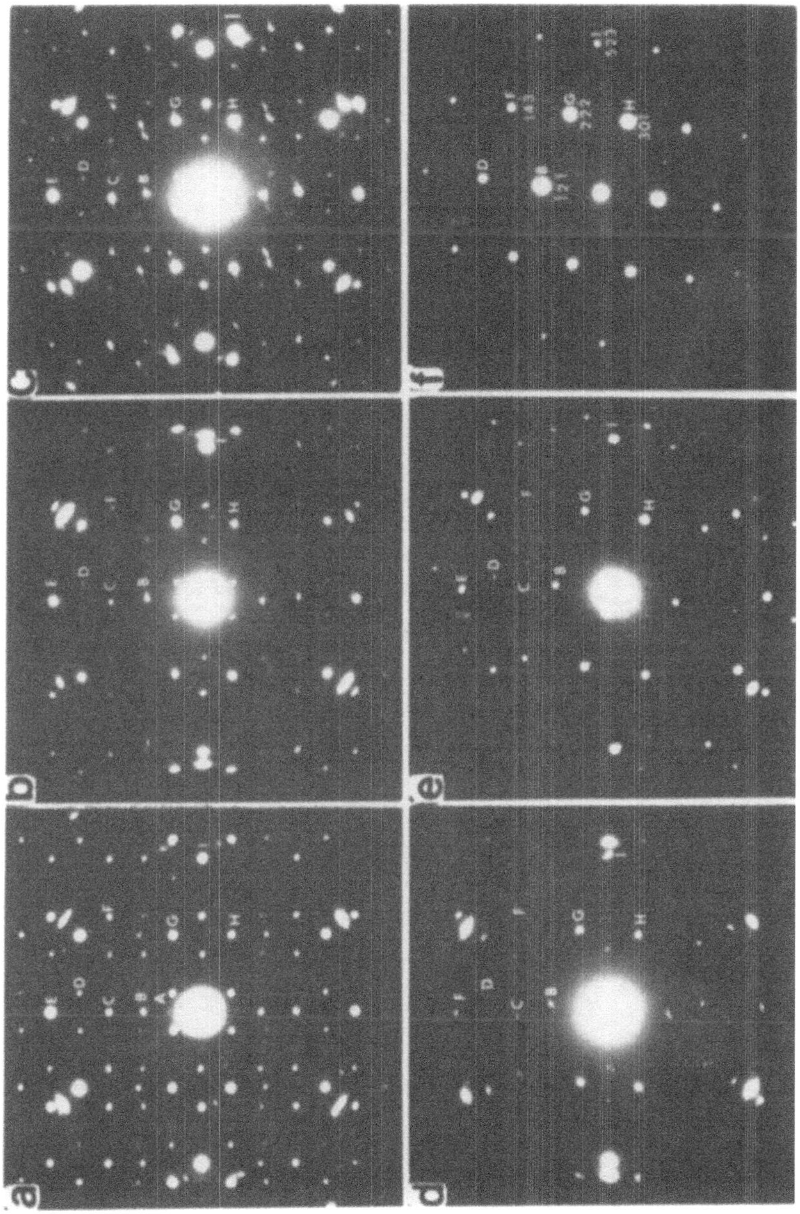

Figure 9. Series of EDP's of Al-Cu-Li alloy taken along the twofold axis showing the continuous transformation from I phase to b.c.c. phase..

diffraction spots belonging to the zero-order zone in figure 9(e) is larger than that of reciprocal-lattice nodes in figure 10(d). This is because the incident electron beam deviates slightly from the twofold axis of T2 phase such that it is in between the twofold axis of T2 phase and the [123] direction of R phase. A further adjustment of the orientation of the sample can make the incident beam parallel to the [123] direction of R phase and the EDP is shown in figure 9(f).

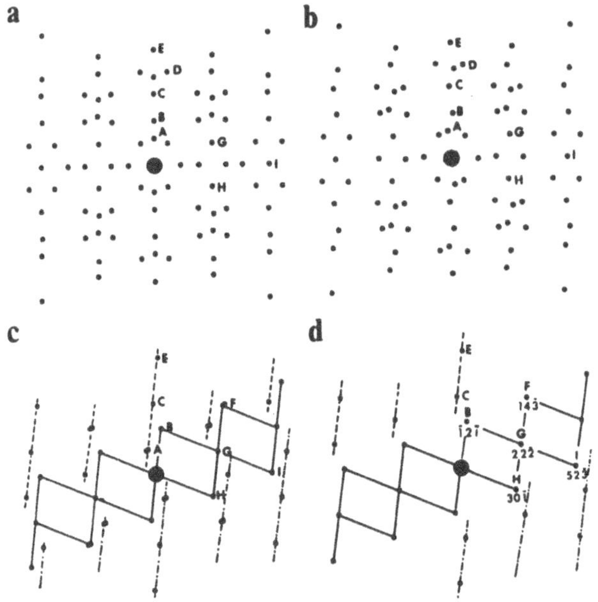

Figure 10. Calculated reciprocal-quasilattice planes for interpreting figure 9(see text).

The above experimental results indicate that the structure of T2 phase which is distorted by the linear phason strain corresponding to equation (4-2) may occupy a wide range continuously from the structure of perfect T2 phase to R phase. In this sense R phase can be treated as a phason defected T2 phase.

5. 6D CRYSTAL RELATED TO BOTH I QUASICRYSTAL AND ITS (1/1) CUBIC APPROXIMANT

5.1. Principle of constructing 6D crystal based on the b.c.c. crystal [33]

The discovery of intermediate states occupying a wide range between I quasicrystals and their (1/1) cubic approximants indicates the possibility to construct a HD crystal related to both the quasicrystal and its approximant. The section of the 6D crystal with the $(\tau/1)$ 3D hyperplane gives the I phase, and the section with the (1/1) 3D hyperplane gives the b.c.c. or almost b.c.c. phase.

The principle of deriving the 6D crystal is easier to be understood by the projection description. In this case the I phase and its (1/1) cubic approximant are obtained by projecting the 6D crystal with the $(\tau/1)$ and (1/1) projection strip respectively[34]. Matrix T given in equation (3-2) is also the projection matrix in the projection method for generating the 3DPT in $E_{\|}$ described by the Cartesian-coordinate system[29,30]. A b.c.c. lattice can also be generated by using the same projection matrix T if the $(\tau/1)$ projection strip is replaced by the (1/1) strip [34]. The relation between the lattice constant of the b.c.c. lattice and that of the 6D hypercubic lattice is given by equation (3-11). Therefore, the 6D structure and the b.c.c. structure can be connected by a matrix T':

$$T' = \frac{\sqrt{2+2\tau^2}}{2\tau^2}T, \qquad (5-1)$$

and

$$\begin{pmatrix} x_1 \\ x_2 \\ x_3 \\ x_4 \\ x_5 \\ x_6 \end{pmatrix} = \frac{1}{2\tau^2}\begin{pmatrix} \tau & 0 & 1 & -1 & 0 & \tau \\ \tau & 0 & -1 & -1 & 0 & -\tau \\ 0 & 1 & -\tau & 0 & \tau & 1 \\ -1 & \tau & 0 & -\tau & -1 & 0 \\ 0 & 1 & \tau & 0 & \tau & -1 \\ 1 & \tau & 0 & \tau & -1 & 0 \end{pmatrix}\begin{pmatrix} x_{\|} \\ y_{\|} \\ z_{\|} \\ x_{\perp} \\ y_{\perp} \\ z_{\perp} \end{pmatrix}. \qquad (5-2)$$

Here $x_{\|}, y_{\|}, z_{\|}$ and $x_1, x_2, x_3, x_4, x_5, x_6$ represent the atomic coordinates in the b.c.c. unit cell and the 6D unit cell, respectively; $x_{\perp}, y_{\perp}$ and $z_{\perp}$ can be recognized as atomic coordinates of a fictitious 3D cubic structure in $E_{\perp}$. The problem in deriving the 6D structure from a b.c.c. structure is in the multisolution of equation (5-2), because parameters $x_{\perp}, y_{\perp}$ and $z_{\perp}$ are arbitrary. Henley and Elser reported that in the rhombic tiling of a b.c.c.

structure, each unit cell is divided into eight prolate rhombohedra (PR) and six rhombic dodecahedra (RD) ,and each RD consists two PR and two OR, and that the coordinate parameters of all atoms in $(Al, Zn)_{49}Mg_{32}$ can be approximately expressed as three kinds of special positions in PR and OR [35]. These special positions are vertices, edge centers of the two kinds of rhombohedra, and positions on the long body diagonal of PR. Hence, it is recommended to define the fictitious 3D structure in $E_{\perp}$ such that all vertices and edge centers of the two kinds of rhombohedra are obtained by projecting the vertices and the edge centers in the 6D unit cell, respectively, onto $E_{\parallel}$, and all long body diagonal of PR are by projecting the 3D hyperplane diagonals in the 6D unit cell. Thus a single solution of the 6D structure can be obtained from equation (5-2). The 6D crystal structure derived for Al-Cu-Li alloy is described briefly in the following. The detailed procedure is written in reference [33].

5.2. 6D crystal related to T2 and R phases in Al-Cu-Li alloy[33]

The 6D structure is very simple. There are altogether twenty seven atoms per unit cell. One (Al,Cu) atom is located at the origin, six (Al,Cu) atoms are at edge centers and twenty Li atoms are on 3D hyperplane diagonals. Every two Li atoms divide a 3D hyperplane diagonal into three segments of length ratio $\tau^{-2} : \tau^{-3} : \tau^{-2}$. Among eighty 3D hyperplane diagonals inside the 6D unit cell only ten are occupied. They are parallel to directions [110001], [011001], [001101], [000111], [100011], [$\bar{1}$01100], [0$\bar{1}$0110], [10$\bar{1}$010], [110$\bar{1}$00], [0110$\bar{1}$0]. The other seventy 3D hyperplane diagonals are empty so that their projections in $E_{\parallel}$ — all body diagonals in the two kinds of rhombohedra except the long ones in PR — are empty. The number of independent atoms is three. The 6D space group is $Pm\bar{3}\bar{5}$.

5.3. I quasicrystal structure obtained by projecting the 6D structure

The problem of obtaining the I quasicrystal structure by projecting the 6D structure with the $(\tau/1)$ strip is in the determination of projection window. In general different kinds of atoms have different projection windows. Principally, the projection windows can be determined from the experimental diffraction data[36].

The structure of T2-Al_6CuLi_3 obtained by cutting the 6D structure with the projection window determined by the method of trial-and-error is identical to a decorated 3DPT which consists of two kinds of rhombohedra. (Al,Cu) atoms are located at vertices and edge centers of PR and OR. Li atoms are at long diagonals of PR. Each long diagonal of PR is divided by two Li atoms into segments of length ratio $\tau^{-2} : \tau^{-3} : \tau^{-2}$.

The structure model obtained is similar to that reported by Elswijk et al [37].

6. HREM IMAGE SIMULATION FOR PERFECT AND PHASON DEFECTED QUASICRYSTALS BASED ON HD CRYSTAL

There are two approaches to high resolution electron microscope (HREM) image simulation for quasicrystals. One is based on the decoration of atoms in the 3D space, In this case it is necessary to construct a fictitious crystal with a huge unit cell which is identical to a part of the quasicrystal. Another approach is based on the HD structure. The latter is more convenient for simulating images of phason defected quasicrystals. Evidently, the former approach leads to a tedious work, even if when the quasicrystal is perfect. For quasicrystals with phason strain it is much more difficult to construct the corresponding periodic 3D crystal. In the following it will be shown that the image simulation for both perfect and phason defected quasicrystals by means of the HD crystal is straightforward.

6.1. Image simulation under weak phase object approximation

In section 2 it is shown that the structure factor of a quasicrystal equals the structure factor of the corresponding HD crystal modulated by function $s(\mathbf{G}_\perp)$. When all HD atoms are of the same shape and size, the structure factors for perfect and phason defected quasicrystals are given in equations (2-20) and (2-22), respectively, and for HD atoms with different shape and size they are given in equations (2-20') and (2-22'), respectively. Principally, the HD hyper atoms are in the shape of a polyhedron of which the symmetry is consistent with that of the quasicrystal. For example, in order to obtain the 3D Penrose tiling [28], all HD lattice nodes should be in the shape of a unit triacontahedron. However, for simplicity the HD atoms can be approximated to a sphere or a spherical shell.

The image intensity for quasicrystals under the weak phase object approximation is expressed as

$$I = 1 - 2\sigma\,\mathcal{F}^{-1}[F(\xi,\eta)W(\xi,\eta)], \qquad (6-1)$$

where $W(\xi,\eta)$ denotes the contrast transfer function and $\mathcal{F}^{-1}$ is the operator of inverse Fourier transform. Equation (6-1) is the same as that for crystals except that here the parameters ξ and η are not integers.

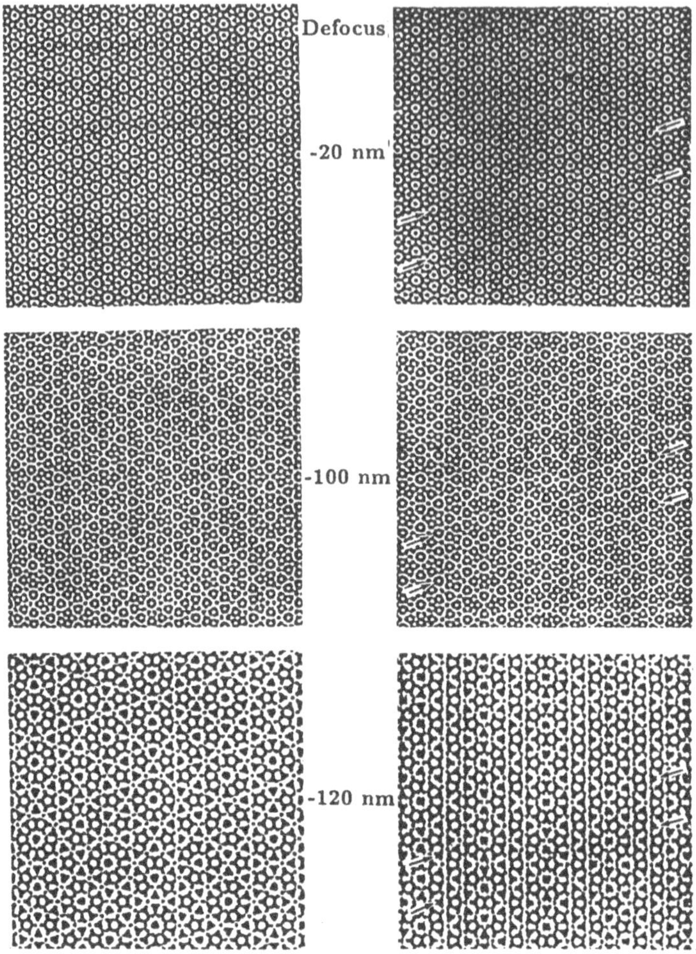

Figure 11. Simulated images of Al-Cu-Li icosahedral quasicrystal. The incident beam is parallel to the fivefold axis. The left column is for perfect quasicrystal and the right column is for phason defected quasicrystal with m=-0.1. Jags are pointed by arrows.

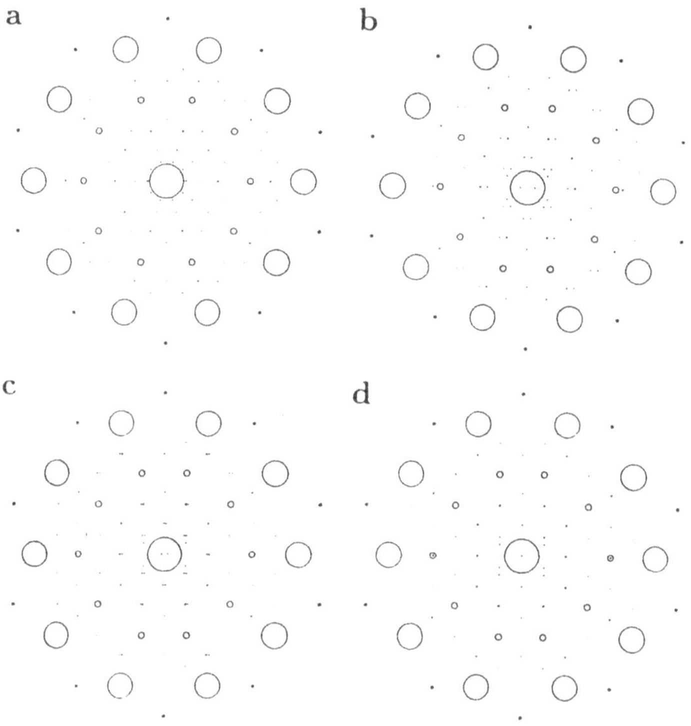

Figure 12. Calculated EDP's of Al-Cu-Li quasicrystal with different values of m. The incident beam is parallel to the fivefold axis.

6.2. Image simulation for Al-Cu-Li quasicrystal [39]

The image simulation is based on the above-mentioned six-dimensional crystal structure model which corresponds to both the icosahedral quasicrystal $T2 - Al_6CuLi_3$ and the b.c.c. crystal $R - Al_5CuLi_3$. The shape of all 6D atoms are approximated to be a sphere extended in the pseudo space. The radius of the sphere is in 30% smaller than the radius of the sphere of which the volume equals to that of the unit triacontahedron. This is because a good fit between the calculated and observed EDP's of T2-Al_6CuLi_3 was obtained by using such an approximation [33].

The electron-optical parameters are as follows: the accelerating voltage 400 KV, the spherical aberration coefficient 1.0 mm, and the radius of objective lens aperture $(0.55)\text{Å}^{-1}$. The phason matrix is the same as shown in equation (4-2).

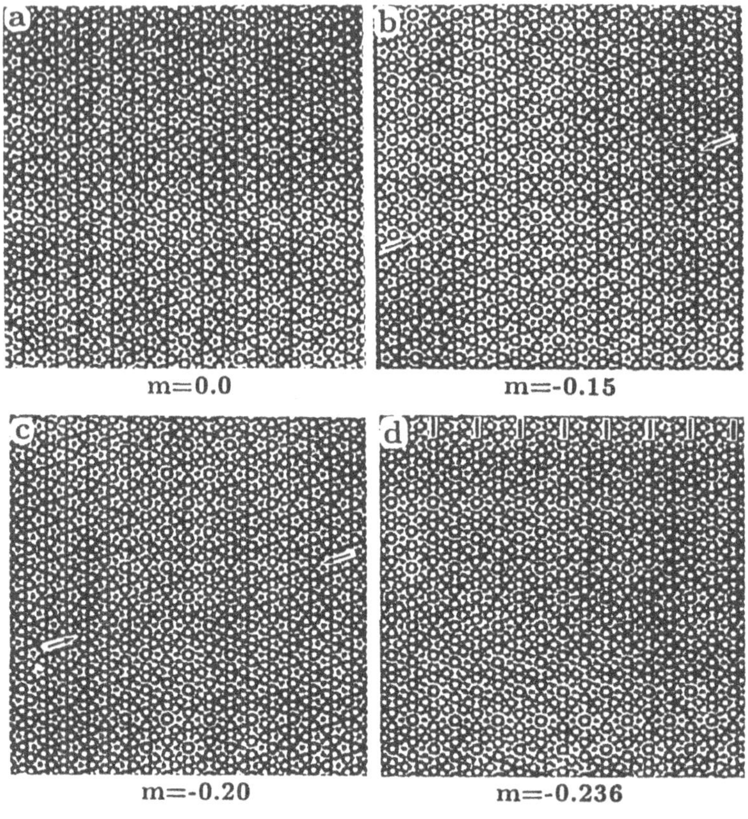

Figure 13. Simulated images of Al-Cu-Li quasicrystal corresponding to EDP's shown in figure 12(a) to 12(d) respectively. Jags in (b) and (c) are pointed by arrows and the 1D periodicity in (d) is indicated by bars.

Figure 11 shows the simulated images for a perfect Al-Cu-Li I quasicrystal projected along the fivefold axis in the left column and for a phason defected quasicrystal with $m = -0.1$ in the right column. The defocus value from the top to the bottom are 20 nm, -1000 nm and -120 nm, respectively. Although the image contrast change evidently with the change of the defocus value, all images in the left column show a perfect fivefold symmetry, while in the right column many jags appear as pointed by arrows, which indicates the degeneration of the fivefold symmetry. Such jags are frequently seen in the published experimental images.

Figure 12(a) to (d) are calculated electron diffraction pattern of Al-Cu-Li quasicrystal with the incident beam parallel to the fivefold axis for $m = 0$, $m = -0.15$, $m = -0.20$ and $m = -0.236$, respectively. The diffraction pattern in figure 12(a) shows a perfect fivefold symmetry. The diffraction spots, especially the weak spots displace along the horizontal gradually with the increase of m from figure 12(a) to (d) , and finally the 1D periodicity is formed in figure 12(d).

Figure 13(a) to (d) are simulated images corresponding to figure 12(a) to (d), respectively. The defocus value is 40 nm for all images, which is close to the Scherzer focus for the present case. The image in figure 13(a) shows a perfect fivefold symmetry. Straight continuous dark bands intersected one another in an angle of 72° can be seen. The distance between adjacent bands obey the golden ratio. When $m = -0.15$ the dark bands are broken and jags appear everywhere (figure 13(b)). But the vertical dark bands are remain to be continuous and some of them are located equidistantly. This indicates the formation of a local one-dimensional (1D) periodicity. With the further increase of m the local 1D periodicity become more clear (figure 13(c)). When $m = -0.236$, white dot strings arranged equidistantly along the vertical become more distinct than the vertical dark bands, and a perfect 1D periodicity along the horizontal as labeled by dark bars can be seen. This indicates that the quasicrystal is transformed into the b.c.c. crystal. The horizontal direction is parallel to one of the basis vectors of the b.c.c. phase. The image shows only the 1D periodicity, because the vertical is not coincident with any zone axis but parallel to the $[01\tau]$ direction of the b.c.c. phase.

Figure 14(a) is an experimental image of Al-Cu-Li quasicrystal taken with the JEM-4000EX electron microscope and the incident beam is parallel to the fivefold axis. It is obvious that the fivefold symmetry is degenerated. For instance, the figure at the center consists of a small white circle surrounded by ten white dots, and around it there are six

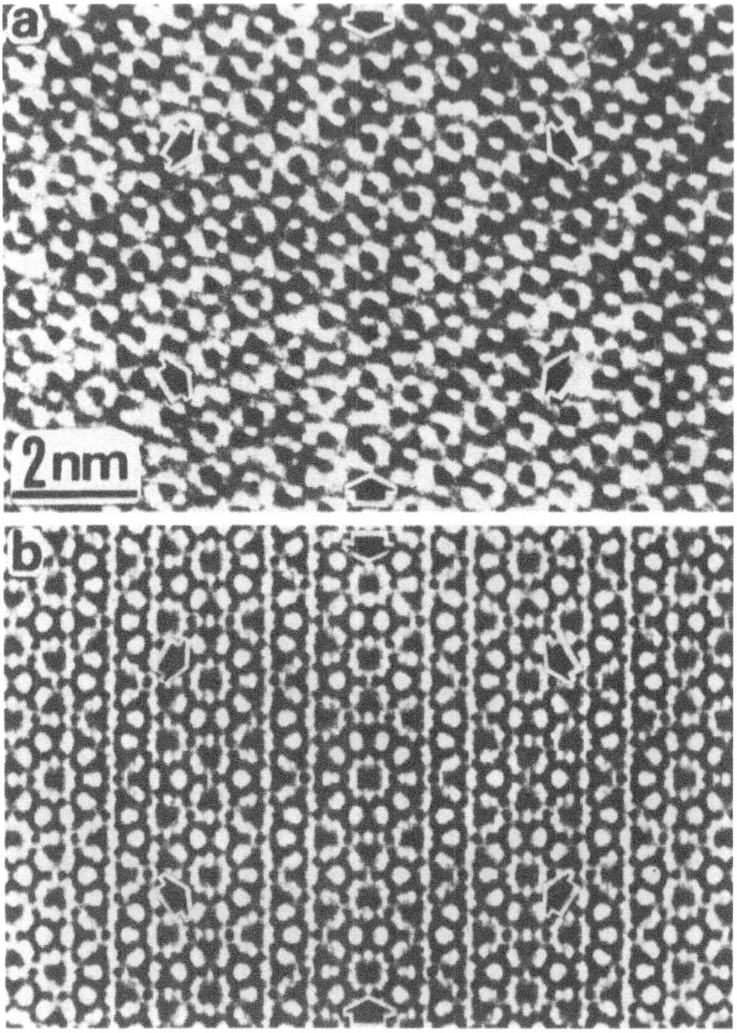

Figure 14. (a) Experimental and (b) simulated image of phason defected Al-Cu-Li quasicrystal. The incident beam is parallel to the degenerated fivefold axis. m=-0.10.

such figures as pointed by arrows instead of five. It is interesting that this image fits well to the simulated image with $m = -0.1$ and underfocus 120 nm(Fig. 3(b)). This can be seen more clearly by referring to the arrows in the two images. Therefore, the local phason strain in a sample can be identified from the HREM image simulation.

7. CONCLUSION

Quasicrystals are closely related with their crystalline approximants. The formers would transform to the latters under the action of a special linear phason strain which is equivalent to the rotation of sectional hyperplane in the cut description and equivalent to that of projection strip in the projection description. In this sense the crystalline approximant can be treated as a phason defected quasicrystal. The quasicrystal can be obtained by cutting a HD crystal with the irrationally orientated 3D hyperplane, and its crystalline approximant is obtained by cutting the same HD crystal with the rationally orientated hyperplane.

Electron diffraction and high resolution electron microscopy study indicates that intermediate states can occupy a wide arrange between the quasicrystal and its crystalline approximant. The structures of intermediate states possess both the structure features of quasicrystal as well as of crystalline approximant. The observation of continuous transformation from I phase to b.c.c. phase implies the close relation between the structure of quasicrystal and that of crystalline approximant. This leads to a new approach to quasicrystal structure determination, which has been proved to be effective.

It is exciting that an idealized HD crystal structure model can be constructed straightforward from the structure of crystalline approximant, which is usually known or easier to be determined than that of the quasicrystal. In such case the problem of quasicrystal structure determination is mainly in the determination of shape and size of HD hyperatoms in the cut description or in the determination of projection windows in the projection description. This can be solved by matching the experimental diffraction intensities with the calculated ones using method of trial and error.

The description of quasicrystal structure in HD space is very beneficial to diffraction pattern and high resolution electron microscope image intensity calculation, especially for quasicrystals with phason defects. The image simulation for phason defected quasicrystal in Al-Cu-Li alloy under the weak phase object approximation and the approximation of

HD atoms to a sphere gives a good fit to the experimental image. Hence, the linear phason strain can be determined averagely from the distortion of EDP's, and locally from high resolution electron microscope images by matching the simulated images based on the 6D structure with the observed images.

Obviously, the above arguments hold for all kinds of quasicrystals.

REFERENCES

1 D. Schechtman, I. Blech, D. Gratias and W. Cahn, Phys. Rev. Lett., 53 (1984) 1951.
2 M. Tanaka, M. Terauchi, K. Hiraga and Hirabayashi, Ultramicroscopy, 17 (1985) 279.
3 D. Schechtman and Blech, Metall. Trans., a16 (1985) 1005.
4 J. D. Budai, J. I. Tischler, A. Habenschuss, G. E. Ice and Elser, Phys. Rev. Lett., 58 (1987) 2304.
5 T. C. Lubensky, J. E. S. Socolar, P. J. Steinhardt, P. A. Bancel and P. A. Heiney, Phys. Rev. Lett., 57 (1986) 1440.
6 A. R. Kortan, H. S. Chen and J. W. Waszczak, J. Mater. Res., 2 (1987) 294.
7 J. E. S. Socolar and D. C. Wright, Phys. Rev. Lett., 59 (1987) 221.
8 Z. Zhang and K. H. Kuo, J. Microsc., 146 (1987) 221.
9 F. H. Li, C. M. Teng, Z. R. Huang, X. C. Chen and X. S. Chen, Phil. Mag. Lett., 2 (1988) 113.
10 Z. H. Mai, S. Z. Tao, B. S. Zhang and L. Z. Zeng, Phys. Rev. B, 38 (1988) 1291.
11 G. Z. Pan and F. H. Li, Proc. 3rd Asia-Pacific Physics Conf., Hong Kong (1988) 116; F.H.Li, G. Z. Pan, S. Z. Tao, M. J. Hui, Z. H. Mai, X. S. Chen and L. Y. Cai, Phil. Mag. B, 29 (1989) 535.
12 Z. H. Mai, S. Z. Tao, B. S. Zhang and L. Z. Zeng, J. Phys.: Condens. Matter, 1 (1989) 2465.
13 F. H. Li and Y. F. Cheng, Acta Cryst., A46 (1990) 142.
14 Z. H. Mai, L. Xu, N. Wang, K. H. Kuo, Z. C. Jin and G. Cheng, Phys. Rev. B, 40 (1989) 12183.
15 H. Zhang and K. H. Kuo, Phys. Rev. B, 41 (1990) 3482.
16 H. Zhang and K. H. Kuo, Phys. Rev. B, 42 (1990) 8907.
17 Z. R. Huang, F. H. Li, C. M. Teng, G. Z. Pan and X. S. Chen, J. Phys.: Condens. Matter, 3 (1991) 2231.
18 Z. R. Huang, K. K. Fung and Y. Q. Zhou, Phil. Mag. B, 63 (1991) 485.
19 Y. Ishi, Phil. Mag. Lett., 62 (1990) 393.

20 P. M. De Wolf, Acta Cryst., A30 (1974) 777.
21 P. Kramer and R. Neri, Acta Cryst. A40 (1984) 580.
22 A. Kalugin, A. Kitaev and L. Levitov, Pis'ma Zh. Eksp. Teor. Fiz., 41 (1985) 119.
23 M. Duneau and A. Katz, Phys. Rev. Lett., 54 2688.
24 V. Elser, Phys. Rev. B, 32 (1985) 4892.
25 V. Elser, Acta Cryst., A42 (1986) 36.
26 A. Yamamoto, and K. Hiraga, Phys. Rev. B, 37 (1988) 6207.
27 K. H. Ishihara, and A. Yamamoto, Acta Cryst., A44 (1988) 508.
28 R. Penrose, Bull. Inst. Math. Appl., 10 (1974) 266.
29 J. W. Cahn, D. Schechtman and D. Gratias, J. Mat. Res., 1 (1986) 13.
30 F. H. Li and L. C. Wang, J. Phys. C, 21 (1988) 495.
31 P. Wainfort, and B. Dubost, J. Phys., Paris, 47 (1986) C3-321.
32 A. Loiseau and G. Lapasset, J. Phys., Paris, 47 (1986) C3-331.
33 G. Z. Pan, Y. F. Cheng and F. H. Li, Phys. Rev. B, 41 (1990) 3401.
34 V. Elser and C. L. Henley, Phys. Rev. Lett., 55 (1985) 2883.
35 C. L. Henley & V. Elser, Phil. Mag.B, 53 (1986) L59.
36 H. F. Fan, Proc. Intern. Workshop on Methods of Structure Analysis of Incommensurate Modulated Structure and Quasicrystals, Bilbao, Spain (1991) 280.
37 M. B. Elswijk and Th. M. De Hosson, Phys. Rev. B, 38 (1988) 1681.
39 F. H. LI, G. Z. Pan H. Hashimoto and Y. Yokota, Ultramicroscopy, in press.

Crystal-Quasicrystal Transitions
M.J. Yacamán and M. Torres (Editors)

Phason defect and phase transition

Zhen-hong MAI   Lu XU

Institute of Physics, Chinese Academy of Sciences, P.O. Box 603, 100080 Beijing, P.R. China

CONTENTS

1. INTRODUCTION

The pioneering work of Shechtman on rapidly quenched Al-Mn alloy[1] opened the door to "forbidden symmetries" in condensed matter physics. Later on it was considered as an example of a new class of materials, "quasicrystalline" phase of solid matter. This discovery has stimulated a flurry of both experimental and theoretical activity. Shortly afterwards other quasiperiodic structures with decagonal orientational order were found [2,3] and there are claims that octagonal[4] and dodecagonal[5] phases also exist. Quasicrystals are solids with long-range quasiperiodic translational order and long-range orientational order. Therefore quasicrystals may exhibit any orientational symmetry including those that are strictly disallowed for periodic crystals. These new phases require the condensed matter physicists to understand a new type of atomic order both on a local and global level[6].

Most of the early investigation focused on the ideal icosahedral quasicrystals. Their diffraction patterns are successfully predicted by the Penrose model[7]. But more detailed analysis of the Bragg peaks in the electron and x-ray diffraction patterns shows , in many cases, distortions in contrast to the perfect diffraction patterns predicted by the Penrose model. a more important sign of disorder is revealed in the x-ray diffraction patterns from icosahedral quasicrystals[8-10]. They show that the peaks are not exact symmetrically arranged —small shifts in peak positions of diffraction spots appear undirectionally. Moreover, some apparent broadening of spots is observed. In some cases other anomalies such as arcs or rings of diffuse intensities are observed. This is evidence of imperfections of quasicrystal structures . We named those quasicrystals as imperfect quasicrystals. To explain the distortions of electron and x-ray diffraction pattern, defect structures [11], phason strains[12] and crystal-chemical model of atomic interactions[13] have been exploited. When we established the paradigm of "quasicrystallography" concerning the possible quasiperiodic structures, a question of whether there is any relation between the structures of quasicrystal and ordinary crystal rose up. Recently the phase transitions from quasicrystalline phase to ordinary crystalline one have been confirmed experimentally and theoretically[14,15]. Moreover, it has been testified that there exist intermediate states between quasicrystals and ordinary crystals. It is now clear that under certain conditions quasicrystals with higher order symmetry can be transformed into ordinary crystals with lower order symmetry.

Among the explanations of quasicrystal structure. Three theories have been advanced: the penrose, glass and random-tiling models. The penrose model suggests that quasicrystals are composed of two or more unit cell that fit together according to specific rules. This model is particularly useful for analyzing some of the basic properties of quasicrystals, such as atomic structure, stoichiometry, diffraction properties. But it has difficulty explaining how these rules might be related to atomic growth processes. The glass model, in contrast, relies on local interactions to join clusters of atoms in a somewhat random way. According to the model, all the clusters have the same orientation, but the structure contains many defects because of random growth. The random-tiling model, which combines some of the best features of its predecessors, considers that there are many nearly degenerate ways of packing structural units, so that a phase with maximum entropy is selected. In this model, the phason degrees of freedom which express fluctuations around an average icosahedral symmetry are expected to have the simple form of gradient-squared elasticity. This leads to Bragg peak in three dimensions and diffuse scattering due to the phason fluctuations.

The purpose of this chapter is reviewing the recent results on the phason defect in quasicrystals and the phase transition from quasicrystals to ordinary crystals. A density-wave description which leads naturally to Landau theory for quasicrystal is handled concisely in Section 2. The distortions in x-ray and electron diffraction patterns observed experimentally in terms of frozen phason strains are presented in Section 3. also the concept of imperfect quasicrystal is introduced. Finally we discuss the phase transformations from quasicrystalline phase to ordinary crystalline phase in Section 4.

2. FREE ENERGY AND PHASON MODE

The problem which we are confronted with when we want to explain the experimental information is to understand the structure of quasicrystal. In periodic crystals the structure is completely specified when both the unit cell and the positions of atoms in this unit cell are determined. But quasiperiodic crystals have actually hidden translational order which can be recovered if the structure is properly specified in a higher dimensional space. For example, icosahedral quasicrystals cannot have 3D periodicity but there are 6D cubic Bravais lattices accepting these symmetries. Recovering periodic schemes at the expense of higher dimensionality allows the tools of crystallography to be used. Numerous techniques have been discussed in the literature for analyzing quasicrystal structure. The approach we used in this section is the phenomenological explanation from the phason strain point of view based on Landau theory.

2.1 Projection method

The construction of quasicrystal structure can be described as the projection of higher dimensional lattice structure[16]. In a higher dimensional space D, a lattice **R** is represented by the density

$$\rho(\mathbf{x}) = \sum_{\mathbf{R}} \delta(\mathbf{x} - \mathbf{R}) \tag{1}$$

where

$$\mathbf{R} = \sum_{i=1}^{N} n_i \mathbf{e}_i \tag{2}$$

$\mathbf{e}_i$ are the mutually orthogonal basic vectors, n_i integers and N the number of dimensions of D. It must be invariant under a higher dimensional reducible representation Γ of the quasicrystal symmetry group. Two mutually orthogonal subspaces of D, $D^{\|}$ and $D^{\perp}$ are constructed. Let $P^{\|}$ and $P^{\perp}$ be two projection operators from D into $D^{\|}$ and $D^{\perp}$ respectively. Therefore , the projections of the basic vectors $\mathbf{e}_i$ in D into $D^{\|}$ and $D^{\perp}$ are $\mathbf{e}_i^{\|} \equiv P^{\|}\mathbf{e}_i$, and $\mathbf{e}_i^{\perp} \equiv P^{\perp}\mathbf{e}_i$. these two sets of vectors $\mathbf{e}_i^{\|}$ and $\mathbf{e}_i^{\perp}$ transform among themselves under two different irreducible representations of the quasicrystal symmetry group, Γ^+ and Γ^- respectively. The Γ^+ is the representation of the real physical space in which the quasicrystal phase exist. The $D^{\|}$ space is often called as a physical space, and the $D^{\perp}$ orthogonal to the physical space is called as a perpendicular space.

However, the sublattice $\mathbf{R}^{\|} = \sum_{i=1}^{N} n_i \mathbf{e}_i^{\|}$ is not the real quasicrystal lattice, one can define a window function $W(\mathbf{x}^{\perp})$

$$W(\mathbf{x}^{\perp}) = \begin{cases} 1, & \textbf{if } \mathbf{x}^{\perp} \in \Omega; \\ 0, & \textbf{otherwise.} \end{cases} \tag{3}$$

where Ω is a volume (or area) in perpendicular space $D^{\perp}$. In general, it is a projection of a hypercubic unit cell in D into $D^{\perp}$. The quasicrystal lattice are represented by the density

$$\rho(\mathbf{x}^{\|}) = \sum_{\mathbf{R}} W(\mathbf{R}^{\perp})\delta(\mathbf{x}^{\|} - \mathbf{R}^{\|}). \tag{4}$$

Similarly, reciprocal lattices $\mathbf{G}^{\|}$ and $\mathbf{G}^{\perp}$, corresponding to lattices $\mathbf{R}^{\|}$ and $\mathbf{R}^{\perp}$ respectively, can be obtained by the projection method.

2.2 Phason variable

Considering a liquid with fully translational and rotational symmetries, this liquid may condense in a possible ordered structure at low temperature. According to the theory of Landau and Lifshitz[17], the condensed phase is described by order parameters which transform as an irreducible representation of the symmetry group of the liquid phase. In the density-wave description for an ordered structure, the mass density $\rho(\mathbf{r})$ can be expanded in a Fourier series

$$\rho(\mathbf{r}) = \sum_{\mathbf{G}\in L_R} \rho_G e^{i2\pi\mathbf{G}\cdot\mathbf{r}} \tag{5}$$

where L_R is the reciprocal lattice associated with the ordered phase, there is a minimal set containing N_R vectors $\mathbf{G}_i$ ($i = 1, \cdots, N_R$) from which L_R is constructed. Each ρ_G is a complex number with an amplitude $|\rho_G|$ and a phase ϕ_G. Since $\rho(\mathbf{r})$ is real, $\rho_G^*=\rho_{-G}$ and hence, $\phi_G=-\phi_{-G}$.

The Landau free energy F of the system can be expanded in a power series in $\rho(\mathbf{r})$. In the expansion of F for an ordered phase, the kth power of $\rho(\dot{\mathbf{r}})$ gives rise to terms of the form

$$\begin{aligned} F^{(k)} &= A_k \sum_{m_1,m_2,\cdots,m_k} \int d\mathbf{r}\rho_{\mathbf{G}_{m_1}}\rho_{\mathbf{G}_{m_2}}\cdots\rho_{\mathbf{G}_{m_k}} exp[\sum_{i=1}^{k}\mathbf{G}_{m_i}\cdot\mathbf{r}] \\ &= VA_k \sum_{\mathbf{G}_m\in l_R} \Delta(\sum_{m=1}^{k}\mathbf{G}_m) cos[\sum_{m=1}^{k}\phi_{\mathbf{G}_m}] \prod_{m=1}^{k} |\rho_{\mathbf{G}_m}| \end{aligned} \tag{6}$$

where V is the volume, the factor $\Delta(x) \equiv \delta_{x,0}$ guarantees that only scalar terms will contribute to the expression for the scalar quantity F. The coefficients A_k depend upon the detail of the physical system (atomic species, interatomic interactions, temperature, pressure, etc.). Moreover the stability of a variety of structures is dependent upon the A_k. However, it has been pointed out by Alexander and McTague[18] that the free energy can often be lowered further by combining density waves with eight vectors forming an octahedron or six vectors forming a tetrahedron, The former represents the reciprocal lattice of bcc structure, and the later that of fcc structure. Comparing the third-order term of free energy between the octahedron and icosahedron structure, they obtained the same conclusion above. Bak[19] discussed the free energy of the icosahedral structure by taking into account of the fifth-order term, which arises from combining sets of five vectors to form regular planar pentagons. He professed that the free energy of icosahedral structure can become favorable compared with the energy of the bcc structure under especial conditions of coefficients. The resulting structure has precisely the symmetry of diffraction spectra observed in icosahedral quasicrystals.

Minimization of the energy F accompanies with constraints among the ϕ_G's. These constraints leave unspecified N_R ϕ_G, which correspond to hydrodynamic variables. When these variables shift uniformly, F remains unchanged. In quasicrystalline phase, the rotational symmetry corresponds to a disallowed crystal symmetry and determines a unique ratio of incommensurate length scales which defines the quasicrystal structure. In the case of periodic crystals, N_R=d, the number of dimensions. In the case of quasicrystals, $N_R=n_i d$, where n_i is the number of relatively incommensurate lengths. For discussing the symmetry of quasicrystal structure, it is useful to consider a larger set of N phase ϕ_n. The phases ϕ_n transform among themselves under the higher dimensional reducible representation Γ of the quasicrystal. The $N - N_R$ phases are the linear combinations of these N_R phases. The N_R hydrodynamic degrees of freedom in phases can be parametrized[20] by two vectors, displacement **u** and phason variable **w**. The phases ϕ_n can be written as

$$\phi_n = \mathbf{G}_n^{\parallel} \cdot \mathbf{u} + \mathbf{G}_n^{\perp} \cdot \mathbf{w} + \phi_n^{\circ} \tag{7}$$

where **u** and **w** would transform under different irreducible representations Γ^+ and Γ^- respectively, as similar with $\mathbf{R}_n^{\parallel}$ and $\mathbf{R}_n^{\perp}$. It can be seen that the higher dimensional displacement $\tilde{\mathbf{u}}$ may be written as $\tilde{\mathbf{u}} = \mathbf{u} \oplus \mathbf{w}$.

2.3 Phason-induced elastic free energy

Note that the uniform **u** and **w** do not change the free energy of the quasicrystal whereas the nonuniform **u** and **w** do. Resembling the case of periodic crystals, the nonuniform **u** represents modulations of positions of all mass, while **w**, similar to incommensurate crystals[21], describes relative rearrangements of atoms. As will discuss in Section 2.4, the **w** causes rearrangements of tilings, leading to break the symmetry of ideal quasicrystals.

If an analytic expansion of the free energy is possible[22], its first term at long wavelength will be quadratic in the spatial gradients of **u** and **w**. The elastic free energy will have the form

$$F_{el} = 1/2 \int d^d\mathbf{r}[K_{ijkl}^{uu}\mathbf{u}_{ij}\mathbf{u}_{kl} + K_{ijkl}^{ww}\nabla_i\mathbf{w}_j\nabla_k\mathbf{w}_l + 2K_{ijkl}^{uw}\mathbf{u}_{ij}\nabla_k\mathbf{w}_l]$$

$$\equiv \int d^d\mathbf{r} f_{el}(\nabla_i\tilde{\mathbf{u}}_\alpha) = F_{\mathbf{uu}} + F_{\mathbf{ww}} + F_{\mathbf{uw}} \tag{8}$$

where i, j, k, l, α are indices of Cartesion coordinates, K_{ijkl}^{uu}, K_{ijkl}^{ww}, and K_{ijkl}^{uw} are generalized elastic constant tensors , and f_{el} is the elastic free energy density. $F_{\mathbf{uu}}$, $F_{\mathbf{ww}}$ and $F_{\mathbf{uw}}$ are the phonon, phason and phonon-phason coupling elastic energies respectively. The $\mathbf{u}_{ij}$ represent the phonon strains, while $\nabla_i\mathbf{w}_j$ the phason strains. Including pure rotations of the system, $\mathbf{u}_{ij}$ could be written as $\mathbf{u}_{ij}=\nabla_i\mathbf{u}_j$. It is clear that ∇_i and u_j transform among themselves under the representation Γ^+ respectively, and $\mathbf{w}_j$ transform under the representation Γ^-. Therefore, one can decomposed the $\mathbf{u}_{ij}$ and $\mathbf{w}_{ij}$ into different types of irreducible strains, each which transforms under a irreducible representation respectively.

F_{el} should be invariant under all of the point group operations of the system. This implies that each individual term in F_{el} must transform under the identity representation of the point group. By these constraints, several independent elements of the elastic tensor exist and be called as elastic constants. These elastic constants can be determined by the identity representations of the direct product representations of the irreducible strains, each one represents an independent elastic constant. Furthermore, the explicit form for the elastic free energy are determined by taking into account of the precise constituent of the irreducible phonon and phason trains.

Another approach to deal with the phason elastic energy of quasicrystal was suggested by Ishii[23]. Expecting a phase transition from quasicrystal phase to a lower symmetry, modulated quasicrystal phase which has a (maximal) subgroup g′ of the quasicrystal point group g. By reducing the phason field to the irreducible representations of g′, the terms of the elastic free energy in Eq.(8) will be modified to transform as the identity representation of g′. The nonvanishing irreducible phason mode can be given by the bases of the identity representation of g′. Since the phason stress is intrinsic to the system of quasicrystal, the symmetry breaking induced by the phason perturbation locked in the nonvanishing irreducible mode corresponding to lower symmetry g′ is spontaneous. Ishii has deduced possible structural modulations induced by the phason degree of freedom for icosahedral and decagonal phases.

By energetic analysis of quasicrystal system using group theory[24] , one can see that unlike the case of ordinary periodic crystal, there exist phason variable **w** in quasicrystal phase, the spatial gradient of **w**, $\nabla_i \mathbf{w}_j$, will result in the elastic free energy in the system. Furthermore, with the help of group theory, the phason mode lock-in will break the system into lower symmetric phase. It has been realized experimentally.

2.4 Phason-effect on quasicrystal

To describe the mass density wave modified by phason field, a linear (uniform) phason strain field is used

$$\mathbf{w}_j = x_i M_{ij} \tag{9}$$

where x_i are Cartesion components of position **x**, $M_{ij} \equiv \nabla_i \mathbf{w}_j$ is a second-rank phason strain tensor. The density wave order parameters $\rho_{\mathbf{G}^{\parallel}}$ in Eq.(5) is modified as

$$\rho_{\mathbf{G}^{\parallel}} = \mid \rho_{\mathbf{G}^{\parallel}} \mid exp[i2\pi \mathbf{G}^{\perp} \cdot \mathbf{R}^{\parallel}\mathbf{M}] \tag{10}$$

The amplitude fluctuation is neglected in Eq. (10).

By placing a window at a position $\mathbf{y}=\mathbf{y}^{\perp} \in D^{\perp}$, the mass density of Eq.(4) is represented by

$$\begin{aligned}\rho(\mathbf{x}^{\parallel}) &= \int d\mathbf{x}^{\parallel} W(\mathbf{x}^{\perp} - \mathbf{y}^{\perp}) \\ &= \sum_{\mathbf{R}} \delta(\mathbf{x}^{\parallel} - \mathbf{R}^{\parallel}) W(\mathbf{R}^{\perp} - \mathbf{y}^{\perp})\end{aligned} \tag{11}$$

Then a series of Fourier transformations of Eq.(11) can be performed for $\mathbf{x}^{\parallel}$, $\mathbf{x}^{\perp}$ and $\mathbf{q}^{\perp}$ sequently. Finally, one obtains

$$\rho(\mathbf{q}^{\parallel}) = \sum_{\mathbf{G}} W^*(\mathbf{G}^{\perp})\delta(\mathbf{g}^{\parallel} - \mathbf{G}^{\parallel}) exp(-i2\pi \mathbf{G}^{\perp} \cdot \mathbf{y}^{\perp}) \tag{12}$$

where $W^*(\mathbf{G}^{\perp})$ is the Fourier transformation of $W(\mathbf{G}^{\perp})$ for $\mathbf{x}^{\perp}$. Comparing Eq.(12) with (10), it is clear that

$$\mathbf{y}^{\perp} = -\mathbf{R}^{\parallel}\mathbf{M} \in D^{\perp} \tag{13}$$

Therefore the density function $\rho(\mathbf{x}^{\parallel})$ in reciprocal space and $\rho(\mathbf{q}^{\parallel})$ in the real space are modified as

$$\rho(\mathbf{x}^{\parallel}) = \sum_{\mathbf{R}} \delta(\mathbf{x}^{\parallel} - \mathbf{R}^{\parallel}) W(\mathbf{R}^{\perp} + \mathbf{R}^{\parallel}\mathbf{M}) \tag{14}$$

$$\rho(\mathbf{q}^{\parallel}) = \sum_{\mathbf{G}} \delta(\mathbf{g}^{\parallel} - (\mathbf{G}^{\parallel} + \mathbf{M}\mathbf{G}^{\perp})) W^{*}(\mathbf{G}^{\perp}) \tag{15}$$

From Eqs.(14) and (15), it is evident that a linear phason strain $\mathbf{w}$ shifts the Bragg peaks $\mathbf{G}^{\parallel}$ to that of $\mathbf{G}^{\parallel} + \Delta\mathbf{G}^{\parallel}$, where $\Delta\mathbf{G}^{\parallel} = \mathbf{M}\mathbf{G}^{\perp}$. That is, the shift of Bragg peaks is controlled by the vector $\mathbf{G}^{\perp}$: (1) the magnitude of the shifts will be proportional to $\mid \mathbf{G}^{\perp} \mid$; (2) since the intensity of Bragg peak decreases roughly with increasing $\mid \mathbf{G}^{\perp} \mid$, the faintest Bragg peaks will be shift the most. If there is also nonlinear phason strain, there will be peak broadening. As a result, phason strain will spoil the symmetry properties of diffraction patterns of ideal quasicrystals. Also the atoms in real space will be rearranged.

3. PHASON DEFECT

Since the discovery of quasicrystal, a lot of work have been devoted to the study of their structure in both experimental and theoretical ways. Later on distortions in the electron diffraction patterns from several icosahedral phase alloys such as Al-Mn, Al-Cu-Li, Al-Mn-Si have been reported by many authors. Lubensky and co-workers[12] argue that this class of distortions indicates the presence of anisotropic strains of the phason variable in the icosahedral phase. Such strains also contribute to the x-ray peak widths and line sharps. Socolar and Wright[25] gave out a formula of the phason field in one domain and calculated the average diffraction intensities of all domains in a specimen, which agrees with the experiment done by Huang et al.[26]. Mai and co-workers investigated the distortions in both x-ray diffraction precession patterns from an Al-Li-Cu large single quasicrystal[9] and electron diffraction patterns from Cr-Ni-Si alloy[27] and derive a detailed expression of the phason strain for explaining the distortions of diffraction patterns. This indicates that the study of quasicrystals has entered a new stage of imperfect quasicrystals.

3.1 Perfect icosahedral quasicrystal

In 1952, Frank[28] pointed out that some metals with moderately complex crystal structures can be supercooled some tens of degrees below the melting point and can exhibit an icosahedral atomic arrangement. In 1983, by extending the concept of hexatic bond-orientational order (BOO) to 3D, Steinhardt et al. reported that under some circumstances glasses might show icosahedral BOO[29]. In 1984, Shechtman et al.[1] reported firstly an Al-Mn supercooled alloy produced electron diffraction patterns with "sharp" spots and exhibits the icosahedral point symmetry (Fig. 1). As is well known, the sharpness of the Bragg peaks means long rang ordering of the atoms, while the icosahedral symmetry is incompatible with any periodic order. The results of Shechtman can be understood in several theoretical contexts. By extending the 2D penrose tiling to 3D space- filling structures, the quasicrystal structure with perfect icosahedral symmetry consists microscopically of two unit cells which are arranged nonperiodically but according to definite mathematical rules. Remarkably, the diffraction pattern calculated from this structure predicts sharp peaks and shows good agreement with the icosahedral diffraction patterns from quenched Al-Mn samples. Bak [19], Levine

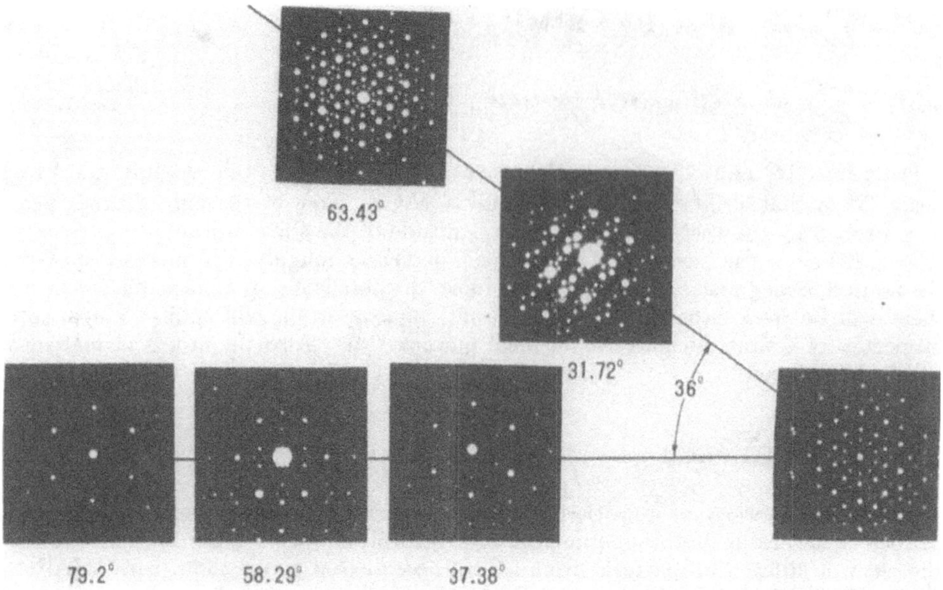

Fig.1 Electron diffraction pattern of icosahedral phase in Al-Mn alloy (Courtesy D.Shechtman).

et al.[20] and Mermin[30] presented phenomenological models based on Landau theory with icosahedral point symmetry and quasiperiodic or incommensurate translational order respectively. Now symmetry behavior of perfect quasicrystals are fully understood.

3.2 Imperfect quasicrystal

Tanaka and co-workers[31] firstly justified that the intensity distribution in convergent-beam electron diffraction (CBED) patterns obtained from icosahedral quasicrystals of a melt-quenched Al-Mn alloy deviates in geometry from the perfect icosahedral symmetry. One can see that reflections in Fig. 2 are deviated in a zig-zag way from the radial line, especially in the directions h and k. In the other directions the deviation is much smaller.

Similar deviation of diffraction patterns were also observed by x-ray diffraction methods. High-resolution x-ray scattering measurements of the icosahedral phase[32] show somewhat broadening and shift of peaks which does not increase uniformly with scattering angle.

To further investigate the distortion patterns, one of the intensive work is to obtain large-sized single quasicrystals which are of major interest in understanding of their structures and physical properties. A stable ternary compound in the Al-Li-Cu system, formerly designated as $T_2 - Al_6Li_3$Cu by Hardy and Silcock[33], has recently been found to exhibit icosahedral quasicrystal structure prepared either by solid state

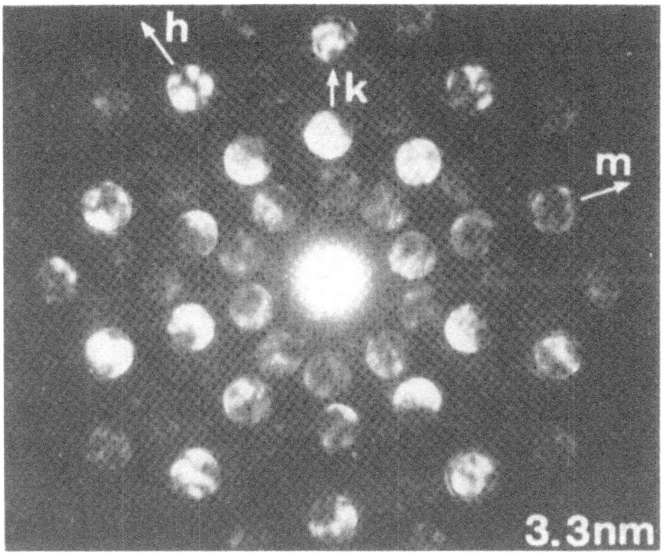

Fig.2 CBED pattern of the quasicrystal taken with the electron incidence parallel to a fivefold rotation axis from 3.3nm diameter areas (Courtesy M.Tanaka).

precipitation[34] or by conventional ingot casting[35]. Using x-ray diffraction precession photography Mai and co-workers[35,9] have observed distortions in the x-ray precession photographs diffracted from large $Al_6 Li_3$Cu single quasicrystals, one of the samples with the size of $4 \times 4 \times 0.25mm^3$. The x-ray precession photographs were taken along all of the fivefold, threefold and twofold axes within an experimental error $30'$ comparing with the axes of the ideal icosahedral symmetry. *Fig*.3 shows the stereographic projection of the symmetry element of the icosahedral group.

The symmetries of x-ray diffraction patterns along six fivefold axes are broken by a mirror and deviated from those of ideal icosahedral quasicrystal (Fig.4(a)). For instance: (1) In a perfect icosahedral quasicrystal(Fig.4(b)) the diffraction spots lie along straight lines, but in Fig. 4(a) they do not, except for those along A-A; (2) Some diffraction spots, for example, denoted c appear in Fig.4(b), but they disappear in Fig.4(a); (3) at the position B in Fig.4(b) there are two separate diffraction spots, but in Fig.4(a) there is only one.

For the ten threefold axes, four of them (marked A in Fig.3) contain the three fold symmetry (Fig.5(a)). And symmetry of diffraction patterns taken alone the rest threefold axes is broken by a mirror plane (Fig.5(c)).

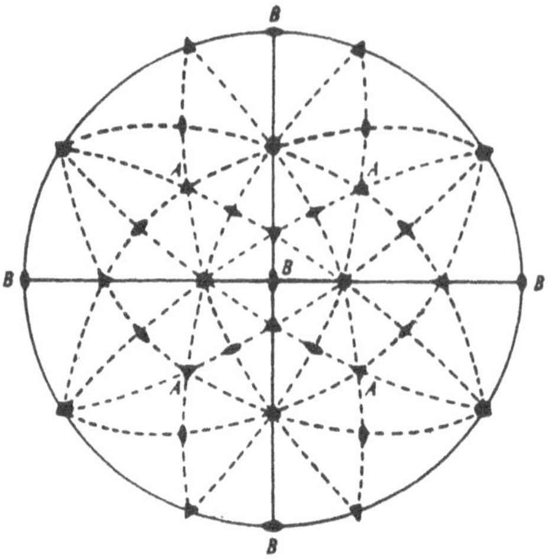

Fig.3 Stereographic projection of the symmetry element of the icosahedral group.

For the fifteen diffraction precession photographs taken along twofold axes, only three of them coposses a symmetry of mirror plane (Fig.6(a)). They are marked by B in Fig.3. Fig.6(c) is the diffraction pattern along the other twelve twofold axes. It is obvious that it is not of mirror symmetry.

The distortions of diffraction patterns are similar for all samples, although the magnitude of the distortion varies from sample to sample. Therefore we reasonably affirm that imperfect quasicrystals whose structures slightly deviate from those of ideal quasicrystals exist in the nature.

Edagawa[36] has investigated phason strains quenched through the solidification processes of $Al_{75}Si_6Mn_{21}$, $Al_{65}Cu_{20}Fe_{15}$ and $Al_{60}Li_{30}Cu_{10}$ quasicrystalline alloys by examining the shapes and widths of x-ray diffraction peaks and compared the experimental results with the theoretical models of linear phason. Most of the peaks for melt-spun Al-Si-Mn, Al-Cu-Fe and Al-Li-Cu have a shoulder or tail which can be attributed from linear phason strain (Fig.7(a)-(c)). For example, the fivefold (211111) peaks have symmetric shape and the narrowest width of all the peaks, and the threefold (111000) ones have a shoulder on the left-hand side. Regarding the twofold peaks, the (332002) ones are symmetric, (221001) have a tail on the left-hand side and (111100) have a tail on the right-hand side. The peaks for Bridgman-grown Al-Li-Cu (Fig.7(d)) are more symmetric as compared with these for melt-spun Al-Li-Cu, indicating that there is little linear phason strain in the Bridgman-grown sample.

By measuring the monotonic increases of the full-width at half-maximal (FWHM) of the peaks with $\mathbf{G}^{\perp}$, it suggests that the behavior of the dependences of FWHM of peaks for the melt-spun samples is regarded as the effect of the anisotropic linear phason. In contrast, only the effect of the anisotropic random phason is seen for Bridgman-grown Al-Li-Cu. These results are suggestive of strong correlation between the linear phason strain and the growth process.

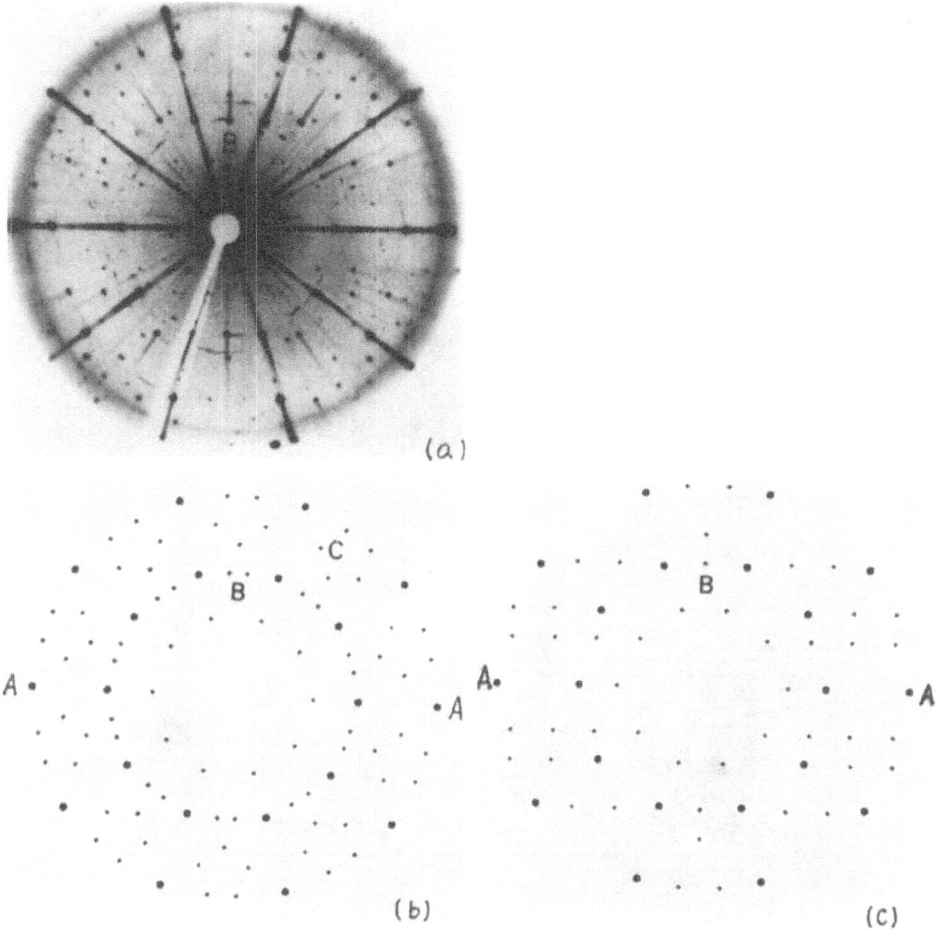

Fig.4 Zero-level x-ray diffraction precession photographs along the fivefold axis: (a) Experiment result, (b) Computer simulation pattern of perfect icosahedral phase and (c) Computer simulation pattern under an action of phason strain ($\alpha = 0.22$).

The distortions in the electron diffraction patterns of octagonal and decagonal quasicrystals have also been observed. Fig.8 shows the electron diffraction pattern of an octagonal quasicrystal in Mn-Fe-Si alloy [37] with a broken eightfold symmetry. The weak spots along the 45° and 135° directions are more or less periodic (shown by the arrows) displaying a pseudo-fourfold symmetry. As mentioned in Section 2.4, the lower the intensity, the greater the peak shift; this means that the shift is proportional to $|\mathbf{G}_\perp|$ for linear phason strains.

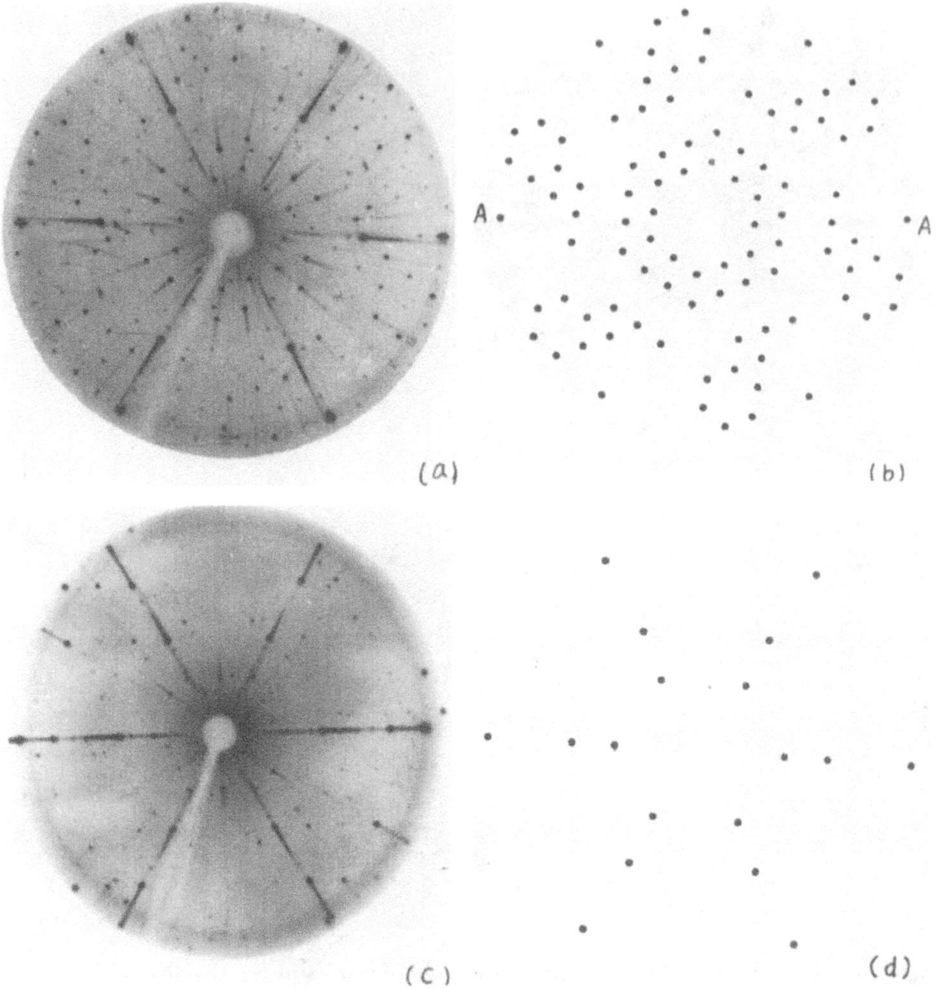

Fig.5 Zero-level x-ray diffraction precession photographs: (a) Experiment result along the threefold axis marked A in Fig.3, (b) Computer simulation along the threefold axis with(a) , (c) Experiment result along the other threefold axis in Fig.3 and (d) Computer simulation along the other threefold axis with (c).

3.3 Phason defect

As described in Section 2. the elastic energy of the icosahedral quasicrystals has terms coupling gradients in **u** to gradients in **w**. In principle, there should be shifts and broadening of diffraction peaks arising from the induced variation in **u** whose

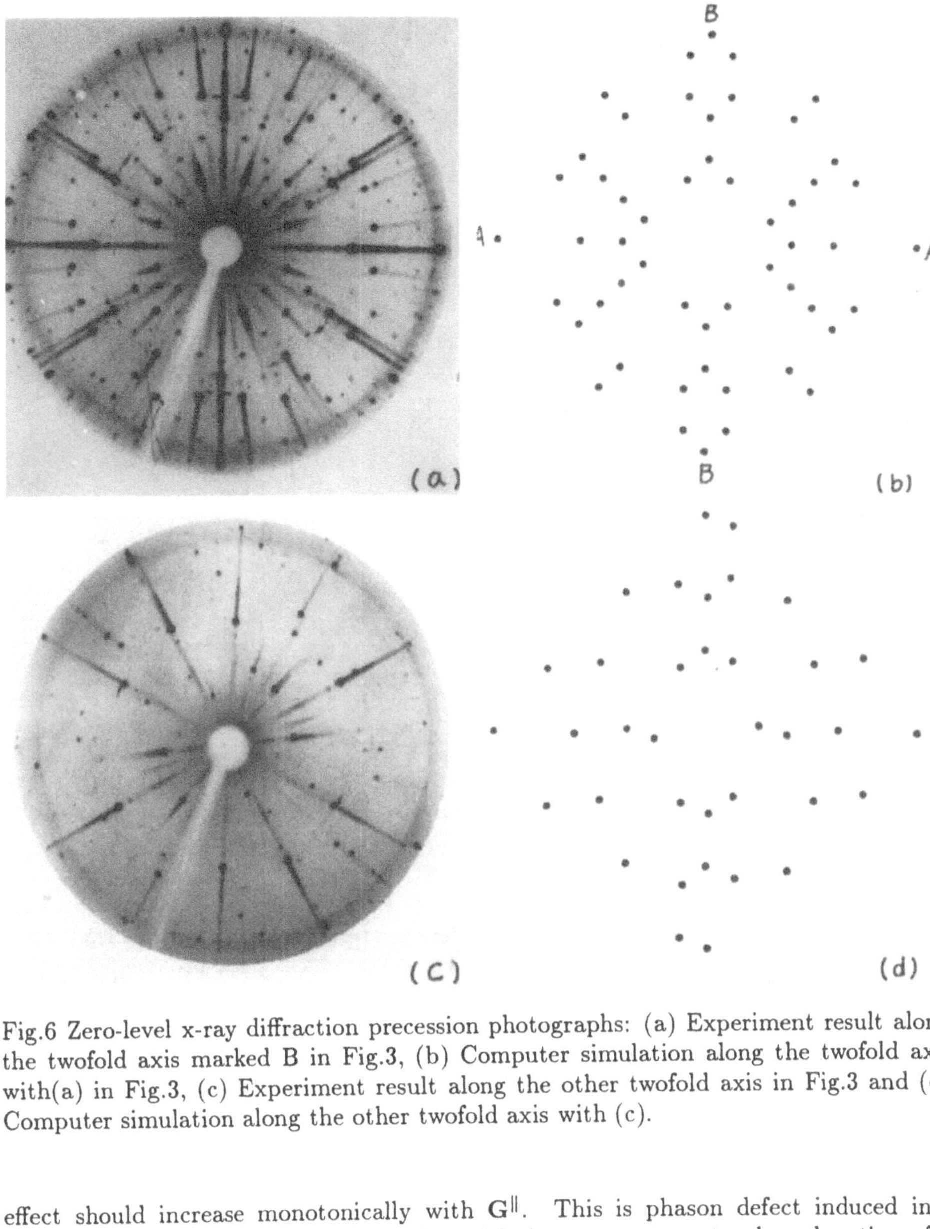

Fig.6 Zero-level x-ray diffraction precession photographs: (a) Experiment result along the twofold axis marked B in Fig.3, (b) Computer simulation along the twofold axis with(a) in Fig.3, (c) Experiment result along the other twofold axis in Fig.3 and (d) Computer simulation along the other twofold axis with (c).

effect should increase monotonically with $\mathbf{G}^{\|}$. This is phason defect induced into quasicrystal. Within the quasicrystal model there are two natural explanations for the finite peak widths found in the high-resolution x-ray diffraction measurements: (1) random isotropic quenched **w** strains (or dislocations), and (2) anisotropic quenched **w**

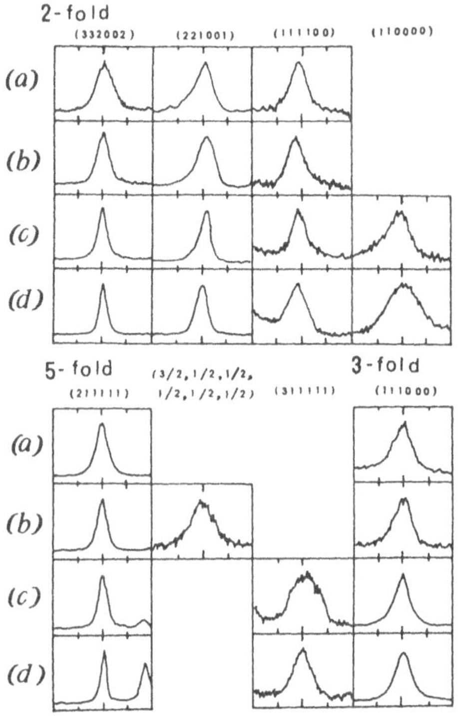

Fig.7 The twofold,threefold and fivefold peaks for melt-spun (a) Al-Si-Mn, (b) Al-Cu-Fe, (c) Al-Li-Cu and (d) Bridgman-grown Al-Li-Cu. The peaks are arranged in order of increasing **G** . The unit in the abscissa is 0.01Å (Courtesy K.Edagawa).

strains. Effect (1) produces an intrinsic broadening of every peak, and thus leads to a broadening of the x-ray diffraction peaks. Effect (2) produces peak broadening in two ways. First, to the extent that it produces an intrinsic anisotropic broadening of each peak. Secondly, anisotropic **w** strains shift peak anisotropically in different directions. We should also emphasize that any other effects producing quenched variations in **u**, such as variations in composition, can also produce $\mathbf{G}^{\perp}$-dependent shifts and broadening. Careful measurements of the line shape and the variation of peak width with $\mathbf{G}^{\perp}$ may lead to an understanding of the nature and source of **w** strains. We also note that

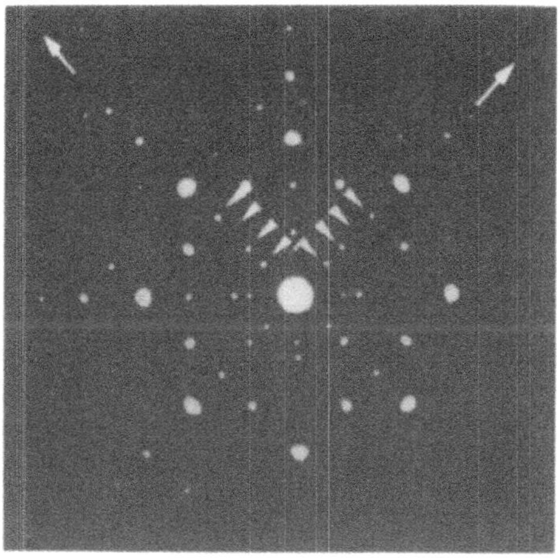

Fig.8 Electron diffraction pattern of a rapidly solidified Mn-Fe-Si alloy with a broken eightfold symmetry (Courtesy Z.M.Wang).

the effect (2) can lead to asymmetrical broadening of the peaks in the x-ray powder diffraction experiments.

The distortions of diffraction precession patterns in Section 3.2 can be explained using the soft linear phason mode. All the x-ray diffraction spots can be indexed following Elser's method[16] using six independent vectors directed to the vertices of an icosahedron:

$$e_1^{\|} = A\begin{pmatrix}0\\1\\\tau\end{pmatrix} \qquad e_2^{\|} = A\begin{pmatrix}0\\-1\\\tau\end{pmatrix} \qquad e_3^{\|} = A\begin{pmatrix}\tau\\0\\1\end{pmatrix}$$

$$e_4^{\|} = A\begin{pmatrix}1\\\tau\\0\end{pmatrix} \qquad e_5^{\|} = A\begin{pmatrix}-1\\\tau\\0\end{pmatrix} \qquad e_6^{\|} = A\begin{pmatrix}-\tau\\0\\1\end{pmatrix}$$

there $A = \frac{1}{(1+\tau^2)^{\frac{1}{2}}}$, and τ is the golden meaning, $\frac{1+\sqrt{5}}{2}$. A general Bragg vector is expressed as

$$\mathbf{G}^{\|} = \frac{\pi}{a}\sum_{i=1}^{6} n_i e_i^{\|} \tag{16}$$

where n_i are integers and might be thought of as the indices of diffraction spots, and a is a quasilattice parameter which is equal to 5.045Åin reference [9]. For each $\mathbf{G}^{\|}$, the associated vector $\mathbf{G}^{\perp}$ is defined

$$\mathbf{G}^{\perp} = \frac{\pi}{a}\sum_{i=1}^{6} n_i \mathbf{e}_i^{\parallel} \tag{17}$$

where

$$\mathbf{e}_1^{\perp} = -\mathbf{e}_1^{\parallel} \qquad \mathbf{e}_2^{\perp} = \mathbf{e}_2^{\parallel} \qquad \mathbf{e}_3^{\perp} = \mathbf{e}_4^{\parallel}$$

$$\mathbf{e}_4^{\perp} = \mathbf{e}_6^{\parallel} \qquad \mathbf{e}_5^{\perp} = \mathbf{e}_3^{\parallel} \qquad \mathbf{e}_6^{\perp} = \mathbf{e}_5^{\parallel}$$

According to the Section 2.4, in 3D icosahedral quasicrystals there are six broken-symmetry hydrodynamic variables, three of them are described as phonon modes $\mathbf{u}$, while the other three modes as phason modes, $\mathbf{w}$. If the spatial variations in $\mathbf{w}$ are approximately linear over a given microregion, then the phason field could be given by $\mathbf{w}(\mathbf{x}) = \mathbf{w}(\mathbf{0}) + \mathbf{x} \cdot \mathbf{M}$. Let the x, y, z axes correspond to the three twofold axes (marked B in Fig.3). In this case, the $\mathbf{M}$ is given by

$$\mathbf{M} = \alpha \begin{pmatrix} -1 & 0 & 0 \\ 0 & 0 & -1 \\ 0 & 1 & 0 \end{pmatrix} \tag{18}$$

where α is a constant dependent on the magnitude of the phason strain. Consequently it is possible to simulate the x-ray precession photography by computer. Figs.4(c), 5(b) and 5(d), 6(b) and 6(d) are the simulations of x-ray precession photographs taken along fivefold,threefold(A), threefold, twofold (B) and twofold axes respectively. For Fig.4(c) the x-ray intensity was conserved in the simulation; it is very satisfying to see that the simulations fit the experimental results well. Comparing Fig.4(a), 4(b) and 4(c), it is easy to understand the deviation of x-ray diffraction precession patterns of imperfect quasicrystals from those of ideal quasicrystal: (1) some diffraction spots, for example the spots denoted c in Fig.4(b), have a large distortion along the direction of the fivefold axis under the action of phason strain; they then shift beyond the range in which diffraction spots are selected in the reciprocal space. Therefor they disappear in the zero-level x-ray precession patterns (Fig.4(c)). (2) in Fig.4(a) the diffraction spots do not lie along straight lines, except for those along A-A. The reason for this distortion is that the shift direction of the spots along A-A are in the A-A line, these spots are still in a straight line. (3) At the position B in Fig.4(b) there are two separate diffraction spots. Under the action of the phason strain they shift face to face and overlap each other, so that at the position B in Fig.4(c) only one spot could be observed; this is in good agreement with the experiment result shown in Fig.4(a).

From the results above, one can see that the symmetries of the distorted quasicrystals deviated from those of ideal icosahedral quasicrystals. The phason strain in $T_2 - Al_6Li_3Cu$ leads to the degeneration of the icosahedral group m$\bar{3}\bar{5}$ into the group m$\bar{3}$ to which $R - Al_5Li_3Cu$ belongs.

We now understand that the phason strain introduces shifts and broadenings of diffraction spots. That is, the diffraction patterns of quasicrystals with phason defects deviate from those of ideal quasicrystals. Such quasicrystals are recognized as imperfect quasicrystals. Since each degree of phason strain corresponds to a state of imperfect quasicrystals, it is reasonably predicted that there exist several intermediate states from quasicrystal to ordinary crystal structure. It will be discussed in Section 4.

3.4 Dislocation in quasicrystal

With the further study of quasicrystals, a question that whether the paradigm in crystallography can be used in quasicrystallography was opened. Working on hydrodynamic theories, Levine[20] predicted that dislocations should exist in quasicrystals. The relation between dislocations and the tiling pictures has been clarified by Socolar[38]. Since the quasicrystal lattice is aperiodic dislocations in quasicrystals cannot be interpreted as insertions or removals of half-planes of atoms. This means that dislocations in quasicrystals have to be characterized by displacements not only in the translational (phonon) degree of freedom but also in the phason variable. The Burgers vector of a dislocation for icosahedral quasicrystals in 6D space can be written as $\mathbf{b} = (\mathbf{b}_u, \mathbf{b}_w)$, where the 3D vectors $\mathbf{b}_u$ and $\mathbf{b}_w$ are the "phonon" parts and "phason" ones respectively. No dislocation lies in the $\mathbf{b}_u$=0 or $\mathbf{b}_w$=0 subspaces. Thus $\mathbf{b}_u$, the purely translational part of the Burgers vector is an incomplete characterization of a dislocation in a quasicrystal. The $\mathbf{G} \cdot \mathbf{b} = 0$ invisibility criterion has to be modified for quasicrystals.

In Section 2 we have a expression of the phase of density-wave of quasicrystal[Eq.(7)]:

$$\phi_n = \mathbf{G}_n^{\parallel} \cdot \mathbf{u} + \mathbf{G}_n^{\perp} \cdot \mathbf{w} + \phi_n^{\circ}$$

where n=1,....,6. Hatwalne et al [39] have deduced the $\mathbf{u}$ and $\mathbf{w}$ in terms of the phases ϕ_n

$$\mathbf{u} = \mathbf{G}^{-2} \sum_{n=1}^{6} \mathbf{G}_n^{\parallel} \phi_n \tag{19-a}$$

$$\mathbf{w} = \mathbf{G}^{-2} \sum_{n=1}^{6} \mathbf{G}_n^{\perp} \phi_n . \tag{19-b}$$

As well know, dislocation are line defects which leave the phases invariant modulo 2π: a dislocation in the ith phase can be expressed by

$$\int_{\Gamma} d\phi_i = 2\pi N_i \tag{20}$$

for any loop Γ surrounding the dislocation line, where N_i can be thought of as the number of extra half-planes of constant phase inserted into the ith density wave. Therefore one have

$$\int_{\Gamma} d\mathbf{u} = \int_{\Gamma} G^{-2} \sum_{n=1}^{6} \mathbf{G}_n^{\parallel} d\phi_n = \mathbf{b}_u \tag{21-a}$$

$$\int_{\Gamma} d\mathbf{w} = \int_{\Gamma} G^{-2} \sum_{n=1}^{6} \mathbf{G}_n^{\perp} d\phi_n = \mathbf{b}_w \tag{21-b}$$

Hence, the invisibility criterion for a dislocation with Burgers vector$(\mathbf{b}_u, \mathbf{b}_w)$ is proposed as

$$\mathbf{G}^{\parallel} \cdot \mathbf{b}_u + \mathbf{G}^{\perp} \cdot \mathbf{b}_w = 0 \tag{22}$$

From Eqs.7 and 20, it is evident that the number of extra half-layers inserted in the ith density wave by a dislocation with Burgers vector($\mathbf{b}_u$,$\mathbf{b}_w$) is

$$N_i = \frac{1}{2\pi}(\mathbf{G}_i^{\parallel} \cdot \mathbf{b}_u + \mathbf{G}_i^{\perp} \cdot \mathbf{b}_w)$$

If the invisibility condition Eq.22 holds for a given $\mathbf{G}_i$, the corresponding n_i is 0. Hence that density-wave component may be expected to be least distorted and therefore to show the least contrast.

Recent observations of dislocations and grain boundaries in quasicrystals have been reported [40]. Devaud - Rzepski[41] observed dislocations in $Al_{65}Cu_{20}Fe_5$ icosahedral quasicrystals. They deduced from their analysis of Fourier-filtered lattice fringe pictures a direction of the Burgers vector in 6D space parallel to a [1,p,q,0,-(1+p),(2+q)] direction. Since their picture were taken along a single direction only, the parameters p and q could not be determined directly. Assuming that, on energetic grounds, the Burgers vector should have minimum length in 6D space, they concluded that p = q = -1 leading to [1, -1, -1,0,0,1]. Zhang et al. [42] also studied dislocations in icosahedral $Al_{65}Cu_{20}Fe_{15}$ by electron diffraction contrast analysis and high-resolution electron microscopy. The results provide direct evidence of a Burgers vector in 3D space parallel to a twofold axis of the icosahedral phase,[0/1,1/0,1/1]. These two results coincide with each other.

4. PHASE TRANSITION FROM QUASICRYSTAL TO ORDINARY CRYSTAL

We have discussed the distortions in the x-ray and the electron diffraction patterns of quasicrystals and explained the shift and the broadening of diffraction spots in terms of phason strain in Section 3. However, the questions of whether the distorted quasicrystal phase is an intermediate phase between quasicrystal and ordinary crystal and of how they are connected together are still opened. Recently many authors have reported the experimental observations of phase transition from quasicrystals to ordinary crystals[14,43], even from 2D decagonal quasicrystal to 1D quasicrystal [44] or reversible transitions between a microcrystalline rhombohedral phase and a perfect icosahedral phase[45]. Mai et al. [15,27] have treated a continuous quasicrystalline-to-crystalline transition by successive increase of linear phason strain in a quasicrystal. In this section we discuss how far the imperfect quasicrystalline phase can move towards the crystalline phase and what extent the phason strain can be used to explain such phenomena.

4.1 Icosahedral-cubic transition

Based on the experimental observation of distortions in electron and x-ray diffraction patterns, Mai et al.[15] have theoretically deduced how icosahedral symmetries of quasicrystals could be connected with m3m or m3 symmetries of a ordinary crystal by an action of soft phason strain using group theory and projection method.

In Section 3.3, we have the expressions of Bragg vector $\mathbf{G}^{\parallel}$ and $\mathbf{G}^{\perp}$ (Eqs.(16) and (17)). In general, the symmetry operations in the 6D space produce a 6D representation of the icosahedral group I, which can be decomposed into irreducible representations of I. The character table of the icosahedral group I is shown in Table 1. It is clear that

$$\Delta = \Gamma_3 + \Gamma_{3'}$$

Table 1
Character table of the icosahedral point group [$\tau = (1+\sqrt{5})/2$]

I	E	$12C_5$	$12C_5^2$	$20C_3$	$15C_2$
Γ_1	1	1	1	1	1
Γ_3	3	τ	$1-\tau$	0	-1
$\Gamma_{3'}$	3	$1-\tau$	τ	0	-1
Γ_4	4	-1	-1	1	0
Γ_5	5	0	0	-1	1
Δ	6	1	1	0	-2

Therefore the 6D space can be decomposed into two 3D-invariant subspaces. The process of projecting a vector of 6D space into the 3D hyperplane $Z_3^{\parallel}$ can be represented by the projection matrix $\mathbf{P}^{\parallel}$ and into the 3D space $Z_3^{\perp}$ by $\mathbf{P}^{\perp}$. The projection matrices $\mathbf{P}^{\parallel}$ and $\mathbf{P}^{\perp}$ are given by

$$\mathbf{P}^{\parallel} = A\begin{pmatrix} 0 & 0 & \tau & 1 & -1 & -\tau \\ 1 & -1 & 0 & \tau & \tau & 0 \\ \tau & \tau & 1 & 0 & 0 & 1 \end{pmatrix} \tag{23-a}$$

$$\mathbf{P}^{\perp} = A\begin{pmatrix} 0 & 0 & 1 & -\tau & \tau & -1 \\ -1 & -1 & \tau & 0 & 0 & \tau \\ -\tau & \tau & 0 & 1 & 1 & 0 \end{pmatrix} \tag{23-b}$$

where $A = \frac{1}{\sqrt{1+\tau^2}}$, consequently $\mathbf{e}_i^{\parallel} = \mathbf{P}^{\parallel}\mathbf{e}_i$ and $\mathbf{e}_i^{\perp} = \mathbf{P}^{\perp}\mathbf{e}_i$. If we project a vector $\mathbf{L}=(\mathbf{n}_1, \ldots\ldots, \mathbf{n}_6)$ from the Z_6-space into the 3D spaces $Z_3^{\parallel}$ and $Z_3^{\perp}$, the wave vector $\mathbf{G}'^{\parallel}$ is given by :

$$\begin{aligned} \mathbf{G}'^{\parallel} &= \mathbf{G}^{\parallel} + \mathbf{M}\mathbf{G}^{\perp} \\ &= \frac{\pi}{a}A\begin{pmatrix} (\tau-\alpha)(n_3-n_6)+(1+\alpha\tau)(n_4-n_5) \\ (\tau-\alpha)(n_4+n_5)+(1+\alpha\tau)(n_4-n_2) \\ (\tau-\alpha)(n_1+n_2)+(1+\alpha\tau)(n_3+n_6) \end{pmatrix} = \frac{\pi}{a}\mathbf{P}_{\alpha}^{\parallel}\mathbf{L} \end{aligned}$$

where $\mathbf{M}$ is the same as used in Section 3.3. So

$$\mathbf{P}_{\alpha}^{\parallel} = A\begin{pmatrix} 0 & 0 & \tau-\alpha & 1+\alpha\tau & -1-\alpha\tau & -\tau+\alpha \\ 1+\alpha\tau & -1-\alpha\tau & 0 & \tau-\alpha & \tau-\alpha & 0 \\ \tau-\alpha & \tau-\alpha & 1+\tau\alpha & 0 & 0 & 1+\tau\alpha \end{pmatrix}$$

This means that under the action of phason strain, the hyperplane $Z_3^{\parallel}$ has been transformed into another hyperplane $Z_{\alpha 3}^{\parallel}$. The six basis vectors in Z_6 are projected into $Z_{\alpha 3}^{\parallel}$, one can obtain

$$\mathbf{e}_{\alpha 1}^{\parallel} = \mathbf{P}_{\alpha}^{\parallel}\mathbf{e}_1 = A\begin{pmatrix} 0 \\ 1+\alpha\tau \\ \tau-\alpha \end{pmatrix} \qquad \mathbf{e}_{\alpha 2}^{\parallel} = \mathbf{P}_{\alpha}^{\parallel}\mathbf{e}_2 = A\begin{pmatrix} 0 \\ -1-\alpha\tau \\ \tau-\alpha \end{pmatrix}$$

$$\mathbf{e}^{\|}_{\alpha 3} = \mathbf{P}^{\|}_{\alpha}\mathbf{e}_3 = A\begin{pmatrix}\tau-\alpha\\ 0\\ 1+\tau\alpha\end{pmatrix} \qquad \mathbf{e}^{\|}_{\alpha 4} = \mathbf{P}^{\|}_{\alpha}\mathbf{e}_4 = A\begin{pmatrix}1+\alpha\tau\\ \tau-\alpha\\ 0\end{pmatrix}$$

$$\mathbf{e}^{\|}_{\alpha 5} = \mathbf{P}^{\|}_{\alpha}\mathbf{e}_5 = A\begin{pmatrix}-1-\alpha\tau\\ \tau-\alpha\\ 0\end{pmatrix} \qquad \mathbf{e}^{\|}_{\alpha 6} = \mathbf{P}^{\|}_{\alpha}\mathbf{e}_6 = A\begin{pmatrix}-\tau+\alpha\\ 0\\ 1+\alpha\tau\end{pmatrix}$$

From these expressions of basis vectors $\mathbf{e}^{\|}_{\alpha i}$, the following results can be obtained:
(1) When $\alpha = 0$, the six vectors $\mathbf{e}^{\|}_{\alpha i}$ are the six linear independent basis vectors of a perfect icosahedron which are the same as those used in Section 3.3.
(2) When $\alpha = (\tau - 1)/(\tau + 1) = 0.236$

$$\mathbf{e}^{\|}_{\alpha 1} = B\begin{pmatrix}0\\1\\1\end{pmatrix} \qquad \mathbf{e}^{\|}_{\alpha 2} = B\begin{pmatrix}0\\-1\\1\end{pmatrix} \qquad \mathbf{e}^{\|}_{\alpha 3} = B\begin{pmatrix}1\\0\\1\end{pmatrix}$$

$$\mathbf{e}^{\|}_{\alpha 4} = B\begin{pmatrix}1\\1\\0\end{pmatrix} \qquad \mathbf{e}^{\|}_{\alpha 5} = B\begin{pmatrix}-1\\1\\0\end{pmatrix} \qquad \mathbf{e}^{\|}_{\alpha 6} = B\begin{pmatrix}-1\\0\\1\end{pmatrix}$$

where $B = \frac{1}{\sqrt{1+\tau^2}}\frac{\tau+2}{\tau+1}$. It is clear that there are only three linearly independent vectors $\mathbf{e}^{\|}_{\alpha 1}$, $\mathbf{e}^{\|}_{\alpha 3}$ and $\mathbf{e}^{\|}_{\alpha 4}$. They are the basis vectors of the fcc lattice. Then the reciprocal lattice of an icosahedron is transformed into one possessing m3m point group symmetry under the action of the phason field.
(3) When $\alpha = \frac{(k\tau-1)}{(\tau+k)}$

$$\mathbf{e}^{\|}_{\alpha 1} = C\begin{pmatrix}0\\K\\1\end{pmatrix} \qquad \mathbf{e}^{\|}_{\alpha 2} = C\begin{pmatrix}0\\-K\\1\end{pmatrix} \qquad \mathbf{e}^{\|}_{\alpha 3} = C\begin{pmatrix}1\\0\\K\end{pmatrix}$$

$$\mathbf{e}^{\|}_{\alpha 4} = C\begin{pmatrix}K\\1\\0\end{pmatrix} \qquad \mathbf{e}^{\|}_{\alpha 5} = C\begin{pmatrix}-K\\1\\0\end{pmatrix} \qquad \mathbf{e}^{\|}_{\alpha 6} = C\begin{pmatrix}-1\\0\\K\end{pmatrix}$$

where $K \neq 1$ and $C = \frac{1}{\sqrt{1+\tau^2}}\frac{\tau+2}{\tau+k}$

Then the reciprocal lattice possesses m3 point group symmetry. Obviously, when K is a rational number, the $\mathbf{e}^{\|}_{\alpha i}$ define a reciprocal lattice of crystal. When K is an irrational number the $\mathbf{e}^{\|}_{\alpha i}$ define a 3D incommensurate phase.
(4) When $\alpha = \tau$

$$\mathbf{e}^{\|}_{\alpha 1} = \frac{1}{A}\begin{pmatrix}0\\1\\0\end{pmatrix} \qquad \mathbf{e}^{\|}_{\alpha 2} = \frac{1}{A}\begin{pmatrix}0\\-1\\0\end{pmatrix} \qquad \mathbf{e}^{\|}_{\alpha 3} = \frac{1}{A}\begin{pmatrix}0\\0\\1\end{pmatrix}$$

$$\mathbf{e}^{\|}_{\alpha 4} = \frac{1}{A}\begin{pmatrix}1\\0\\0\end{pmatrix} \qquad \mathbf{e}^{\|}_{\alpha 5} = \frac{1}{A}\begin{pmatrix}-1\\0\\0\end{pmatrix} \qquad \mathbf{e}^{\|}_{\alpha 6} = \frac{1}{A}\begin{pmatrix}0\\0\\1\end{pmatrix}$$

In this case only three of the $\mathbf{e}^{\|}_{\alpha i}$ are linearly independent, $\mathbf{e}^{\|}_{\alpha 1}$, $\mathbf{e}^{\|}_{\alpha 3}$, and $\mathbf{e}^{\|}_{\alpha 4}$. It is evident that they are the basis vectors of the simple cubic lattice possessing m3m point group symmetry.

From the results obtained above, one could see that firstly, under an action of phason strain, the transformation from an icosahedral quasicrystal into a crystal can take place. Secondly, the effect of phason strain on the hyperplane $Z^{\|}_3$ is to change its position in 6D space, besides the origin of 6D space. Each value of α corresponds to one hyperplane, so that there are several intermediate states between icosahedral quasicrystal and cubic crystal. In fact, these theoretical results are consistent with the studies of Al-Li-Cu quasicrystals by both the x-ray diffraction technique[9] and the electron diffraction method[14]. In Chapter 5 Li has shown a series of electron diffraction patterns of $T_2 - Al_6Li_3Cu$ with various degrees of distortion. It confirms that a certain quenched linear phason strain introduces imperfections into icosahedral quasicrystals, even phase transition from an icosahedral quasicrystalline phase to a ordinary crystalline one. Ishii[10] suggested that the transformation of icosahedral Al-Li-Cu to the bcc R-phase is expected to be a second -order transition.

It should be noted that the icosahedral quasicrystal can transform either directly to a related crystalline phase or through a 2D decagonal quasicrystal as an intermediate state[47]. Recently Kim et al.[48] have investigated the solid state transformation of icosahedral phase by *in situ* heating experiments in TEM. As-rapidly-solidified Al-20at% Mn consists mainly of a dendritic icosahedral phase with a small amount of interdendritic fcc α-Al. During subsequent heat treatment at temperature below 500°C the dendritic icosahedral phase grows and consumes the interdendritica α-Al. At about 500°C the decagonal phase nucleates near icosahedral dendrites and grain boundaries. Then it grows into the icosahedral matrix by lateral motion of ledges 10-20nm high across facet planes normal to the twofold symmetry axes. At about 600°C the decagonal phase transforms into a crystalline phase , probably Al_4Mn. This result may suggest that solid-state decomposition of the icosahedral phase is a mechanism of decagonal phase formation in as-rapidly-solidified Al-Mn alloys. But Dong et al.[49] claimed that the decagonal phase is stable in the temperature range between 973 and 1350K and grows directly from the liquid and no other crystalline phase is involved in the growth process in the light of study on the phase transformation in $Al_{65}Cu_{17.5}Co_{17.5}$ alloy by neutron diffraction.

4.2 Icosahedral-rhombohedral transition

Quasicrystalline phase in an alloy of approximate composition $Al_{65}Cu_{20}Fe_{15}$ was firstly reported by Tsai et al.[50]. A nearly single icosahedral phase is now known to be formed in Al-Cu-Fe alloy either by rapid quenching or by conventional solidification. Experimental results indicate that the icosahedral phase in Al-Cu-Fe system is stable only within a certain temperature and concentration range. Outside this range the icosahedral phase is either metastable or in a supersaturated state and will decompose into stable phase during annealing. The phase transition of icosahedral quasicrystals in Al-Cu-Fe depends on the alloy composition and annealing temperature[51−53].

Liu[51] reported that two modes of decomposition of the icosahedral phase, discontinuous or continuous reaction, were observed in rapidly solidified $Al - Cu - Fe$ alloys. The discontinuous decomposition is characterized by the existence of a definite reaction front separating the icosahedral phase from the resulting crystalline one and its kinetics is controlled by the migration of the reaction front into the icosahedral phase. A peritectoid reaction as observed in $Al_{70}Cu_{20}Fe_{10}$ or precipitation as in $Al_{65}Cu_{20}Fe_{15}$

are typical for this kind of decomposition. In contrast, Continuous decomposition can occur without any definite reaction front. As observed in $Al_{60}Cu_{30}Fe_{10}$ during annealing at about 620° for 55h, the icosahedral phase transforms continuously and homogeneously into a face-centered rhombohedral structure R$\bar{3}$m with a=37.71 Å and α=63.43°. In $Al_{65}Cu_{20}Fe_{15}$ alloy annealing at 540° leads to discontinuous decomposition of icosahedral phase by precipitation of $Al_{13}F_4$ face-centered rhombohedral structure with a=61.02Å, $\alpha = 63.43°$.

Audier et al.[45] directly confirmed by TEM *in situ* thermal treatment. Al-Cu-Fe samples annealed at lower temperatures or slowly solidified exhibit a periodic microcrystalline structures in which the symmetry of individual grains is found to be rhombohedral. A unit cell of this crystalline phase can be represented as the Ammann proplate rhombohedron with $a \approx 37.7\AA = 2\tau^3 \times 4.45\AA$ where $\tau = (\sqrt{5}+1)/2$. This is the first observation of structural transformation between a stable quasicrystal and its modulated phase.

Recently Liu et al.[54] present evidences found by TEM which show that the continuous decomposition of the icosahedral quasicrystal proceeds through a series of intermediate modulated structures before reaching the rational approximants of a rhombohedral structure. Fig.9 shows a complex modulated structure intermediate between the icosahedral and the rhombohedral structures. In region A some prolate and oblate rhombi are outlined which are the same unit cells as those for constructing a Penrose tiling[7], but their arrangement is well modulated from the icosahedral one. Some oblate rhombi have become to distribute periodically, implying the formation of rhombohedral microdomains. Such microdomains can also see in some other regions in Fig.9. Since there is no definite interface between the rhombohedral microdomains and the modulated structure, it is reliable that the modulated structure evolves coherently and continuously into the rhombohedral one. Moreover, differently oriented rhombohedral microdomains, as shown in region B, can result from this evolution because of the group-subgroup relation between the icosahedral m$\bar{3}\bar{5}$ and the rhombohedral $\bar{3}$m structures.

It is known[15] that successive increasing the linear phason strain in a quasicrystal can lead to a series of intermediate modulated structures between quasicrystal and related ordinary crystal. Ishii[46] has discussed the icosahedral-rhombohedral structure transformation in Al-Cu-Fe alloy by locking the magnitude of the phason strains to a special value giving rational approximations to irrational ratios inherent in quasicrystals. The resulting rhombohedral unit cell structure is essentially the same as that proposed by Audier[45].

The structural transition between the icosahedral and the rhombohedral phases is expected to be a first-order transition because we have third-order term in an expansion of the elastic free energy in a series of phonon (elastic) and phason strain fields. By examining the elastic free energy, it would be found that the rhombohedral phase might be stabilized compared with the icosahedral phase in a significant region in the parameter space of the phason 'elastic' constants even if the icosahedral phase is locally stable[23]. If the effective phason stiffness is temperature dependent and becomes negative at lower temperatures, as suggested in the entropic stabilization mechanism[55], a weakly first-order transition to the trigonal (rhombohedral) phase may be plausible. However, the icosahedral- rhombohedral structure transition is still controversial at present.

It should be pointed out that reversible transitions from rhombohedral microcrystals to an icosahedral quasicrystal have been recently reported by Janot et al.[56] Samples of $Al_{63.5}Cu_{24.5}Fe_{12}$ were examined by *in situ* temperature dependence of electron

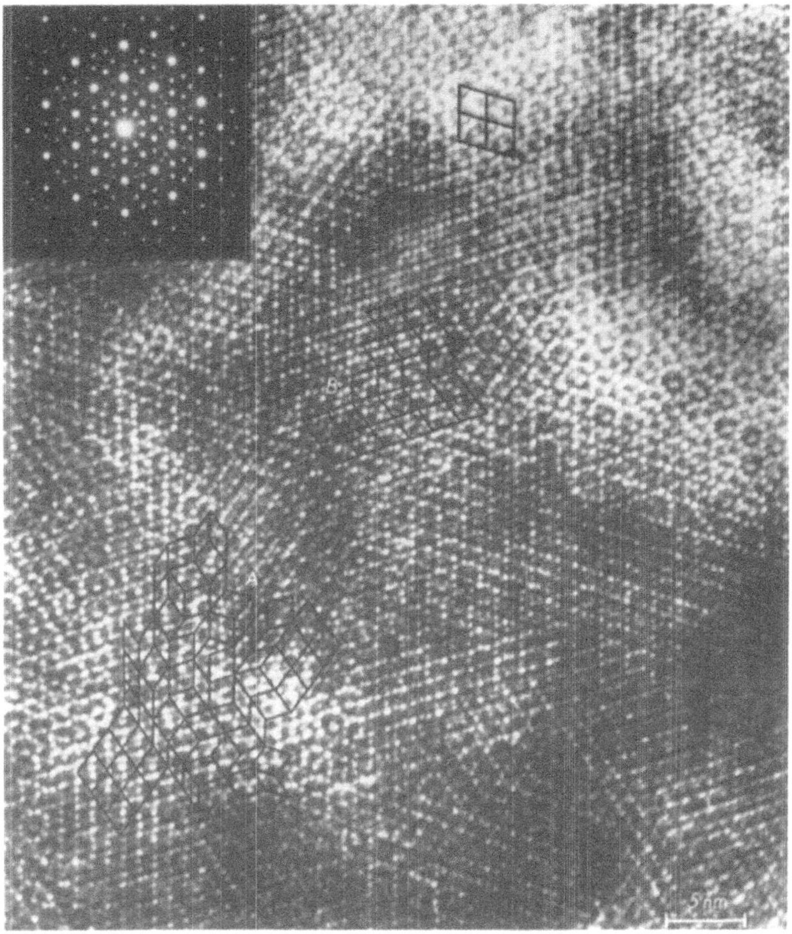

Fig.9 HREM image of an intermediate modulated structure during the continuous decomposition in $Al_{60}Cu_{30}Fe_{10}$ (Courtesy W.Liu).

diffraction, HREM image and *in situ* neutron power diffraction. Experimental results demonstrate that the equilibrium phases of $Al_{63.5}Cu_{24.5}Fe_{12}$ samples are rhombohedral microcrystals up to 650°C and perfect icosahedral quasicrystal only between about 750° and 820°C, with reversible transition from one to the other through an intermediary modulated icosahedral structure, finally turn into a mixture of cubic and monoclinic crystal above 865°C before total melting.

Bancel[57] has also discussed the structural transition of $Al_{65}Cu_{23}Fe_{12}$ in terms of a temperature dependent phason elastic constant derived from a configurational entropy term in the elastic free energy. Measuring the temperature dependence of x-ray peak intensities of icosahedral $Al_{65}Cu_{23}Fe_{12}$, he found that below 670°C the peak broadening suggests the onset of the structural transition. These data were explained by softening phason mode which drives a structural transition to a lower-symmetry phase. Above 670°C phasons equilibrate rapidly and the mode softening decrease the Debye - Waller factor of single grains of i-AlCuFe.

It is fair to mention that the transition from rhombohedral microcrystal at room temperature to perfect quasicrystal at higher temperature via modulated quasicrystal is not yet a universally accepted phenomenon.

4.3 Octagonal-β-Mn-type structure transition

In 1987, Wang et al.[4] reported the discovery of the octagonal quasicrystal in rapidly solidified Cr-Ni-Si and V-Ni-Si alloy. Somewhat later this phase also observed in rapidly solidified Mn-Si[58] and Mn-Si-Al[59]. The diffraction patterns of the new structure show a 2D quasiperiodicity with eightfold rotational symmetry and 1D periodicity along the eightfold axis. HREM observation shows that 45° twins of β-Mn-type structure (space group $P4_132$, a=0.62nm) frequently coexist with the octagonal phase in the Cr-Ni-Si alloy. Moreover, both the β-Mn-type structure and the octagonal quasicrystal can be described as composed of squares and 45° rhombi, in the former in a periodic and in the latter in a quasiperiodic arrangement. It is suggestive that these structures are closely related. Using the linear phason strain theory, Mai et al. have studied the process of transition from an octagonal quasicrystal to a crystalline phase with β-Mn-type structure in Cr-Ni-Si alloy[27].

Based on the knowledge described in Section 2, one can obtain the real-space tilings and the diffraction patterns of the quasicrystal respectively by calculating Eqs.(15) and (14). Since the octagonal quasicrystal is a 2D one, the higher- dimensional hypercubic lattice corresponding to the octagonal quasicrystal is a 4D lattice, the four basis vectors of $D^{\parallel}$ are

$$\mathbf{e}_1^{\parallel} = \frac{1}{\sqrt{2}}\begin{pmatrix}1\\0\\0\\0\end{pmatrix} \quad \mathbf{e}_2^{\parallel} = \frac{1}{\sqrt{2}}\begin{pmatrix}\frac{1}{\sqrt{2}}\\\frac{1}{\sqrt{2}}\\0\\0\end{pmatrix} \quad \mathbf{e}_3^{\parallel} = \frac{1}{\sqrt{2}}\begin{pmatrix}0\\1\\0\\0\end{pmatrix} \quad \mathbf{e}_4^{\parallel} = \frac{1}{\sqrt{2}}\begin{pmatrix}\frac{-1}{\sqrt{2}}\\\frac{1}{\sqrt{2}}\\0\\0\end{pmatrix}$$

and those of the $D^{\perp}$ are

$$\mathbf{e}_1^{\perp} = \frac{1}{\sqrt{2}}\begin{pmatrix}0\\0\\1\\0\end{pmatrix} \quad \mathbf{e}_2^{\perp} = \frac{1}{\sqrt{2}}\begin{pmatrix}0\\0\\\frac{-1}{\sqrt{2}}\\\frac{1}{\sqrt{2}}\end{pmatrix} \quad \mathbf{e}_3^{\perp} = \frac{1}{\sqrt{2}}\begin{pmatrix}0\\0\\0\\-1\end{pmatrix} \quad \mathbf{e}_4^{\perp} = \frac{1}{\sqrt{2}}\begin{pmatrix}0\\0\\\frac{1}{\sqrt{2}}\\\frac{1}{\sqrt{2}}\end{pmatrix}$$

the basis vectors of the reciprocal lattice are the same as those of the real lattice, namely, $\mathbf{e}^{*\parallel} = \mathbf{e}_i^{\parallel}, \mathbf{e}^{*\perp} = \mathbf{e}_i^{\perp}$, i=1,......,4. Let $\mathbf{M}$ be

$$\mathbf{M} = \alpha \begin{pmatrix}0 & 0 & 0 & 1\\0 & 0 & 1 & 0\\0 & 1 & 0 & 0\\1 & 0 & 0 & 0\end{pmatrix}$$

where α is a parameter which represents the strength of the linear phason strain.

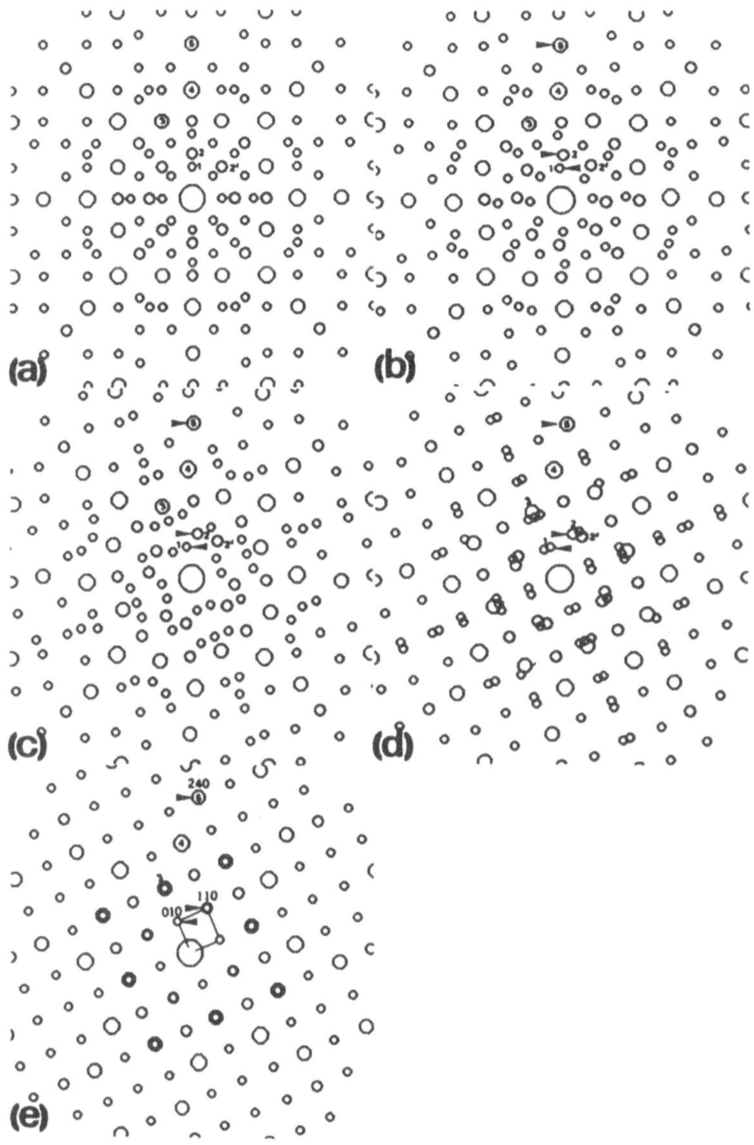

Fig.10 Computer simulation of electron diffraction patterns along the eightfold axis with different magnitude of phason strain: (a) α=0, (b) α=0.05, (c)α=0.15, (d) α=0.3 and (e) $\alpha = \sqrt{2} - 1$.

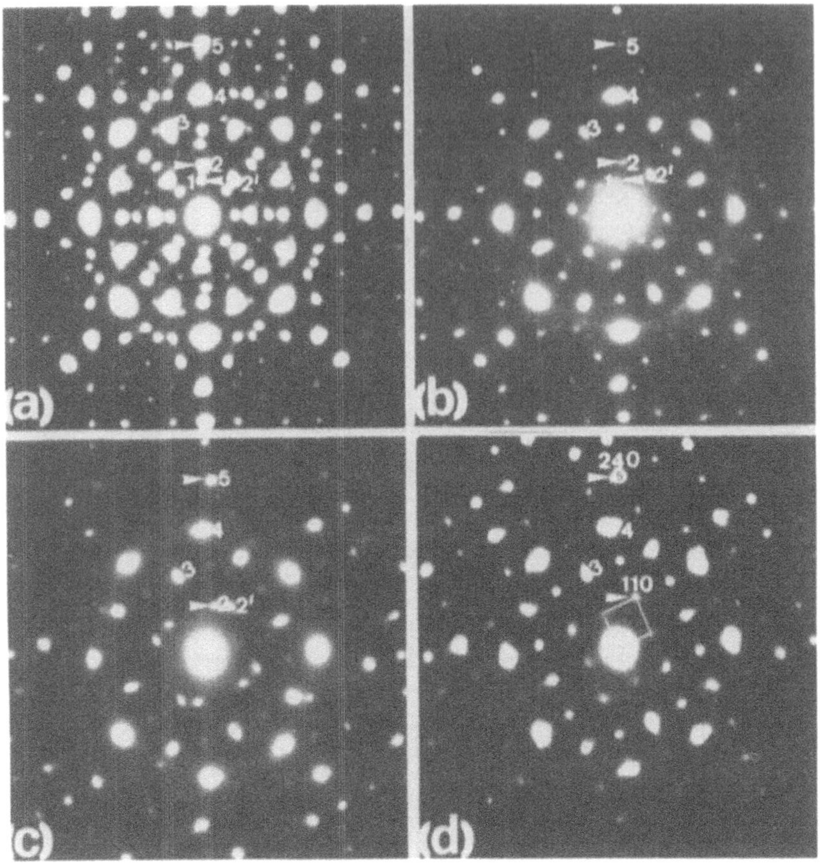

Fig.11 Electron diffraction pattern of Cr-Ni-Si alloy (a)–(c) the intermediate phase and (d) a β-Mn-type structure.

Consequently it is possible to study the structural change of octagonal quasicrystals under different strength of phason strain. Fig.10 shows the computer simulation of the diffraction patterns with different degree of phason strain field Fig.10 (a) is a diffraction pattern of perfect octagonal quasicrystal. One can see that due to the action of phason strain, the diffraction spots calculated shift (indicated by arrows) and rotate around the ordinary eightfold axis (Fig.10(b)). Comparing Fig.10(a) with Fig.10(b), it is clear that the eightfold symmetry of the perfect octagonal quasicrystal is broken into a fourfold symmetry, but there is no translational periodicity. With the increase of the value of α, the diffraction spots calculated shift further, and tend to aggregate into clusters (Figs.10(c) and (d)). When the value of α reaches $\sqrt{2}$-1, the spots in each cluster

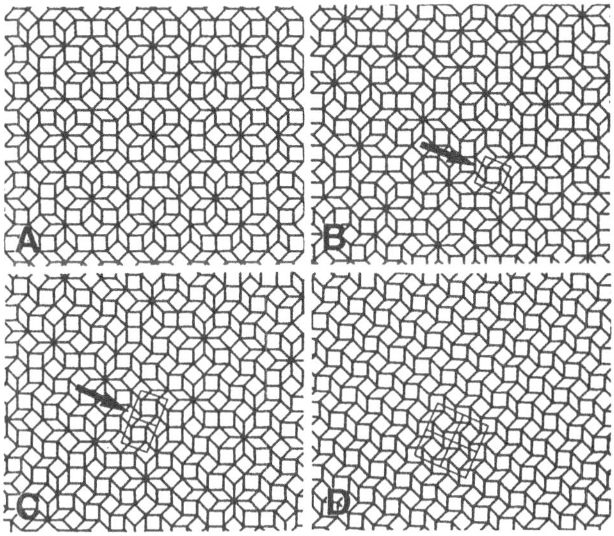

Fig.12 Computer simulations of real-space tiling (a) $\alpha = 0$, (b) $\alpha = 0.1$, (c) $\alpha = 0.2$ and (d) $\alpha = \sqrt{2} - 1$.

are superposed upon a single point, so that the diffraction pattern has a translational periodicity (Fig.10(e)). It is obvious that the results of the simulated diffraction patterns show the process of transition from an octagonal quasicrystal to a crystalline structure, i.e., from a quasiperiodicity with eightfold rotation symmetry to a periodicity with fourfold rotation symmetry.

This results have been verified experimentally. Fig.11 is the electron diffraction patterns of quasicrystals in rapidly quenched Cr-Ni-Si alloy taken one by one moving the selected area aperture or specimen. It shows the process of the transition from the octagonal quasicrystal to the β-Mn-type structure. Comparing Fig.10 with Fig.11, it is obvious that the computer simulation patterns are in good agreement with the experimental ones. For example, spots 2 and $2'$ approach each other in Fig.11(b) and (c), and eventually become the 110 diffraction spot of the β-Mn-type structure.

Based on the Eq.(15) the real-space tiling consisting of squares and 45° rhombi can be constructed. Fig.12 shows the tilings with different values of α. Fig.12(a) is the real-space tiling of perfect octagonal quasicrystal. From Figs.12(b) and (c) one can see that local areas having translational periodicity (indicated by arrows) appear.

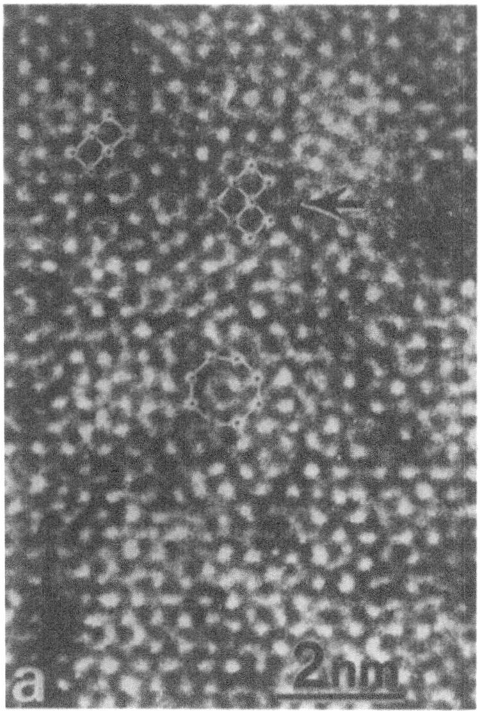

Fig.13 HREM image of Cr-Ni-Si quasicrystal (one octagon unit is outlined) with many nuclei of the β-Mn-type structure (unit cell outlined) (Courtesy N.Wang).

More surprising is that the crystallographic structure in these areas is the same as that of the β-Mn-type structure[60]. As a result, with the increase of value of α, the embryos of the β-Mn-type structure grow up. When the value of α reaches $\sqrt{2}$-1, these periodic structures grow arbitrarily large (Fig.12(d)). Note that the value $\alpha = \sqrt{2} - 1$ is also used in Fig.10(e). Alternately, under the action of phason strain field with the strength $\alpha = \sqrt{2} - 1$, the octagonal quasicrystal is transformed completely into a crystal with the β-Mn-type structure. This theoretical result has been confirmed by HREM image[60]. Fig.13 is a HREM image taken from a Cr-Ni-Si quasicrystal. It is evident that in addition to octagons consisting of eight bright dots, some short-range periodic arrangements of bright dots resembling those corresponding to β-Mn-type structure can also be recognized. It is reasonable to interpret the continuous transformation of the octagonal quasicrystal to the β-Mn-type structure as the result of the gradual introduction of short-range translational periodicity in the quasicrystal.

4.4 2D decagonal–1D quasicrystal transition

After the discovery of the icosahedrally related decagonal quasicrystal in Al-Mn

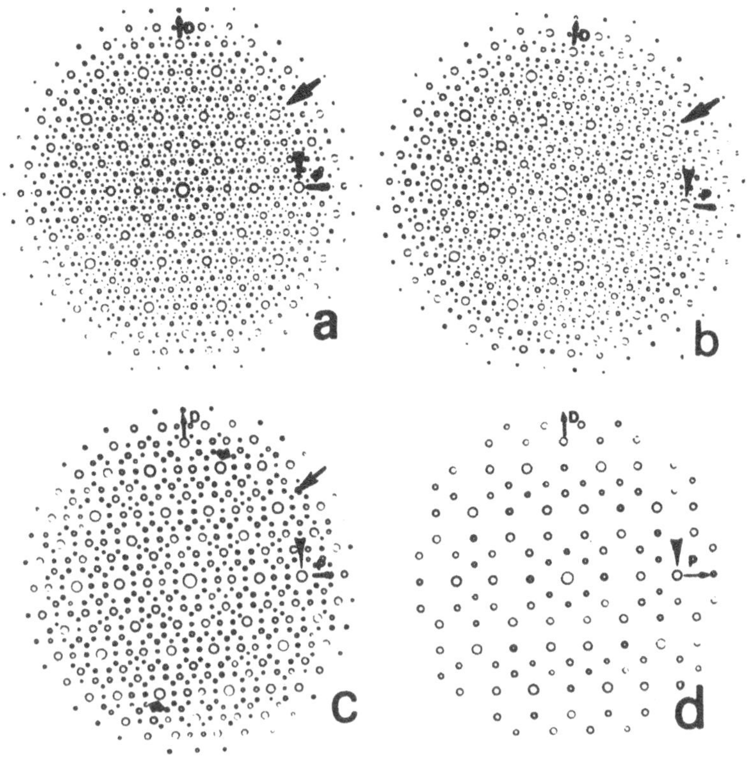

Fig.14 Simulated electron diffraction patterns of 1D quasicrystals: (a) $\alpha = 0$, (b) $\alpha = -0.02$, (c) $\alpha = 0.055$ and (d) $\alpha = 0.38$ (courtesy H.Zhang).

alloys, many new ones have been found in other Al-M alloys (M is transition metals including the platinum group metals). The decagonal quasicrystal is also a 2D one which displays tenfold rotational symmetry in a quasiperiodic plane and periodic along the tenfold axis perpendicular to this aperiodic plane. If one of these two quasiperiodic directions becomes periodic, an 1D quasicrystal, namely, periodic in two directions with the third one remaining aperiodic, will be resulted. The 1D quasicrystal has been

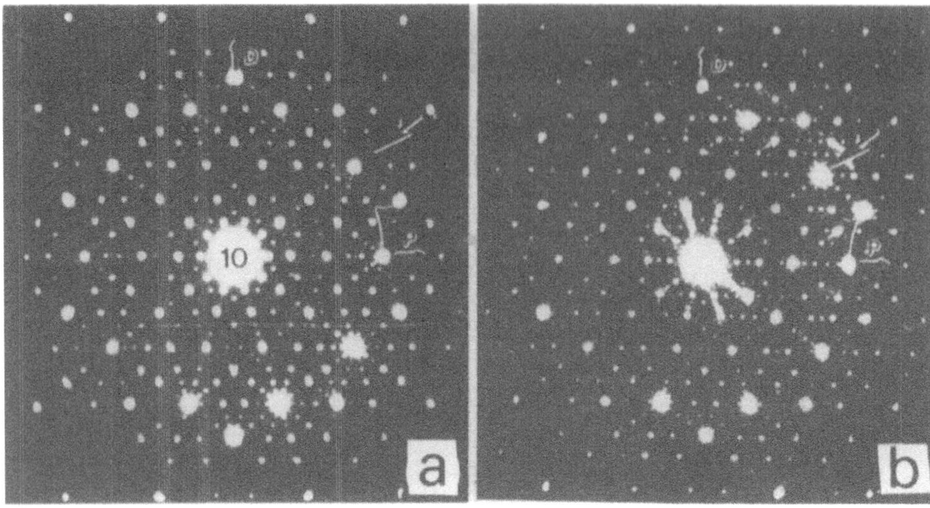

Fig.15 Electron diffraction patterns (a) 2D quasicrystal and (b) 1D quasicrystal (Courtesy H.Zhang).

observed in rapidly solidified Al-Ni-Si, Al-Cu-Mn, and Al-Cu-Co alloys[61]. Following Mai's work[9,27], Zhang et al.[31] have deal with the transformation of the 2D decagonal quasicrystal into the 1D quasicrystal from the phason strain point of view.

For 2D quasicrystal, the second-rank tensor **M** is given by

$$\mathbf{M} = \alpha \begin{pmatrix} 1 & 0 \\ 0 & 0 \end{pmatrix}$$

Fig.14(a) is a simulated diffraction pattern with α=0. One can see that the intervals between the two neighboring spots along one of the ten twofold P directions in Fig.14(a) are in the ratio of $1{:}\tau{:}\tau^2$. Then the phason strain is introduced gradually into the 2D decagonal quasicrystal. Fig.14(b) is the simulated diffraction pattern with α=-0.02. The weak ten spots (indicated by arrow) are shifted to lie on an ellipse with the long axis in the direction marked by D while the ten strong spots are more or less still lying on concentric circles. Moreover, the spots along the horizontal twofold direction (indicated by P) are periodic giving a periodicity of $13 \times a$, where a is the lattice parameter. This means that 2D simulated quasicrystal now transforms into 1D quasicrystal. With further increase of the phason strain the shifts of spots proceed further. When $\alpha = -0.145$, another 1D quasicrystal with a periodicity of $5 \times a$ in the P direction is obtained. Only five equally spaced strong spots remain in the P direction; all weak spots either coincide with these strong spots or disappear in the pattern.

If the sign of α is positive, the spots will shift in the reverse direction. Fig.14(c) is the simulated electron diffraction pattern with α=0.055. Again it shows an 1D quasicrystal with a periodicity of $8 \times a$ along the P direction. The ten weak spots (indicated by arrow) lie on an ellipse similar to those in fig.14(b), but the shifts of spots are now more obvious. The spots do not lie on a straight line along a twofold direction except the P direction. Further increasing the phason strain to $\alpha = 0.38$ results in the 1D quasicrystal with a periodicity of $3 \times a$ (Fig.14(d)). In this case the ellipse consisting of weak spots no longer exists. But sets of ten strong spots are lying on an ellipse, implying that the presence of severe phason strains produces signicant shifts of all spots, including the strong ones.

The theoretical inference has been verified experimentally. Fig.15(a) and (b) are the electron diffraction patterns of 2D and 1D quasicrystals in Al-Cu-Co alloy. Fig.15(b) was taken after the specimens was undergone a heat treatment at 800°C for 40h. One can see sharp diffraction spots with tenfold rotation symmetry in Fig.15(a) while a periodicity of $13 \times a$ along P direction in Fig.15(b). Comparing Fig.14(a) and (b) with Fig.15(a) and (b), the match is reasonably good.

From the results discussed above, it is clear that (1) 2D quasicrystals can be transformed into 1D quasicrystals in certain conditions. (2) The periodicities in 1D quasicrystal are the various approximants of the Fibonacci series (1,2,3,5,8,13,...).

5. SUMMARY

In the case of ordinary crystal, the intrinsic soft phonon model can be applied to understand some kinds of the phase transitions from crystal to crystal, the later is a modulated phase induced by the phonon field. In the case of quasicrystal, due to its incommensurate characteristic, it is predicted that the phason elastic free energy emerges during the quasicrystal growth process. Moreover, the phason field relaxes more slowly than that of the phonon in quasicrystals. The phason field induce strains, leading to defects in quasicrystal.

Numerous experiments affirm that phason- induced modulated quasicrystalline phases exist. These phases can be recognized as intermediate phases between perfect quasicrystals and ordinary crystals. The phase transitions from quasicrystal to ordinary crystal are now proposed through two way, continuous or discontinuous. However, some experiments based on the result of thermal treatments claim that there exist reversible phase transitions from modulated quasicrystal to perfect quasicrystal. It is supposed that the quasicrystalline phase is a stable one whose energy state is higher than that of crystalline phase, but lower than that of the instable state perturbed by phason fields. Therefore a phase located in the instable state would transforms into a phase with lower energy state, either crystalline phase through intermediate modulated phases or quasicrystalline phase.

It is sure that the further study of the rule of phason behavior in quasicrystals will promote the more deeply understanding of the nature of quasicrystals, *vice versa*.

Authors would like to thank Chao-Ying Wang and Lan-Sheng Wu for their kind collaboration.

6. REFERENCES

[1] D.Shechtman, I.Blech, D.Gatias and J.W.Cahn, Phys. Rev. Lett., **53**, 1951 (1984).
[2] L.A.Bendersky, Phys. Rev. Lett., **55**, 1461 (1985).
[3] K.Chattopadhyay, S.Lete, R.Prasad, S.Ranganathan, G.N.Subbanna and N.Thangaraj, Scripta Metall., **19**, 1331 (1985).
[4] N.Wang, H.Chen and K.H.Kuo, Phys. Rev. Lett., **59**, 1010 (1987).
[5] T.Ishimasa, H-U.Nissen and Y.Fukano, Phys. Rev. Lett., **55** , 511 (1985).
[6] D.Levine and P.J.Steinhardt, Phys. Rev. Lett., **53** , 2477 (1984).
[7] R.Penrose, Bull. Inst. Math. Appl., **10** , 266 (1974); A.L.Mackay, Physica, **114A** , 609 (1982).
[8] C.L.Henley, Comments Cond. Matter Phys., **13** , 59 (1987).
[9] Zhenhong Mai, Shizhong Tao, Lingzhi Zeng and Baoshan Zhang, Phys. Rev., **B 38**, 12913 (1988).
[10] H.S.Chen, A.R.Kortan and J.M.Parsey, Phys. Rev., **B 38**, 1654 (1988).
[11] L.Pauling, Phys. Rev. Lett., **58** , 365 (1987); Nature, **317**, 512 (1984).
[12] T.C.Lubensky, J.E.S.Socolar, P.J.Steinhardt, P.A.Bancel and P.A.Heiney, Phys. Rev. Lett., **57**, 1440 (1986).
[13] L.A.Aslanov, Acta Cryst., **A 47** , 63 (1991).
[14] F.H.Li, G.Z.Pan, S.Z.Tao, M.J.Hui, Z.H.Mai, X.S.Chen and L.Y.Cai, Phil. Mag., **59**, 535 (1989).
[15] Zhenhong Mai, Shizhong Tao, Baoshan Zhang and Lingzhi Zeng, J. Phys.: Condens. Matter, **1**, 2465 (1989).
[16] V.Elser, Acta Cryst., **A 42** , 36 (1986); Phys. Rev. Lett., **54**, 1730 (1985).
[17] L.D.Landau and I.E.Lifshitz, Statisical Physics, 2nd ed. (Pergamon, New York, 1968), Chap. XIV.
[18] S.Alexander and J.P.McTague, Phys. Rev. Lett., **41**, 702 (1978).
[19] P.Bak, Phys. Rev. Lett., **54** , 1517 (1985); Phys. Rev., **B 32**, 5764 (1985).
[20] D.Levine, T.C.Lubensky, S.Ostlund, S.Ramaswamy, P.J.Steinhardt and J.Toner, Phys. Rev. Lett., **54**, 1520 (1985).
[21] P.Bak, Rep. Prog. Phys., **45** , 587 (1981); V.L.Pokrovsky and A.L.Talopov, Theory of Incommensurate Crystals, in Sov. Science Reviews (Horwood, Zurich, Switzerland, 1985).
[22] T.C.Lubensky, in Introduction to Quasicrystas, edited by M.V.Jarić, (Academic Press, Inc, 1988), p100.
[23] Y.Ishii, Phys. Rev., **B 39** , 11862 (1989); J.Non-cryst. Solids, **117/118**, 840 (1990).
[24] Michael Tinkham, Group Theory and Quantum Mechanics, (McGraw-Hill, New York, 1964).
[25] J.E.S.Socolar and D.C.Wright, Phys. Rev. Lett., **59**, 221 (1987).
[26] Z.R.Huang, G.Z.Pan, D.Y.Yang and X.S.Chen, Solid State Commun., **63**, 951 (1987).
[27] Z.H.Mai, L.Xu, N.Wang, K.H.Kuo, Z.C.Jin and G.Chen, Phys.Rev., **40**, 12183 (1989).
[28] F.C.Frank, Proc. Roy. Soc. (London), **A 215**, 43 (1952).
[29] P.J.Steinhardt, D.R.Nelson and M.Ronchetti, Phys. Rev., **B 28**, 784 (1983).
[30] N.D.Mermin and S.M.Troian, Phys. Rev. Lett., **54**, 1524 (1985).
[31] M.Tanaka, M.Terauchi, K.Hiraga and M.Hirabayashi, UI-Tramicroscopy, **17**, 279 (1985).

[32] P.Bancel, P.A.Heiney, P.W.Stephens, A.I.Goldman and P.M.Horn, Phys. Rev. Lett., **54**, 2422 (1985).
[33] H.K.Hardy and J.M.Silcock, J of the Institute of Metals, **84**, 423 (1955-56).
[34] W.A.Cassada, G.J.Shiffet and S.J.Poon, Phys. Rev. Lett., **56**, 2276 (1986).
[35] Zhenhong Mai, Baoshan Zhang, Mengjun Hui, Zhaorong Huang and Xishen Chen, Materials Science Forum, **22-24**, 591 (1987).
[36] K.Edagawa, Phil. Mag. Lett., **61**, 107 (1990).
[37] Z.M.Wang and K.H.Kuo, Acta Crystall., **A 44**, 857 (1988).
[38] J.E.S.Socolar, T.C.Lubensky and P.J.Steinhardt, Phys. Rev., **B 34**, 3345 (1986).
[39] Y.Hatwalne and S.Ramaswamy, Phil. Mag. Lett., **61**, 169 (1990).
[40] C.H.Chen, J.P. Remeika, G.P.Espinosa and A.S.Cooper, Phys. Rev., **B 35**, 7737 (1987).
[41] J.Devaud-Rzepski, M.Cornier-Quiquandon and D.Gratias, Proc. of the 3rd International Conference on Quasicrystals, 1989 Vista Hermonza, Mexico.
[42] Z.Zhang, M.Wollgarten and K.Urban, Phil. Mag. Lett., **61** , 125 (1990).
[43] Z.Zhang, N.C.Li,and K.Urban, J. Mater. Res., **6** , 366 (1991).
[44] H.Zhang and K.H.Kuo, Phys. Rev., **B 41** , 3482 (1990).
[45] M.Audier and P.Guyot , Proc. of the Anniversary Adriatic Research Conference on Quasicrystals,1989 Trieste, ed M.V.Jarić and S.Lundquist (Singapore:World Scientific) p74.
[46] Y.Ishii, Phil. Mag. Lett., **62** , 393 (1990).
[47] R.J.Schaefer and L.Bendersky, Scr. Metall., **20** , 745 (1986).
[48] D.H.Kim, K.Chattopadhyay and B.Cantor. Phil. Mag., **A 62** , 157 (1990).
[49] C.Dong, J.M.Dubois, M.De Boissieu and C.Janot, J. Phys.: Condens. Matter, **3** , 1665 (1991).
[50] A.P.Tsai, A.Inoue and T.Masumoto, Jpn. J. Appl. Phys., **26** , L1505 (1987).
[51] W.Liu and U.Köster, Mater. Sci. Eng., **A 133** , 388 (1991).
[52] S.Ebalard and F.Spaepen, J. Mater. Res., **5** , 62 (1990).
[53] J.Devand-Rzepski, A.Quivy, Y.Calvayrac, M.Cornier-Quiquadon and D.Gratias, Phil. Mag., **B 60** , 855 (1989).
[54] W.Liu, U.Köster and A.Zaluska, Phys. Stat. Sol., (a) **126** , K9 (1991).
[55] M.Widom, Proc. of the Anniversary Adriatic Research Conference on Quasicrystals, 1989 Trieste, ed M.V.Jarić and S.Lundquist (Singapore: World Scientific) p337.
[56] C.Janot, M.Audier. M.De Boissieu and J.M.Dubois, Europhys. Lett., **14** , 355 (1991).
[57] P.Bancel, Phys. Rev. Lett., **63** , 2741 (1989); Ibid., **64** , 496 (1989).
[58] W.Cao, H.Q.Ye and K.H.Kuo, Phys. Stat. Sol., (a) **107** , 551 (1988).
[59] N.Wang, K.K.Fung and K.H.Kuo, Appl. Phys. Lett., **52** , 2120 (1988).
[60] N.Wang and K.H.Kuo, Phil. Mag. Lett., **61** , 63 (1990).
[61] L.X.He, X.Z.Li, Z.Zhang and K.H.Kuo, Phys. Rev. Lett., **61** , 1116 (1988).
[62] H.Zhang and K.H.Kuo, Phys. Rev., **B 41** , 3482 (1990).

Crystal-Quasicrystal Transitions
M.J. Yacamán and M. Torres (Editors)

Twinning of Quasicrystals and Related Crystals

S.Ranganathan[a], Alok Singh[a], R.K.Mandal[b] and S.Lele[b]

[a]Centre for Advanced Study, Department of Metallurgy, Indian Institute of Science, Bangalore 560 012, India.

[b]Centre for Advanced Study, Department of Metallurgical Engineering, Banaras Hindu University, Varanasi 221 005, India

Abstract

The phenomenon of twinned aggregates of crystals with motifs of non- crystallographic symmetry leading to non-crystallographic point group symmetry for the aggregate as a whole is analysed here by adopting the method of coset decomposition based on group theoretical approach to higher dimensional space to see the three-dimensional analogue. This is illustrated with examples of irrational twinnings of rational approximants of the icosahedral and the decagonal quasicrystals. The twinning of the icosahedral quasicrystal is also considered, where the twinned grain mimics the symmetry of the decagonal quasicrystal. This is analysed with the concept of the dichromatic pattern of a bicrystal.

1. INTRODUCTION

Shechtman, Blech, Gratias and Cahn [1] stunned the scientific community by announcing the discovery of a phase with icosahedral symmetry, called a quasicrystal, in a rapidly solidified Al-Mn alloy. Pauling [2] advanced an alternative explanation for the icosahedral symmetry as arising from multiple twinning of a cubic crystal.

Mackay [3] pointed out that while Pauling [2] emphasized the cubic nature of twinning, he gave no guidance about its indefinite continuation. The quasicrystal provides an algorithm for indefinitely great extension. It was known to Euclid (his last theorem) that five cubes could be inscribed in a dodecahedron. Mackay [4] deduced from this relationship that along a five-fold axis, all five cubes have an axis $[\tau 10]$. Along a three-fold axis, two cubes have [111] and three have an axis $[\tau^2 10]$. Along a two-fold axis one cube has [100] and four have $[\tau\ 1\ 1/\tau]$. Though Pauling's description of twinning varied, in his review [5] he drew attention to this relationship between the dodecahedron and the cube (Fig. 1) and commented that this could lead to a synthesis of the twinning and quasicrystal viewpoints.

The experimental demonstration that this postulated twinning relationship exists in polycrystalline aggregates is due to Bendersky et al. [6,7]. They found crystalline aggregates with overall icosahedral symmetry in Al-Mn-Fe-Si alloys, but which were distinctly different from Pauling's [2] twinning model. They also emphasized the need to scrutinize the concept of twinning. Bendersky [8] has discussed in detail the new concept of hypertwinning. These twins, whose twin axis was irrational, were further studied by Mandal et al.[9], and discovered in Al-Fe-V-Si by Srivastava and Ranganathan [10] and in Al-Mn-Ge by Lalla et al.[11].

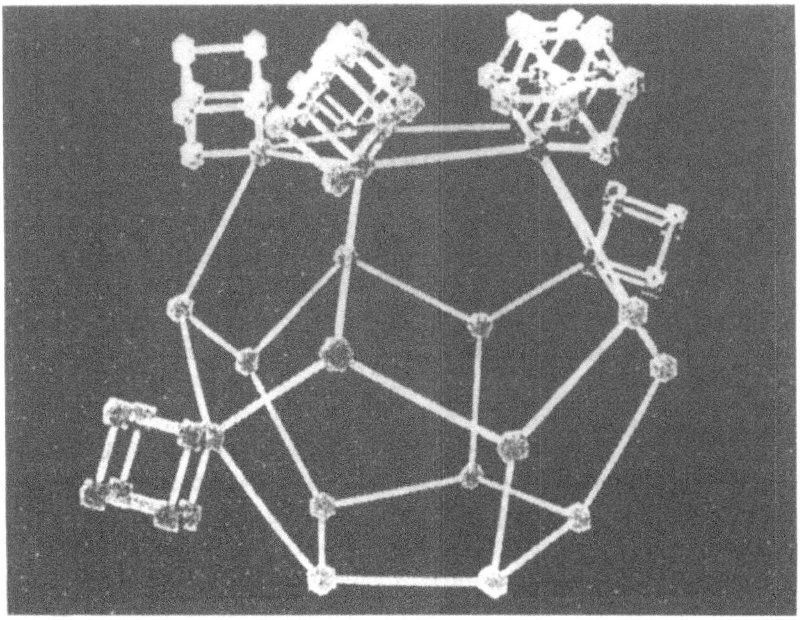

Figure 1. Twinning about an icosahedral seed in such a way that a 3-fold axis of the cubic crystal coincides with a 3-fold axis of the seed (Pauling [5]).

A twin of an icosahedral quasicrystal, with a mirror plane normal to one of the twin axes, was discovered by Koskenmaki et al.[12] in an Al-Mn-Fe alloy. The icosahedral twins were further studied in an Al-Mn alloy by Ranganathan et al.[13] and in Al-Mn-Fe by Singh and Ranganathan [14,15].

The twinning of the icosahedral quasicrystal as well as irrationally twinned crystalline aggregates giving rise to icosahedral and decagonal symmetry are reviewed in this chapter and extends our earlier work [16].

2. TWINNING ELEMENTS

2.1 Definitions

In polycrystalline materials one can usually observe an infinite number of orientation relationships between a pair of grains. If the orientation relationships are governed by a close and finite set of symmetry operations, we say that the different variants are twin related. Otherwise a random aggregate will result with no systematics in the orientation relationships. Symmetry operations which send a single crystal (or a quasicrystal) from one orientation to another are twinning rules or operations and each region having a single orientation is known as a twin variant. The essence of Friedel's attempt to indicate the cause of twinning has now been incorporated in the above discussion. According to Friedel [17], twinning is involved in "homogenous crystalline aggregates in which one observes so large a number of identical mutual orientations between crystals that can rule out random cause". The above definition seems to throw light on the cause of twinning, whereas subsequent advancements have focussed on the geometrical aspects of twinning. The above statement can be supplemented by the following two definitions : Cahn [18] defines a twin "as a polycrystalline edifice built up of two or more homogenous portions of the same crystal species in juxtaposition and oriented with respect to each other according to well defined laws". The International Tables for Crystallography [19] (referred to as IT hereafter) give the following definition for twinning. "A twin consists of two or more single crystals of the same species but in different orientations, its twin components. They are intergrown in such a way that atleast some of their lattice directions are parallel. The twin law describes the geometrical relation between the twin components. It specifies a symmetry operation, the twin operation that brings one of the twin components into parallel orientation with each other. The corresponding symmetry element is called the twin element". One of the important points to highlight in this section is to spell out clearly the synthesis of cause and effect that is Friedel's emphasis on the elimination of "random cause" for orientation relationships and all those subsequent developments with reference to the geometrical aspects of twinning. Twinning in a polycrystal of n dimensions should be considered to arise as a result of the following three governing factors inherently present in its single constituents :

1) the translational invariance of the lattice (translational group under periodic boundary condition)

2) the point like invariance of a crystal (the point group)

3) the crystal invariance (the space group)

All the three leave specific signatures in the reciprocal space and also in the grain boundary. Both features can be studied through the transmission electron microscope by utilising conventional as well as high resolution microscopy but not by X-ray diffractometry. Based on the above discussion, we give the following definition: "When two or more single crystals of same species grow in a finite number of different orientations owing to geometrical and compositional constraints present in them, we say they are the twin related components of a polycrystalline aggregate and the phenomenon is called twinning".

The geometrical constraints imply certain restrictions on the way a motif or basis group of atoms continue across the twin interface. One can visualise two essentially different situations both of which preserve the number of bonds for all atoms including those at the interface. In the first case, there may be a small distortion or a change in the type of the bonds. This preserves the motif orientation approximately. In the second case, one can have a common coherent planar interface such that the number of bonds of the interfacial atoms is again preserved but their orientations are not. This can occur if the twinning operation is a mirror plane at the twin interface, a rotation axis perpendicular to the interface or a 2-fold rotation axis in the twin interface.

An alternative but equivalent classification of twinning follows from symmetrization and dissymmetrization operations [20]. In the first case, one introduces the concept of prototype symmetry which is the symmetry of the highest supergroup attainable by, or conceivable for, the point group of the motif through a small structural distortion, that is, without a reconstruction of bonds. The prototype point group symmetries are $m\overline{35}$ or 10/mmm in all cases considered here. These are compatible respectively with 6d cubic lattices (F,P,I) and an orthogonal lattice (having 5 equal and 1 different lattice parameter).

When the symmetry of a given structure belongs to prototype symmetry (i.e. highest symmetry possible in a given crystal hierarchy), then the twinning can occur according to the two steps mentioned below. First refers to dissymmetrization of the twin aggregates by which they lower their symmetry with respect to the parent one. The symmetrization, then, proceeds by the twinning operation not present previously. This is analogous to the description of a dichromatic pattern (DCP) in bicrystallography [21].

The n dimensional hypertwinning can be defined by qualifying the remark of finite orientations. If these are created due to a t-fold rotation incompatible with n periodic translations, one is discussing n dimensional hypertwin aggregate whose rank in indexing dimension is greater than n. The t-fold rotation gets applied to n dimensional crystal due to the presence of atomic clusters (contributing a majority of atoms in a crystal) having t-fold symmetry. A reduction in the number of parameters in indexing dimension can be achieved by making an appeal to a hypercrystal approach thus revealing the hidden symmetry of n dimensional polycrystalline aggregate in a crystal of m ($m > n$) dimension, m then becomes rank of indexing dimension.

Various types of twinning possible in quasicrystals are considered here. Since we consider the motif to play a dominant role in twinning, it is clear that a similar twinning would occur in a periodic crystal which has motifs similar to those in quasicrystals.

Twinning in such related crystals is also considered here. Owing to the presence of 3d non- crystallographic symmetry in quasicrystals, there is a need to consider higher dimensional descriptions of such structures so that the 3d non-crystallographic symmetry becomes compatible with more than 3 independent periodic translations.

2.2 Prototype Twinning in Crystals and Quasicrystals

The observation of non-crystallographic point group symmetry of the twinning aggregates of crystals with motifs of non- crystallographic geometry (known as Rational Approximant Structure (RAS) in quasicrystalline literature) is a new phenomenon. This is achieved by the method of coset decompositions based on group theoretical approach available in literature but adapting them to higher dimensional space to see the 3d analogue more clearly. Twinning of various phases with space group symmetries G, which are a subgroup of G_o is considered to arise due to a real or virtual phase transition from the latter to the former. If the point group of G is different from that of G_o but their 6d lattices are identical then the change is translation (t-) equivalent. On the other hand, if there is only a change of translational symmetry in going from G_o to G, then the change is class (k-) equivalent [22]. Finally, there could be a change in the lattice as well as point group. If the index of G in G_o is n, then G_o can be expressed as follows [21,23,24]

$$G_o = \phi_1 G + \phi_2 G + \ldots + \phi_n G \qquad (1)$$

where the ϕ_i's (i=1 to n) are a representative set of twinning operations. The n left cosets in Eq. 1 have a one-to-one correspondence with twins in n possible orientations. Clearly, the symmetry of the aggregate of n twins or equivalently their diffraction symmetry is that of G_o [21-26]. Further, there is no orientational change in the transformation from the prototype phase to any of the observed twin variants. This results in the preservation of bond orientational order (BOO) across a twin interface as mentioned earlier. This rules out the necessity for a planar interface. Now we briefly consider examples of t- and k- equivalent cases.

For cases where the point symmetries of a phase and its prototype phase are the same, the twinning operations (ϕ_i) are translations. Hence twins arising in such situations may be designated as translational twins. It is known that there are only three 6d cubic lattices with point symmetry $m\bar{3}\bar{5}$, namely face centred cubic (F), primitive (P) and body centred cubic (I). The translational groups [27] F(2a) and P(a) are subgroups of index 2 in P(a) and I(a) respectively. The symbols a or 2a in the paranthesis indicate the lattice parameters of the unit cells. Thus 2 or 4 twin variants of F will mimic the symmetry of P and I respectively, while 2 twins of P will display the symmetry of I. An example of the latter situation is provided by RAS α-AlFeSi phase. The structure has the space group $Pm\bar{3}$, however, owing to the presence of twins, the diffraction symmetry becomes $Im\bar{3}$ [28]. The twinning operation in 6d is a translation by half the body diagonal of the cube. It follows that the Fourier transforms (or diffraction patterns) of the individual twin variants are identical, since the only difference between the two is a phase shift which does not affect intensities. The effect on the 3d structure obtained by taking a rational cut is remarkable. One obtains two distinct structures which are homometric and are related to each other by a centre of inversion at 1/2 1/2 1/2. Cooper [28] has designated these cell I and cell II. This simple interpretation as translational twins is only possible by adopting the 6d point of view [29]. The twin variants are analogous to antiphase domains in the usual terminology. The second example of the above is offered by the two translational twins observed in a rapidly solidified Al-Cu-Fe alloy [27]. However, unlike the previous case, each of the variants displays icosahedral $Fm\overline{35}$ symmetry, which gives rise to an irrational structure in 3d physical space. The two domains are related by a shift of <1/2 00000>. The presence of such domains have been experimentally confirmed by electron microscopy [27].

When the point symmetries of a phase and its prototype are different but the lattices are identical, one obtains rotational twins. The largest subgroup of the icosahedral group (order 120) is cubic point group $m\overline{3}$ (order 24) with index 5. Thus 5 twin variants are possible. The twinning operation is a 5 fold rotation around a 6d hypercubic axis. Such an operation leaves the 6d lattice invariant but flips the lower symetry motif from one orientation to another. The hypertwins observed in α-AlMnFeSi [6] correspond precisely to this situation. Table 1 shows the number of rotational twin variants for a number of observed icosahedrally related phases. The reduction in the number of parameters required to specify the twinned aggregate in terms of higher dimensional approach is also included. Examples drawn from our work are presented in section 4.

Next we consider the general case where there is a lowering of lattice as well as motif symmetry for an observed phase from the prototype. We restrict ourselves to orthogonal (non-hyper cubic) 6d lattices [29-32] whose centering characteristics are unchanged, that is, for example, a P lattice remains P. 2d pentagonal quasicrystals [30] (PQC) have $\overline{5}m$ point group symmetry and can be obtained by a cut from a 6d orthogonal lattice with 5 mutually equal (a_1 to a_5) and one different lattice parameter (a_6). 5 fold symmetry exists along the axis corresponding to a_6. In the process of lowering the symmetry, the PQC and the respective 6d lattice lose all the 5 fold axes except the one retained and all 2 fold axes which are not normal to the retained 5 fold axis. The index of $\overline{5}m$ in the icosahedral group is 6 and the corresponding 6 rotational twin variants can be obtained by rotation around any of the lost 3 fold and 2 fold axes of symmetry. Various types of the phases of this type are given in Table 1.

Table 1 Phases giving rise to icosahedral symmetry with rank 6 in aggregate form

Name of the Phase	Symmetry	Order of Group	No. of Variants	Reduction	Examples
Cubic	$m\bar{3}$	24	5	9	Al-Mn-Si Al-Mn-Fe-Si Al-Mn-Ge Al-Mn-Cr-Si Al-Fe-V-Si Al-Cu-Li
Rhombohedral	$\bar{3}m$	12	10	24	Al-Cu-Li Al-Cu-Fe
Orthorhombic	mmm	8	15	39	
Monoclinic	2/m	4	30	84	
Orthorhombic	222	4	30	84	
2D QC	$\bar{5}m$	20	6	24	
1D QC	222	4	30	114	
2D QC	$10/mmm^{+}$	40	6	24	Al-Mn-Ni

+ does not belong to the sub-group of icosahedral point group

Analogous to those discussed above in respect of icosahedral symmetry, one may encounter examples of prototype twinning of crystals and quasicrystals leading to 8/mmm, 10/mmm and 12/mmm symmetry. Solids with above point group symmetries show quasiperiodicity in two dimensions and periodicity along the symmetry axis in contrast to the icosahedral quasicrystal with 3d quasiperiodicity. Their rank in indexing dimension is, therefore, five. Wang and Kuo [33] have observed 45° rotational variants and octagonal quasicrystals in Cr-Ni-Si and Mn-Si alloys. The aggregate symmetry in the former is 8/mmm with each of the variants displaying cubic symmetry analogous to β-Mn prototype structure. Kulkarni et al. [34] have found a structural relationship between an octagonal quasicrystal and β-Mn twins using strip and projection method. Ishimasa et al. [35] have reported 30° twins in Ni-Cr systems. The twinning is based on a tetragonal (σ-phase) and an orthorhombic (H-phase) cell. The aggregate symmetry resembles that of 12/mmm. Various types of prototype twinning are possible in many alloy systems resulting in 10/mmm symmetry. The different cases have been listed in Table 2. Illustrations drawn from our work are presented in section 5.

Table 2 Phases displaying diffraction symmetry 10/mmm in aggregate form with rank 5

Name of the Phase	Symmetry	Order of Group	No. of Variants	Reduction	Examples
2D QC	$\bar{5}$m	20	2	5	
1D QC	222	4	10	35	Al-Cu-Co Al-Ni-Si
RAS	222	4	10	25	Al-Fe
RAS	2/m	4	10	25	Al-Fe
RAS	mmm	8	5	10	Al-Mn-Ni Al-Mn-Cu Al-Mn Al-Pd

2.3 Twinning Based on Dichromatic Pattern Concept

It follows from the above discussion of twinning based on prototype symmetry that twinning in any crystal having the full holohedral symmetry (Im$\overline{35}$) of the 6d cubic class cannot be so described because a supergroup prototype symmetry does not exist for such crystals. A second type of twinning is based on the concept of the dichromatic pattern [17,18,20] of a bicrystal and is particularly relevant for the above types of 6d crystals. Consider two twin variants which are semi-infinite and share a planar interface. Two kinds of symmetry operations are present in this situation: those which map each twin into itself, and those which interchange one variant (say, white) into the other (say, black). The latter are the twinning operations and lead to a higher symmetry for the twin pair than that given by the intersection of the symmetry groups of the black and white twin variants. The interpenetrating lattices of the white and black variants then together define a lattice called the dichromatic pattern (DCP). The most frequently observed twinning operations are 2 fold rotations along an n fold symmetry axis (with n odd) or a mirror plane perpendicular to it. However, the two operations are equivalent for centrocemetric crystals. The DCP approach involves first a dissymmetrization operation owing to the lowering of symmetry between the two twin variants with symmetry G and ϕG (ϕ is the twinning operation and G is the space group of the phase). The resulting symmetry of the aggregates, however, is not (G $\cap\phi$G) but [(G $\cap\phi$G) $\cup\phi$] where ϕ also acts as a symmetrization operation [20]. Based on this we now propose to consider some examples related to quasicrystalline phases.

The intersection symmetry group for two icosahedral quasicrystal variants rotated relative to each other by 180° around the 5 fold axis is $\bar{5}$m while the symmetry of the aggregate (and the diffraction pattern) is 10/mmm. Thus this icosahedral twin pair mimics the symmetry of the 2d quasicrystalline decagonal phase, although structurally the two phases are distinct. Such a twin has been observed in Al-Mn alloys and Al-Mn-Fe alloys [12-15] and described in section 3. A similar twin pair can form by 180° rotation around a three fold axis. The symmetry of the aggregate in this case is 6/mmm. The two kinds of DCP lattices discussed above can be understood in terms of twinning of 6d hypercube around the five and three fold axes respectively. The twinning plane is a 5d hyperplane with $\bar{5}m$ and $\bar{3}m$ symmetry. The corresponding planar interface will accordingly display the two respective 2d point group symmetries in the 3d physical space. The introduction of a 5d mirror hyperplane does not permit the bond orientational order of the hypercrystal to propagate across the boundary unlike the prototypic case discussed earlier. The trace of the interface in the hypercrystal in 3d physical space will also display a similar behaviour. Unlike twins in 3d crystals, for icosahedral quasicrystals there is a unique plane passing through the inversion centre of a triacontahedral atomic motifs, which can act as an atomic composition plane.

3. ICOSAHEDRAL TWIN

Koskenmaki et al.[12] showed that the microstructure of rapidly solidified $Al_{86}Mn_7Fe_7$ consists of rosette shaped icosahedral phase grains; in numerous instances these grains were twinned with a mirror plane normal to one of the five-fold axes. Ranganathan et al.[13] studied these icosahedral twins in an Al-10%Mn alloy and showed that the overall symmetry of the grain is similar to that of the decagonal phase. Singh and Ranganathan [14,15] have studied these icosahedral twins in an Al-Mn-Fe alloy and established the orientation relationship between the twinned parts of the quasicrystal.

The microstructure of a melt-spun Al-5at%Mn-5at%Fe alloy shows icosahedral quasicrystal dendrites in an aluminium matrix. The morphology of the icosahedral particles suggests the growth direction to be along five-fold symmetry directions although the growth direction in an Al-Mn icosahedral phase is along the three-fold symmetry directions [36,37]. It was found that a large number of such particles had a curved boundary separating arms with different contrasts. By tilting experiments it was found that in all these instances the grains were twinned.

Figure 2a shows a typical icosahedral grain of melt-spun Al-5Mn-5Fe alloy in a two-fold zone axis orientation. While the orientation of the two-fold zone axis diffraction pattern from the arms marked A is shown in Figure 2b, Figure 2c shows the orientation for the other two arms. The composite diffraction pattern of Figure 2d shows that the two patterns have a five- fold direction in common, marked in the figure.

The orientation relationship between the twinned and the untwinned parts of the icosahedral grains is shown in Figure 2e as two superimposed stereograms of icosahedral phase rotated by 36° around a common five-fold axis in the centre. The five-fold twin axis then becomes a ten-fold. Several twinned grains of the icosahedral phase were tilted along two fold axes common to the twinned and untwinned parts of the grain to determine the exact orientation relationship between them. They were all found to bear the same orientation relationship as in the stereogram of Figure 2e. On tilting 45° from the the common five fold axis (i.e., the twin axis) along a two-fold vector, the zone axis B for the untwinned (denoted by BU along one of the two-fold vector traces in the stereogram) and K for the twinned (denoted by KT) sector is arrived at (zone axis notations of Singh

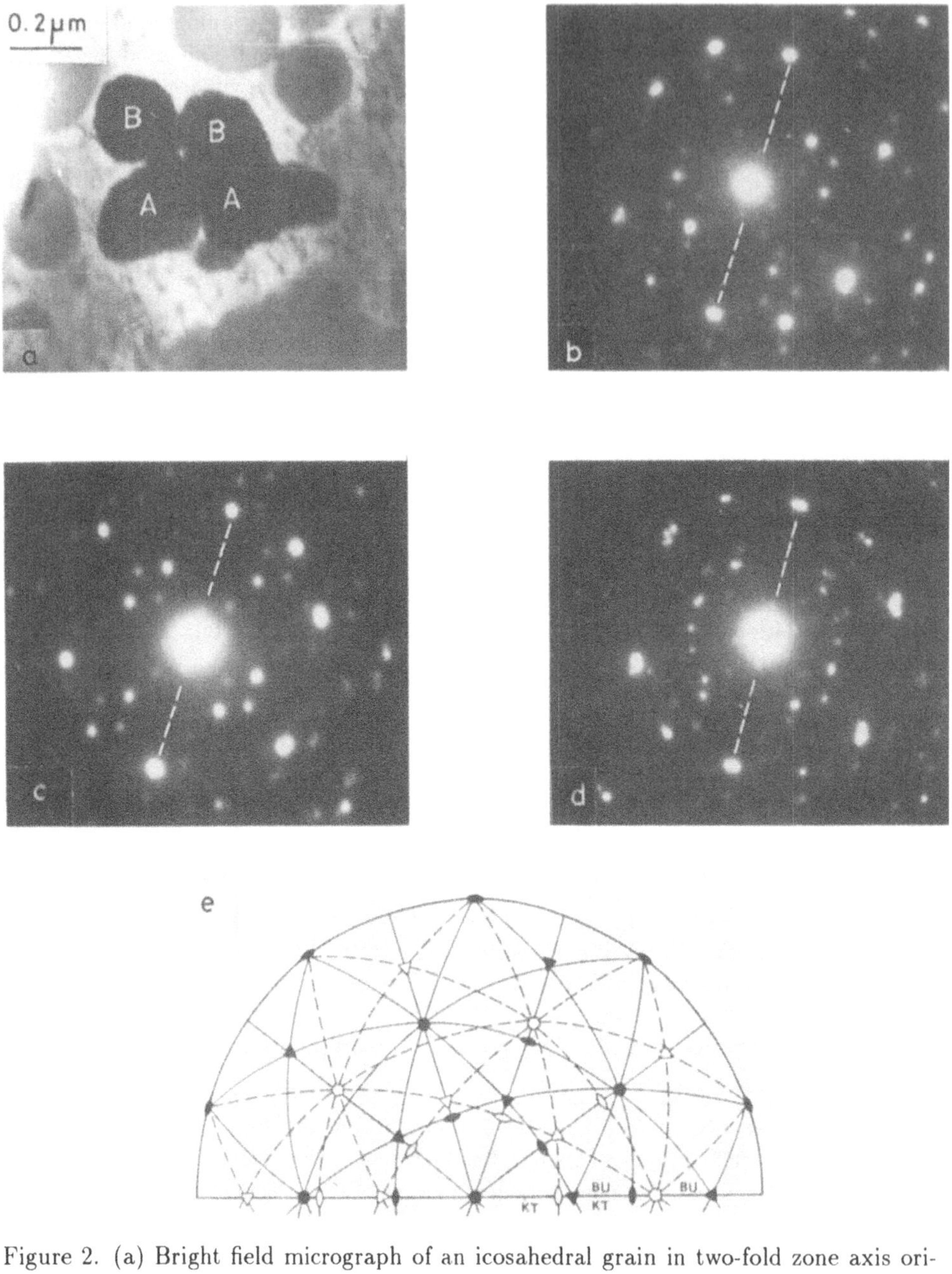

Figure 2. (a) Bright field micrograph of an icosahedral grain in two-fold zone axis orientation in melt-spun Al-5Mn-5Fe alloy. (b) Diffraction pattern marked A in (a), (c) diffraction pattern marked B in (a), (d) a composite diffraction pattern from parts A and B, and (e) a stereogram showing relationship between A and B.

and Ranganathan [38]).

As the growth direction here was shown to be the five-fold direction, a 36° misorientation during the growth along this axis would produce these twins. As observed by Koskenmaki et al.[12], the twin boundaries are curved, thus resembling more to a grain boundary.

4. HYPERTWINS WITH ICOSAHEDRAL SYMMETRY

Rational Approximant Structures can be generated in the projection formalism by changing the orientation of the three dimensional slice related to τ (τ=1.618...) by F_{n+1}/F_n, where F_n is an nth generation number in a Fibonacci series [39,40]. The crystalline phases α-AlMnSi and $Mg_{32}(Al,Zn)_{49}$ are considered to be the 1/1 cubic rational approximant structures of the icosahedral phase [41].

As indicated earlier, Bendersky et al. [6] found aggregates with overall icosahedral symmetry in rapidly solidified Al-Mn-Fe- Si alloys and pointed out that the orientation relationship between crystals is such that icosahedral motifs in all the crystals are parallel. The cubic axes undergo a five-fold rotation about irrational axis $< 1\tau 0 >$, but only five orientations occur among hundreds of crystals. These twins are distinctly different from Pauling's [2] "icosatwins" which are generated by a rotation of <110> by 70.53°. These irrational twins were further studied in Al-Mn-Fe-Si alloy by Mandal et al. [9]. Srivastava and Ranganathan [10] obtained a microstructure of icosahedral phase surrounded by crystalline rings of silicide particles of cubic structure in an Al-Fe-V-Si alloy. The polycrystalline aggregate of cubic crystals with irrational twinning leads to an icosahedral symmetry. Lalla et al. [11] obtained an icosahedral, a decagonal and a new metastable cubic phase in a twinned polycrystalline aggregate configuration giving rise to an icosahedral symmetry in an Al-Mn-Ge alloy.

Electron microscopy of the melt-spun ribbons from $Al_{75}Mn_{10}Cr_5Si_{10}$ reveals a dendritic morphology of the icosahedral phase. Fig. 3 shows bright field image of icosahedral phase, with grains of the order of 2.5 μm, and electron diffraction patterns from it along 5-fold, 3- fold and 2-fold axis respectively.

Apart from the icosahedral phase, the melt-spun foils contained aggregates of twinned crystals. These crystallites are so twinned that the aggregate exhibits icosahedral symmetry. A bright field micrograph of an aggregate and pseudo five-fold, three-fold and two-fold diffraction patterns from it are shown in Fig.4. The bright field micrograph shows a number of moire fringes due to overlapping of crystals of size 0.3 μm to 0.6 μm, identified to be the cubic α-AlMnSi phase. The resemblance of these diffraction patterns to the corresponding ones from the icosahedral phase (Fig. 3) is striking. The twin axis of the rational approximant phases is the irrational [$1\tau 0$] axis [6].

The electron diffraction patterns from the cubic α phase show intensity modulation of diffracted spots due to an icosahedral motif. The [100] pattern can be recognised to be

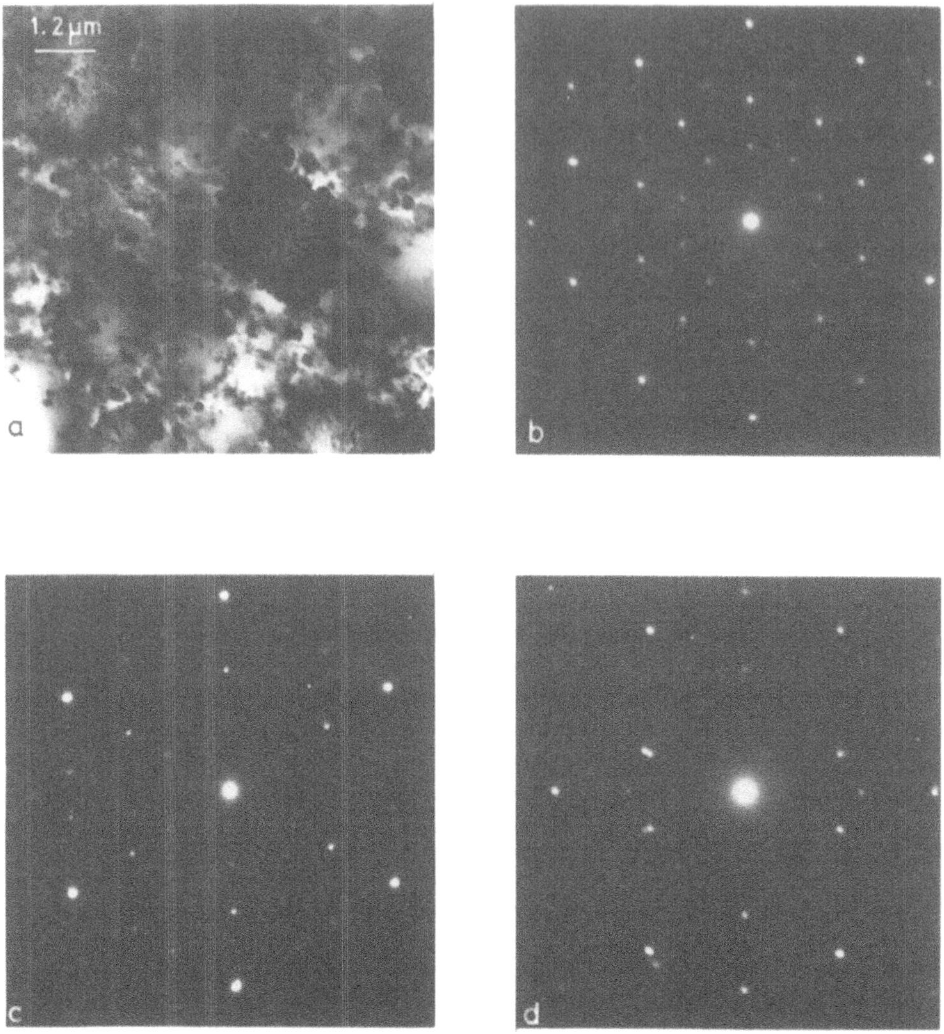

Figure 3. Icosahedral phase in melt-spun Al-Mn-Cr-Si alloy. (a) bright field micrograph, (b) five-fold, (c) three-fold and (d) two-fold diffraction patterns.

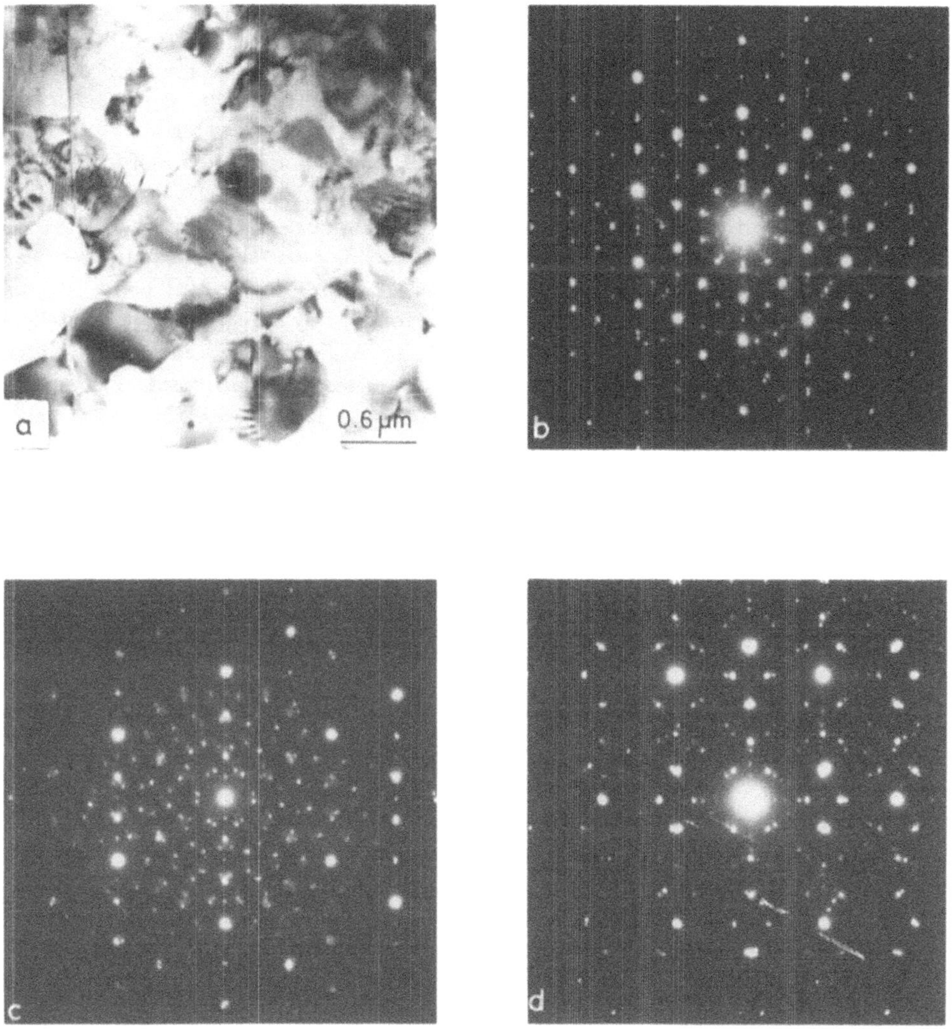

Figure 4. Twinned aggregates of α-AlMnSi phase in melt-spun Al-Mn-Cr-Si alloy. (a) bright field micrograph, (b) pseudo five-fold, (c) pseudo three-fold and (d) pseudo two-fold diffraction patterns.

related to the two-fold pattern of an icosahedral phase and [111] pattern to the three-fold.

Bendersky et al. [6] have demonstrated that five variants of cubic crystals (symmetry $Pm\bar{3}$) rotated 72° along irrational axis $[1\tau 0]$ result in icosahedral point group symmetry $(m\bar{3}\bar{5})$. They have argued that the formation of the polycrystalline aggregate is due to the crystallization of icosahedral phase during growth, but the orientational order of these crystals remains the same as in the icosahedral phase. In another observation Mandal et al. [9] have noted that the icosahedral phase is not always necessary for the nucleation of irrationally twinned sites. But at the prevailing cooling rates, the icosahedral seed can act as a favourable nucleant site for the twinned structure. In a recent paper [10], an intriguing microstructure of icosahedral phase surrounded by an aggregate of crystalline particles of cubic structure has been reported in an Al-Fe-V-Si and an irrational twin relationship of crystalline particles among themselves leading to icosahedral symmetry has been elucidated. It has also been postulated that in this case the formation of crystalline aggregates was due to the decomposition of the quasicrystalline phase [10].

Just as cubic rational approximants can twin to give rise to icosahedral symmetry, rhombohedral rational approximants can twin and give rise to the same symmetry. Audier and Guyot [42] have shown that a twenty-fold twinning of a large unit cell $R\bar{3}m$ crystal of pseudo-icosahedral symmetry may explain various diffraction effects of Al_6CuLi_3 structures.

5. HYPERTWINS WITH DECAGONAL SYMMETRY

Rational approximant structures of the decagonal phase can be twinned leading to decagonal symmetry, as indicated in Table 3. Kumar [43] was the first to point out the close similarity between the crystal structure of monoclinic $Al_{13}Fe_4$ phase and the decagonal phase, making this phase a candidate for the rational approximant. Van Tendeloo et al. [44] established the link between $Al_{60}Mn_{11}Ni_4$ and the decagonal phase. Following this Ranganathan and Chattopadhyay [45] pointed out that the isostructural $Al_{20}Mn_3Cu_2$ and $Al_{24}Mn_5Zn$ were also rational approximants to the decagonal phase. Li and Kuo [46] have shown that the same crystalline structure occurs in binary Al-Mn alloys and christened it as π-phase. A structure closely related to the π- phase, known as Y phase, occurs in Al-Mn-Cu [47,48]. Li et al. [49] and Daulton and Kelton [50] have shown that this phase exists in binary Al-Mn alloys.

A fascinating aspect is the coexistence of Y and π-phases. The patterns from Al_3Mn first presented by FitzGerald et al. [51] were interpreted on this basis by Van Tendeloo et al. [47]. This was confirmed by Daulton and Kelton [50]. Kou's recent set of papers [46,48,49] extend this idea further by highlighting the structural similarity between these two phases.

The twin rolled Al-15.5Cu-8.5Mn alloy shows grains of about half to a micron size with a striated contrast radiating from the centre, giving rise to diffraction patterns resembling those from a decagonal phase [47,52]. Fig.5a is a diffraction pattern along the

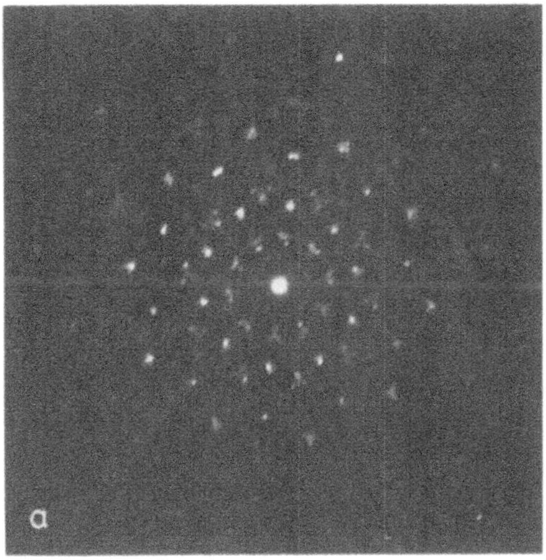

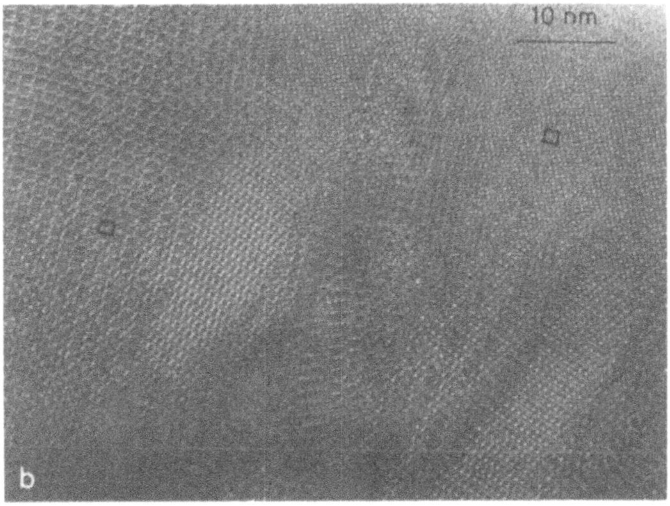

Figure 5. (a) Electron diffraction pattern taken along [010] axis of the Al_3Mn phase showing a pseudo ten-fold symmetry due to twinning in the twin rolled Al-Mn-Cu alloy. (b) A high resolution electron micrograph taken with b axis of Al_3Mn along the electron beam showing microdomains in a grain in twin rolled Al-Mn-Cu alloy.

[010] axis of the Al_3Mn phase. This pattern has a pseudo ten-fold symmetry and, due to the presence of streaking, is clearly distinguished from a real ten-fold. It was found that the Al_3Mn and $Al_{20}Mn_3Cu_2$ coexist in the grains as microdomains with their b axes parallel to each other. The adjacent domains of the same phase were rotated with respect to each other by multiples of 36°.

Fig 5b shows a high resolution electron micrograph taken with the b axis of Al_3Mn and $Al_{20}Mn_3Cu_2$ along the electron beam. The regions showing a rectangular arrangement of spots are the Al_3Mn domains. The regions with a different arrangement of spots are the domains of the B-centred $Al_{20}Mn_3Cu_2$ phase.

6. CONCLUSIONS

Twinning of the icosahedral phase and twinned crystalline aggregates giving rise to quasicrystalline symmetry are analysed as dichromatic pattern of a bicrystal and coset decomposition based on group theoretical approach, respectively, adopted to higher dimensional space.

The twinning of the icosahedral phase with one of the five- fold axis as the twin axis has been illustrated by electron microscopic studies on Al-Mn-Fe alloys and discussed. The icosahedral twin pair mimics the symmetry of the decagonal phase.

The twinned aggregates of the rational approximants of the icosahedral and decagonal quasicrystals are illustrated by the Al-Mn-Cr-Si and Al-Mn-Cu alloys respectively.

7. ACKNOWLEDGMENTS

The authors are grateful to Prof. K. Chattopadhyay for many stimulating discussions. Financial support from the Department of Science and Technology, New Delhi, and the Office of Naval Research Under the Indo-US cooperative project is gratefully acknowledged. One of the authors (SR) is also grateful to the Karnataka State Industrial Investment Development Corporation endowed chair.

8. REFERENCES

1. D.Shechtman, I.Blech, D.Gratias and J.W.Cahn, Phys.Rev. Lett., 53 (1984) 1951.
2. L. Pauling, Nature, 317 (1985) 512
3. A. L. Mackay, Nature, 319 (1986) 103
4. A. L. Mackay, J. Microsc., 146 (1987) 233.
5. L. Pauling, in Aperiodicity and Order, vol. 3, Ed. M. V. Jaric and D. Gratias, Academic Press, New York, 1989, p.137
6. L. A. Bendersky, J. W. Cahn and D. Gratias, Phil. Mag. B, 60, (1986) 837.

7. L. A. Bendersky, J. W. Cahn and D. Gratias, Quasicrystals and Incommensurate Structures in Condensed Matter (eds. M. Jose Yacaman, D. Romeu, V. Castano and A. Gomez), World Sientific, Singapore, 1990, p337

8. L. A. Bendersky, Interfaces: Structure and Properties (ed. S. Ranganathan, C. S. Pande, B. B. Rath and D. A. Smith) Oxford & IBH, New Delhi, 1992, p119

9. R.K. Mandal, G.V.S. Sastry, S. Lele and S. Ranganathan, Scripta Metall. Mater. 25 (1991) 1477.

10. A.K. Srivastava and S. Ranganathan, Scripta Metall. Mater. 27 (1992) 1241.

11. N.P. Lalla, R.S. Tiwari and O.N. Srivastava, J. Mater. Res., 7 (1992) 53.

12. D.C.Koskenmaki, H.S.Chen and K.V.Rao, Scripta Metall. 20 (1986) 1631.

13. S.Ranganathan, R.Prasad and N.K.Mukhopadhyay, Phil.Mag.Lett. 59 (1989) 257.

14. Alok Singh and S. Ranganathan, J. Non-Crystalline Solids (in press)

15. Alok Singh and S. Ranganathan, Acta Metall. Mater. (submitted)

16. R. K. Mandal, S. Lele and S. Ranganathan, Phil. Mag. Lett. (1993) (in press)

17. G. Friedel, Lecons de Crystallographie, Paris, 1926 (2nd ed. 1964 Libraire Sci. A Blanchard, Paris)

18. R. W. Cahn, Adv. Phys., 3 (1954) 363.

19. A. J. C. Wilson (Ed.), International Tables for Crystallography, Vol. C, Kluwer Academic Publishers, London, 1992.

20. A. V. Shubnikov and V. A. Kopstick, Symmetry in Science and Art, Plenum, New York, 1974.

21. V.K. Wadhawan, Phase Transitions 9 (1987) 297.

22. J.W. Cahn and G. Kalonji, Solid-Solid Phase Transformations (Eds.) H.I.Aaronson, D.E.Laughlin, R.F.Sekerka and C.M.Wayman, AIME, USA, 1981, p3.

23. B. K. Vainshtein, V. M. Fridkin and V. L. Indembom, Modern Crystallography II, Springer-Verlag, New York (1982)

24. R. C. Pond and D. S. Vlachavas, Proc. Roy. Soc. London A 386 (1983) 95.

25. R. Portier and D. Gratias, J. de. Physique (France) C4, (1982) 17.

26. T. Hahn (Ed.), International Tables for Crystallography, Vol. A, Kluwer Academic Publishers, London, 1989.

27. J. Devaud-Ryespski, A. Quivy, Y. Calvyrac, M. Cornier- Quiquandon and D. Gratias, Phil. Mag. B 60 (1989) 855.

28. M. Cooper, Acta Cryst. 23 (1967) 1106.

29. R.K. Mandal, Ph.D. Thesis, Banaras Hindu University (1990).

30. R.K. Mandal and S. Lele, Phys. Rev. Lett. 62 (1989) 2695.

31. R. K. Mandal and S. Lele, Phil. Mag. B 63 (1991) 513.

32. R.K. Mandal and S. Lele, Phil. Mag. B, (in Press).

33. N. Wang and K. H. Kuo, Phil. Mag. B 60 (1989) 387.

34. U. D. Kulkarni, S. Banerjee and S. Ranganathan, Scripta Metall. Mater., 25 (1991) 529.

35. T. Ishimasa, H. U. Nissen and Y. Fukano, Phys. Rev. Lett. 55 (1985) 511.

36. R.J.Schaefer, L.A.Bendersky, D.Shechtman, W.J.Boettinger and F.S.Biancaniello, Metall.Trans.A 17A (1986) 2117.

37. H.-W.Nissen, R.Wessicken, C.Beeli and A.Csanady, Phil.Mag.B 57 (1988) 587.

38. Alok Singh and S.Ranganathan, Scripta Metall.Mater. 25 (1991) 409.

39. K. N. Ishihara, Mater. Sci. Forum 22-24 (1987) 223.

40. D. Gratias and J. W. Cahn, Scripta Metall. 20 (1986) 1193.

41. V. Elser and C. L. Henley, Phys. Rev. Lett. 55 (1985) 2883.

42. M. Audier and P. Guyot, Acta Metall., 36 (1988) 1321.

43. V.Kumar, Mater.Sci.Forum 22-24 (1987) 283.

44. G. Van Tendeloo, T. Van Landuyt, S. Amelinckx and S. Ranganathan, J. Microsc. 149 (1988) 1.

45. S. Ranganathan and K. Chattopadhayay, Phase Transitions, 16/17 (1989) 67.

46. X. Z. Li and K. H. Kuo, Phil. Mag. B 65 (1992) 525.

47. G. Van Tendeloo, Alok Singh and S. Ranganathan, Phil. Mag. A 64 (1991) 413.

48. X. Z. Li and K. H. Kuo, Phil. Mag. B 66 (1992) 117.

49. X. Z. Li, D. Shi and K. H. Kuo, Phil. Mag. B, 66 (1992) 331.

50. T. L. Daulton, K. F. Kelton and P. C. Gibbons, Phil. Mag. B 63 (1991) 687.

51. J. D. Fitz Gerald, R. L. Withers, A. M. Stewart and A. Calka, Phil. Mag. B 58 (1988) 15.

52. Alok Singh, A.K. Srivastava and S. Ranganathan, Proceedings of Symposium on Microstructure of Materials, Berkeley, 1992 (in press)

Crystal-Quasicrystal Transitions
M.J. Yacamán and M. Torres (Editors)

THE STRUCTURAL RELATIONS BETWEEN AMORPHOUS, ICOSAHEDRAL, AND CRYSTALLINE PHASES

J. C. Holzer[a] and K. F. Kelton[b]

[a]W. M. Keck Laboratory of Engineering, California Institute of Technology, Pasadena, California 91125

[b]Department of Physics, Washington University, St. Louis, Missouri 63130

1. INTRODUCTION

The tetrahedron constitutes the densest packing of four hard spheres, or equivalently the most stable configuration of four bodies interacting through a pairwise central potential. As illustrated in Fig. 1, however, undistorted assemblies of tetrahedra cannot fill space. Twenty tetrahedra can pack around a common vertex to form a compact structure only if the tetrahedral edges on the surface are lengthened by 5%. The resultant structure, characterized by five-fold, three-fold, and two-fold rotational symmetries, is the icosahedron (Fig. 2), first enumerated by Plato as one of the divine geometrical forms.[1] Structures with icosahedral symmetry abound in nature, particularly in biological systems. This symmetry is conspicuously absent in condensed matter structures, however, due to its fundamental incompatibility with translational periodicity.[2,3]

Because of its energetic stability, the icosahedron dominates the local symmetry in many large unit cell crystalline phases and is postulated to occur frequently in liquids and glasses. Diffraction patterns from these structures, however, show expected crystalline or amorphous features. Evidence for the local icosahedral order is inferred from the observed intensity modulations of the diffraction patterns,[4] the existence of a barrier to nucleation of the stable crystal phase,[5] and computer simulations.[6-9]

In 1984 Shechtman, Blech, Gratias, and Cahn announced the remarkable discovery of an alloy of aluminum and manganese that produced crystalline-like diffraction patterns with an icosahedral point group symmetry.[10] This 'icosahedral phase' (i-phase) has now been found in many metallic alloys. The vast majority of the reports of i-phase formation have been in binary, ternary, and quaternary aluminum and titanium alloys, with at least one of the alloy components being a transition metal. While i-phase formation has been observed in other alloys, they do not fall into any specific categories. A partial list of i-phase-forming alloys can be found in a recent review by Kelton.[11] The report by Shechtman *et al.* was the first of a growing class of condensed matter phases termed quasicrystals, which exhibit crystalline-like diffraction patterns with a noncrystallographic point group symmetry.

While conventional explanations for these apparently noncrystallographic phases have been sought in large unit cell crystals or multiply twinned structures,[12-16] those approaches are incapable of explaining the observed diffraction data. Evidence from dark field images,[17] high resolution transmission electron microscopy,[18-23] convergent beam studies,[24,25] x-ray diffraction,[26] and field ion microscopy[27,28] agree that the i-phase is a new class of condensed matter characterized by a noncrystallographic orientational symmetry, and cannot be explained by conventional crystallographic models.

In most cases the i-phase is metastable relative to competing crystalline phases. Just as for any metastable phase, its occurrence requires that the formation of more thermodynamically stable phases be kinetically inhibited. The standard technique for achieving this condition is rapid quenching from the liquid phase. While most reports of i-phase formation have been in these rapidly quenched alloys, it has also been shown to form by a variety of other processes. These include devitrification of metallic glasses,[29-33] vapor condensation,[34,35] solidification under high pressure,[36] electrodeposition,[37,38] solid state precipitation,[39,40] interdiffusion of multilayers,[41] mechanical alloying,[42] and ion implantation.[43] The formation of the i-phase by a crystal to icosahedral transformation (solid state precipitation, interdiffusion of multilayers, and mechanical alloying) is interesting in that it demonstrates that the i-phase can nucleate in preference to more stable crystalline phases, even though the initial material may not have icosahedral short-range order.

Figure 1. Twenty regular tetrahedra sharing a common vertex. Note the gaps. [From J. F. Sadoc and R. Mosseri, J. de Physique, Coll. C8, 421, (1985).]

Figure 2. Regular icosahedron formed by twenty distorted tetrahedra. [From D. R. Nelson and F. Spaepen, in *Solid State Physics*, (H. Ehrenreich and D. Turnbull, eds.) vol. 42, p. 1, Academic Press, N.Y., 1989.]

In those systems where the i-phase is metastable, it will transform to the stable crystalline phase (or phase mixture) by thermal annealing. Although this route to crystallization is most often studied,[44,45] crystallization can also be induced by ion beam bombardment,[46] application of high pressures,[36,47] and mechanical alloying.[48]

It is also possible to vitrify the i-phase in some cases. Icosahedral Al-Mn[49] and Al-V[50] transforms to an amorphous phase under low temperature electron irradiation. 'Spontaneous vitrification', vitrification of a metastable crystalline phase by low temperature annealing, has been reported for an i-phase in Fe-Cu produced by ion beam mixing.[51] Finally, Eckert *et al.*[48] have shown that an icosahedral to amorphous transition can be induced by ball milling in the Al-Cu-Mn system. In fact, they found that depending on the experimental conditions, an amorphous, icosahedral, or crystalline phase could be obtained by mechanically alloying the pure elements. Furthermore, the three phases could be transformed into to each other reversibly by further milling at the proper intensity.

These qualitative studies of phase transformations to and from the i-phase demonstrate that in most cases the i-phase is intermediate in free energy between crystalline and amorphous phases. However, evidence suggests that icosahedral AlLiCu,[52] GaMgZn,[53] AlCuFe,[54] AlCuRu,[55] AlCuOs,[56] and AlPdMn[57] may be equilibrium phases. The AlCuFe system has been studied extensively, and it has been shown that the i-phase is stable only at high temperatures, transforming to a crystalline approximant at lower temperatures.[58-60] Recent work on AlPdMn indicates that this i-phase is actually stable at room temperature.[61] Much of the current research efforts in this field are aimed at finding and studying these equilibrium quasicrystals.

In this article we will discuss the evidence for short-range polytetrahedral order in glasses, undercooled liquids, and complex crystalline phases and the long-range polytetrahedral order in the i-phase. The majority of models for quasicrystals stress the similarities between their structures and those of related crystalline phases. This is the focus of most of the articles in this book. We will also discuss aspects of this crystal — quasicrystal structural similarity, but particular attention will be given here to the relationships between liquid or glassy structures and the i-phase. Evidence for the similarity of these structures provided by x-ray, electron, and neutron diffraction and by more local probes such as EXAFS and NMR will be examined. We will review work demonstrating that phase transformation studies based on diffraction measurements alone can sometimes be misleading. A quantitative analysis of the mode and kinetics of the transformation can provide additional insight that is often overlooked. Transformation studies and the information they give on structure as well as stability will therefore be discussed at length. Particular attention will be given to recent measurements of the interfacial energy between the i-phase and a related metallic glass.

This review focuses on polytetrahedral order in condensed systems and in detail on the relation between the undercooled liquid, or glass, and the i-phase. Other quasicrystalline phases such as the decagonal phase are only briefly mentioned. Furthermore, no discussion is provided of the crystallography, detailed atomic structure, competing structural models, or physical properties of quasicrystals. The interested reader is referred to other articles in this book or to several review articles[11,62-65] that discuss these points in depth. We will devote much space to discussions of phase transitions to and from the i-phase. Studies of these phase transitions provide insight into the formation and stability of the i-phase. In addition, structural information about the i-phase can be obtained by an examination of the crystalline transformation products. Much attention will be given to the studies of a limited number of metallic glasses that devitrify to i-phases of the same composition. Polytetrahedral order in condensed systems, including clusters, crystalline phases (emphasizing those closely related to the i-phase — the so called crystalline approximants), and liquids and glasses, will be discussed, and the existing experimental evidence for a close structural relation between glassy, quasicrystalline, and crystalline phases will be reviewed. Since polytetrahedral order in the undercooled liquid or glass is assumed to be the origin of the barrier to nucleation of the translationally ordered phase, and since an understanding of nucleation theory is necessary to examine the measurements of the interfacial energy, we will provide a brief review of steady-state and transient nucleation in condensed phases. A thorough review of the amorphous to icosahedral phase transition will then be given. We will explain how a detailed investigation of this transition has allowed us to obtain the first estimate of the interfacial energy between the amorphous and icosahedral phase. We will examine interfaces between the crystal and amorphous phases and speculate on the nature of the interface between the glass and the i-phase. Finally, a summary and concluding remarks are given.

2. POLYTETRAHEDRAL ORDER IN CONDENSED SYSTEMS

A common thread of short-range polytetrahedral order connects liquids, glasses, crystals, and quasicrystals. In quasicrystals this orientational order is maintained over distances comparable to the coherence lengths for translational order in traditional crystals. Before discussing the structural similarities between quasicrystals and these other condensed phases in the next section, it is useful to review the evidence for polytetrahedral order in liquids, glasses and crystals. For more detail, the reader is referred to an excellent review article on this subject.[66]

Attempts to understand polytetrahedral packings date to Frank[5] and Bernal[67,68] in liquids and glasses and to Coxeter,[69,70] a mathematician who introduced a mean field theory of dense-random packing, emphasizing the importance of polytope {3,3,5}, a regular four dimensional solid that allows a tetrahedral packing on the surface, S^3. The surface of a soccer ball, tessellated by pentagons and hexagons, provides a simple illustration of this idea. Attempts to pack these shapes on a flat surface will be thwarted by unavoidable gaps; only by curving the plane will the gaps disappear. This polytope has become a paradigm for the study of the types of disorder generated in three-dimensional structures, arising from the geometrical frustration associated with attempts to maintain extended icosahedral order. It has been used to discuss defects in polytetrahedral crystals,[71] as a model for metallic glasses,[72] and most recently to construct a theory of the statistical mechanics of geometrical frustration.[73-78]

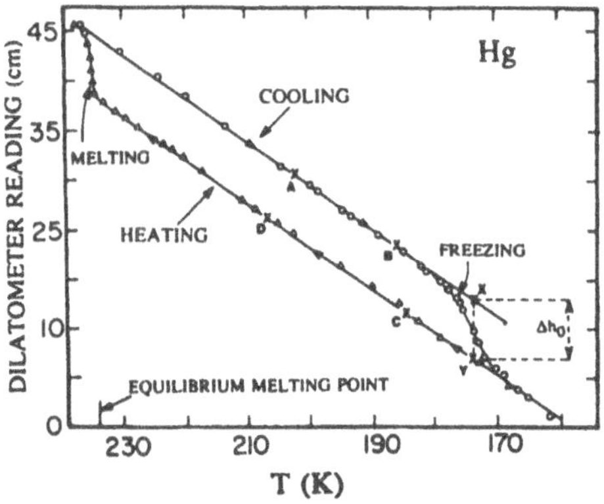

Figure 3. Dilatometric measurements on an mercury emulsion, demonstrating the large undercooling possible in many liquids. [From D. Turnbull, J. Chem. Phys. **20**, 411 (1952).]

2.1. Liquids

Fahrenheit noticed in 1714 that many liquids can be maintained out of equilibrium without crystallizing.[79] The amount of undercooling attainable before solidification can often be quite large, as is illustrated in Fig. 3, showing dilatometric measurements made by Turnbull[80] of liquid and solid mercury emulsions. The classical theory of nucleation, described below, demonstrates that such large undercoolings indicate a large interfacial energy between the undercooled liquid and crystalline phases, in some cases as large as 0.62 times the heat of fusion per atom in the crystal. This is a surprising result given the similarity in the densities of liquids and solids, the average interatomic distances, and the coordination numbers.

In 1952 Frank,[5] considering hard sphere packings, pointed out that there exist three possible arrangements of twelve spheres in contact with a central sphere so that transformations between these configurations requires the breaking of contact with the central sphere (constituting a barrier to transformation), the two close-packed crystalline configurations (fcc and hexagonal) and the noncrystallographic icosahedral packing. For particles interacting via a central potential such as the Lennard-Jones potential, the fully relaxed energy of an icosahedral cluster is lower than the crystalline configurations. Frank observed that the large undercoolings of the liquid suggest a large population of atoms in an icosahedral configuration in the undercooled liquid. To form the crystalline phase, this locally stable polytetrahedral order must be replaced by the energetically more costly tetrahedral and octahedral configurations in the close-packed structures. Such icosahedral order in the undercooled liquid could not be observed directly, by diffraction techniques for example, due to orientational averaging. Although it still remains unobserved, indirect evidence and computer simulations have supported that conclusion.

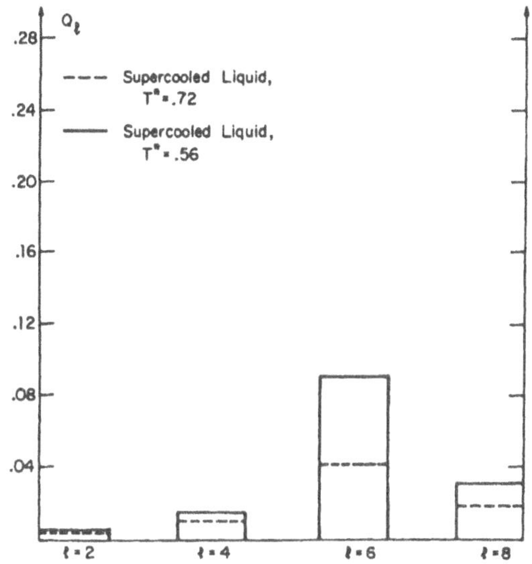

Figure 4. Rotationally invariant bond orientational parameters at two reduced undercooling temperatures. [From D. R. Nelson and F. Spaepen, in *Solid State Physics*, (H. Ehrenreich and D. Turnbull, eds.) vol. 42, p. 1, Academic Press, N.Y., 1989.]

Steinhardt *et al.*[6] made the first quantitative investigation of short-range icosahedral order in undercooled liquids. Particles were taken to interact via the Lennard-Jones pair potential

$$V(r) = 4\varepsilon\left[\left(\frac{\sigma}{r}\right)^{12} - \left(\frac{\sigma}{r}\right)^{6}\right], \quad (1)$$

with periodic boundary conditions. Expanding the density of bonds, ρ, on a unit sphere, surrounding onc of thc atoms at position r, in terms of the spherical harmonics,

$$\rho(r,\Omega) = \sum_{l=0}^{\infty} \sum_{m=-l}^{l} Q_{lm}(r) Y_{lm}(\Omega) , \qquad (2)$$

and defining a set of rotationally invariant bond-orientational order parameters,

$$Q_l = \sqrt{\frac{4\pi}{2l+1} \sum_{m=-l}^{l} |Q_{lm}|^2} , \qquad (3)$$

an increase in the averaged orientational term Q_6 was observed with undercooling (Fig. 4). For icosahedral clusters, only values for $l = 6, 10, 12, \ldots$ are nonvanishing; for fcc clusters, terms with $l = 4, 6, 8, \ldots$ are nonzero. Although this increase in Q_6 could have therefore arisen from either an increase in the number of fcc or icosahedral clusters, an inspection showed that at large undercooling (corresponding to $T^* = .56$ in Fig. 4), the third order cubic invariant[81] was much larger than the value indicative of fcc symmetry.[6,82] Further, the correlation length for icosahedral order grew to three or four interparticle distances with undercooling. Mountain and Brown[83] also reported increases in Q_6 with undercooling in liquids that form a glass.

2.2. Crystals

Some crystal structures contain many atoms that are tetrahedrally coordinated, frequently forming relatively undistorted icosahedra. Often, other polyhedra join the tetrahedrally packed clusters. The bcc structure of $MoAl_{12}$, for example, contains one icosahedral cluster centered at the origin of the unit cell that is linked to the cluster at the body center by an octahedron, sharing opposite faces with both (Fig. 5).[84] The α(AlMnSi)[85] is another example of this type of packing that has proven important for understanding the structure of some i-phases. Larger, 54-atom Mackay icosahedra[86] are packed at the origin and body center as in $MoAl_{12}$, also connected by an octahedron. It is nearly bcc; some of the atom sites exterior but associated with the icosahedron at the body center are missing.

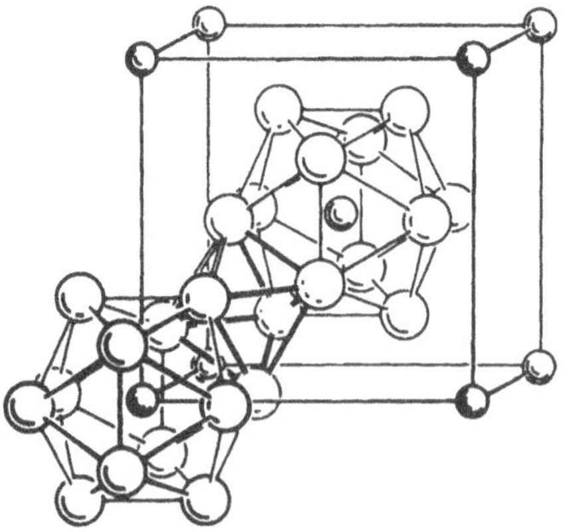

Figure 5. The crystal structure of $MoAl_{12}$. [From L. Pauling, *The Nature of the Chemical Bond, Third Edition*, P. 425, Cornell University Press, 1960.]

The Frank-Kasper[87-89] phases are particularly interesting since they are tetrahedrally close-packed (t.c.p) structures, containing only tetrahedral interstitial sites. Nearly all members of this structural family are alloys of transition metals. The Frank-Kasper phases can be described as layered structures with four layers per lattice repeat along one direction, two main layers and two subsidiary layers of lower atomic density. The main layers consist of pentagons and/or hexagons and triangles; the subsidiary layers consist of squares or rectangles and triangles. In most cases, smaller atoms occupy icosahedral sites and larger atoms sit in sites of non-icosahedral coordination, accommodating the frustration that arises from packing icosahedra. The classic example of polytetrahedral packing is $Mg_{32}(Zn,Al)_{49}$, a large unit cell bcc phase; only one of the constituent polyhedra, the irregular truncated icosahedron, is not a polyhedron with perfect icosahedral symmetry. Other examples of t.c.p. phases include the A15 phases of β-W and Cr_3Si and the σ phase of $Cr_{46}Fe_{54}$ and β-U; a table listing many Frank-Kasper phases with useful structural information is provided in Ref. 66.

Frank and Kasper pointed out that if atoms lying on coordination shells are joined by lines, coordination polyhedra result, dividing space into tetrahedra. Denoting the coordination around a given atom by ZN (the number of vertices on the coordination polyhedra), Frank and Kasper showed that these phases could be built from only four polyhedra, Z12, Z14, Z15, and Z16, shown in Fig. 6. Consistent with the faces being nearly equilateral triangles, emphasizing the tetrahedral symmetry, each polyhedra has 12 vertices with surface coordination of 5 (Z-12), with any additional vertices having coordination 6. None of the 6-coordinated vertices are adjacent. Frank and Kasper showed that connections between the atoms with surface coordination of 6 thread through the crystalline structure, forming a skeleton of lines, disclination lines, containing all the Z14, Z15 and Z16 atoms. Since each of these coordination shells contain at least 2 atoms with sixfold symmetry (there is no Z13 polyhedra), none of these lines can terminate in the crystal structure.

All bonds in polytope {3,3,5} are spindles for fivefold bipyramids, composed of perfect tetrahedra packed around a common bond (Fig. 7a). Since there is a 7.4° mismatch when this structure is projected into 3D space, the tetrahedra must be distorted slightly to ensure closure.If the lengths of the tetrahedra on the edge of the bypramid are lengthened sufficiently, it becomes possible to pack four tetrahedra around the common bond. The coordination of atoms at the end of this bond would then decrease from five to four, corresponding to the loss of one tetrahedron, constituting a +72° disclination line (Fig. 7b). If the lengths along the edge are compressed so that six tetrahedra can pack around the common bond, a -72° disclination line with coordination of six will result (Fig. 7c). Since less distortion is required for the fivefold bipyramid, this bond is taken to be disclination free. Also, as discussed by Frank and Kasper, sixfold bipyramids require less distortion than fourfold ones. As illustrated in Fig. 8, for Z14, the Frank-Kasper polyhedra can be constructed by successive additions of negative disclination lines. A representation of Z14, Z15, and Z16 as nodes in a network of -72° disclination lines is illustrated in Fig. 9 Frank-Kasper phases can therefore be viewed as crystalline structures in which defect lines thread through an otherwise icosahedral medium.

2.3. Glasses

Glasses are liquids that are configurationally frozen upon cooling, so that the amorphous atomic configurations are preserved. For historical reasons, only amorphous solids obtained by cooling the liquid sufficiently fast to kinetically inhibit the formation of crystalline phases are called glasses. Since no differences in structure or properties have been observed between these glasses and amorphous solids of the same composition prepared by other means, such as vapor deposition, mechanical alloying, etc., this technical distinction is now made less often. Cohen and Turnbull[90] predicted that if crystallization could be suppressed, glass formation should be a universal phenomenon. While covalent network-forming glasses (*e.g.* silicates, oxides, etc.) and hydrogen-bonded molecular glasses (*e.g.* glycerine, etc.) are well known, only since Duwez' discovery of Au_4Si[91] has it proven possible to produce glasses from metallic alloys. Subsequently, often based on a search criteria first pointed out by Turnbull and Cohen[92] that

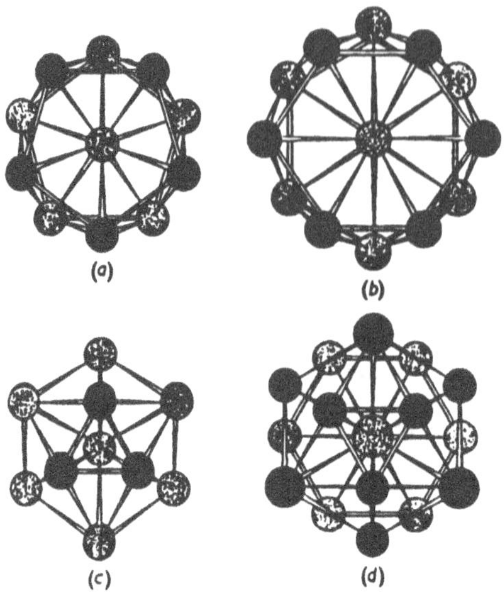

Figure 6. The four Frank-Kasper polyhedra: (a) Icosahedron, Z12, (b) Z14, (c) Z15, (d) Z16. For the icosahedron, two atoms above and below the central atom, along the fivefold axis, are not shown. Similarly, two atoms above and below the central one of Z14, along the sixfold axis, are not shown. For Z16, one atom below the central one is not shown. [From F. C. Frank and J. S. Kasper, Acta Cryst. **11**, 184 (1958).]

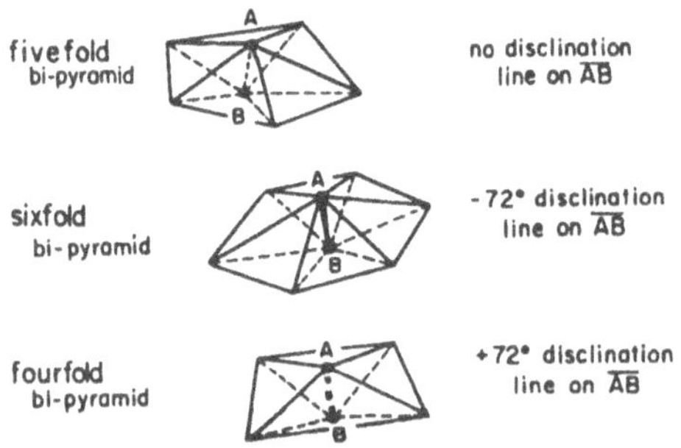

Figure 7. Microscopic definition of defect lines associated with polytetrahedral order. [From D. R. Nelson and F. Spaepen, in *Solid State Physics*, (H. Ehrenreich and D. Turnbull, eds.) vol. 42, p. 1, Academic Press, N.Y., 1989.]

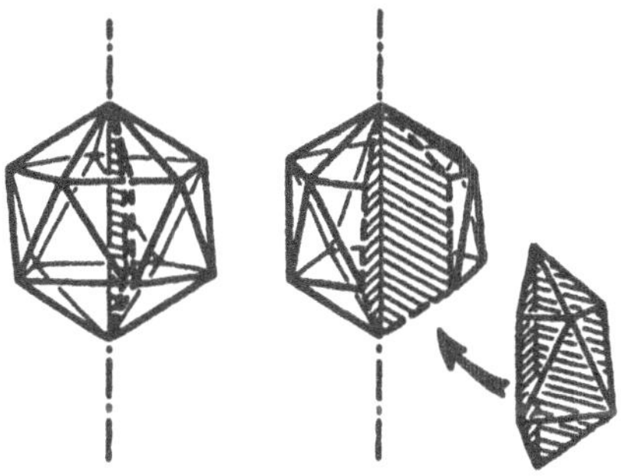

Figure 8. Illustration of the formation of the Z14 Frank-Kasper polyhedra by inserting a disclination wedge into an icosahedron. [From J. F. Sadoc, J. Physique Lett. **44**, L707 (1983).]

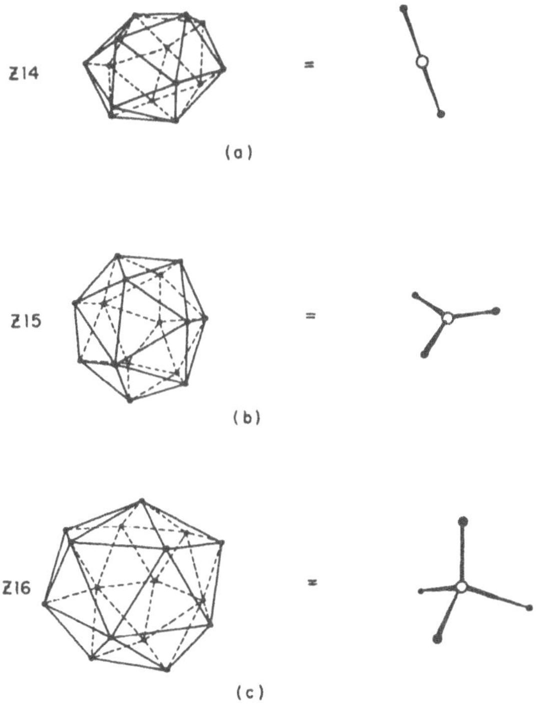

Figure 9. The Frank-Kasper polyhedra represented as nodes in a -72° disclination network. [From D. R. Nelson and F. Spaepen, in *Solid State Physics*, (H. Ehrenreich and D. Turnbull, eds.) vol. 42, p. 1, Academic Press, N.Y., 1989.]

liquids with a composition near a deep eutectic are more likely to form glasses, a large number of glasses have been discovered. Most glasses are metal — metalloid alloys (*e.g.* Au-Si, Ni-P, Pd-Si, Ni-B, etc.), where the late transition or noble metals universally account for approximately 80% of the atoms, or early and late transition metal alloys (*e.g.* Ni-Zr, Cu-Zr, Cu-Ti, etc.). Similarities between the measured structure factors, S(Q), for many amorphous metals and corresponding crystalline alloys suggested that metallic glasses might not be truly amorphous, but are composed of an assembly of microcrystals. Cargill, measuring the x-ray diffraction from amorphous $Ni_{76}P_{24}$, demonstrated that the radial distributions calculated from these microcrystalline models did not fit the experimental data.[93,94] The measured data were instead fit best by the radial distribution function determined by Finney[95] from a dense random packing (DRP) of hard spheres, based on a model proposed originally by Bernal[67] to describe the amorphous structure of monatomic liquids. Treating the hard sphere radius as the only adjustable parameter, the peak positions agree and the splitting of the second neighbor distribution (a particular characteristic of metallic glasses) is reproduced.

Early studies of DRP structures determined the radial distribution by packing ball bearings into rubber bladders, kneading the bladders to maximize the packing fraction, setting the structure in paint or wax, and subsequently taking it apart, recording the location of each sphere by hand.[67,95,96] Bernal showed that the shape of the cavities defined by the nearest neighbors could be described by five polyhedra with equal triangular faces (deltahedral) that are sufficiently small that another sphere of the same size cannot be placed within them (Fig. 10). These polyhedra are the antidefects of the canonical Kasper polyhedra discussed previously, being the nodes of +72° disclination lines.[66] In agreement with Frank's argument for prominent tetrahedral packing in the liquid, 86% of the polyhedra are tetrahedra, while only 6% were octahedra. Studying the Scott model[96] and a larger model (7994 atoms), Finney[95] demonstrated that the coordination polyhedra, determined from the Voronoi cells for each sphere, are dominated by five-sided faces, corresponding to rings of five tetrahedra, the pentagonal bipyramid (Fig. 7a), again in agreement with Frank's proposal for amorphous structures. Even a few cases of perfect pentagonal dodecahedra with perfect icosahedral coordination were found.

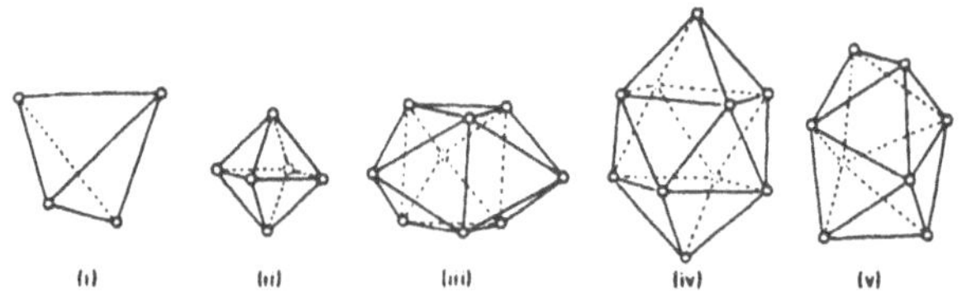

Figure 10. The polyhedra formed by dense random packing of hard spheres of equal size: (i) tetrahedron, (ii) octahedron, (iii) trigonal prisms "capped" with three half octahedra, (iv) Archimedean antiprisms capped with two half octahedra, (v) tetragonal dodecahedron. [From P. H. Gaskell, *Models for the Structure of Amorphous Metals*, in *Glassy Metals II*, Vol. 53, Topics in Applied Physics, Springer-Verlag, Berlin, 1983, pp. 5-47].

All metallic glasses that are deeply metastable near and above room temperature are alloys. Although almost pure glasses have been produced by vapor deposition near liquid helium temperatures,[97] they are unstable to collision-limited crystallization as the temperature is increased. The agreement with Cargill's Ni-P data led Polk[98] to suggest that the smaller metalloid atoms (such as P) occupy the large Bernal holes (Archimedean antiprisms or trigonal prisms), predicting a metalloid concentration of 20 at. %, the observed value. Within this model: (i) metalloid atoms are only surrounded by metal atoms; (ii) the coordination of the metalloid atoms is lower than that of the metal atoms; and (iii) the local environment of the metalloid atoms is similar to that found in crystalline phases.[99] These features have been confirmed by subsequent diffraction studies.[100-102] More recent studies have indicated that the metalloid/metal ratio is too large to accommodate the 20% interstitial metalloids, requiring some distortion of the DRP structure.[103] The qualitative features of the model remain useful, however; in particular predicting the high degree of chemical ordering, also found in analogous crystalline intermetallic compounds.[104] Relaxed binary DRP structures that incorporate this high degree of chemical short-range order, assuming a Lennard-Jones potential, produce extremely good agreement with measured radial distribution functions and structure factors.[105] It is therefore likely that DRP models describe the intrinsic structure of non-network-forming liquids just above the glass transition.

The microscopic defect construction discussed previously for the Frank-Kasper phases can also be used to understand DRP structures.[66] As shown by the DRP model proposed by Ichikawa[97] for amorphous iron films, the unrelaxed structure is dominated by fivefold bipyramids. The number of links of sixfold disclination line are greater than for fourfold, indicating a frustration in 3-D of flat space. Upon relaxation, the number of fivefold bonds increases dramatically, similar to the increase upon undercooling of liquids noted by Steinhardt *et al.*[6]

Nelson[66] argues that the undercooled liquid can be viewed as a tangled collection of $\pm 72^{o}$ disclination lines. In agreement with Ichikawa's results, as the liquid is cooled, some of the plus and minus disclination lines (four- and sixfold bond spindles) annihilate, increasing the degree of short-range icosahedral order. As observed by Ichikawa, however, there remain an excess of sixfold disclination lines, evidence of the geometrical frustration in 3D space. If the cooling is sufficiently slow, these disclination lines will order into a Frank-Kasper phase. As the cooling increases, however, the lines will remain entangled (topological reasons make it difficult for them to cross[66]), unable to kinetically access the more ordered phase, and will drop out of equilibrium - the glass transition. Within this picture, the large atoms in binary alloys of atoms of different sizes relax the frustration by providing preferred sites for the required excess of sixfold bond spindles (typically sitting inside the Z14, Z15 and Z16 coordination shells) and opening up the volume of the structure to fill the cracks in 12 of the coordination shells, resulting in Z12 shells and thus increasing the icosahedral short-range order. This agrees with the results of computer simulations by Jonsson and Andersen.[9]

3. STRUCTURE OF THE ICOSAHEDRAL PHASE

The structure of quasicrystals is frequently viewed as intermediate between the structures of metallic glasses and polytetrahedrally dominated crystalline phases. Like glasses, they have strong short-range order, but lack translational periodicity. Unlike glasses, however, their diffraction patterns are sharply peaked, indicating long-range translational order (assumed to be quasiperiodic), with peaks that reduce to zero at values of Q that are not significantly smaller than those of powdered crystalline samples. Such diffraction patterns and the evidence of strong faceting in some systems[106-109] are more reminiscent of crystalline phases.

Thermodynamically, quasicrytsals are assumed to lie intermediate in energy, higher than crystalline phases but lower than metallic glasses. There is no proof for this belief; it reflects a long-standing prejudice that the lowest free energy state is always that of a periodic system. For at least one alloy, $Al_{65}Cu_{20}Fe_{15}$, there is considerable evidence that the i-phase is more stable than the competing crystalline phase in some temperature ranges.[58-60] This is taken to reflect the additional entropy in the quasicrystal. Also, the spontaneous vitrification of alloys of $Ti_{60}Cr_{40}$ has been reported,[110] though this is in dispute. These results clearly suggest that the most energetically favored structure may not be periodic in all cases.

3.1. Similarity to Crystalline Phases

Similar to the methods of standard crystallography, the atomic structure of quasicrystals is often modeled by first constructing a geometry, and then decorating that geometry with a basis. Several methods exist for generating quasicrystals deterministically, including inflation, tilings with matching rules, and projections from a higher dimensional space. Nondeterministic methods include the icosahedral glass (a random packing of icosahedral clusters, maintaining their orientational correlations) and random tiling methods. These and related methods are reviewed in detail elsewhere.[11,64,65]

There are many examples of the close similarity between quasicrystals and related crystalline phases, demonstrated by electron diffraction. The reader is referred to the contributions in this book by Kuo, Fang-Hua, Ranganathan, and Yacaman, *et al.* for a thorough discussion of these data. The structural similarity between the i-phase and several crystalline approximants in the Ti-Mn-Si system have been discussed by Levine *et al.*[4] They provide evidence that strongly suggests that the i-phase and several related crystalline phases all contain the same cluster of atoms with icosahedral symmetry. The implication is that a common atomic 'building block' is packed differently in the different phases. This is further supported by the NMR measurements of Jeong *et al.*,[111] who found that the ^{55}Mn NMR linewidth and Knight shift of icosahedral $Ti_{63}Mn_{37}$ are identical to those of the related crystalline phases obtained upon annealing.

The notion of a common building block of atoms in the icosahedral and related crystalline phases is also suggested in a variety of other systems. For example, electron diffraction and microscopy,[112,113] EXAFS measurements,[114] and EELS measurements[115] comparing icosahedral Al_6CuLi_3 and cubic Al_5CuLi_3 demonstrate a close similarity in local atomic order. Atomic units with icosahedral symmetry are suggested to be common to both structures.[113]

The i-phase often grows with a fixed orientational relationship to neighboring crystalline phases. Understanding these relationships leads to information about the structural similarities between the phases. It has been demonstrated that the addition of small amounts of Si to icosahedral Al-Mn stabilizes the i-phase against formation of the decagonal phase.[116,117] Based on the close similarity between the <230> and <100> electron diffraction patterns of the cubic α(AlMnSi) phase and the 5-fold and 2-fold patterns of the i-phase, Elser and Henley[118] and Guyot and Audier[119] have suggested that the two phases are closely related, *i.e.* they contain the same atomic building block. Motivated by these observations, Fung and Zhou[120] directly observed the transformation of icosahedral $(Al_6Mn)_{1-x}Si_x$ ($x = 0.02$ and 0.04) into α(AlMnSi) using a heating stage in an electron microscope. They found that the orientational relationship between the transforming i-phase and the α phase is

$$\begin{gathered} i2 // \langle 001\rangle, \langle 132\rangle \\ i5 // \langle 035\rangle \\ i3 // \langle 111\rangle, \langle 052\rangle, \end{gathered}$$

confirming the predictions by Elser and Henley, and Guyot and Audier. The same orientational relationships were observed in as-quenched samples of Al-Mn-Si.[121] In addition, this orientational relationship has been observed between the i-phase and a large unit cell bcc phase in as-quenched samples of $Ti_{73-x}V_{27}Si_x$ ($13 \leq x \leq 25$) (Fig. 11),[122] and for a large unit cell bcc phase nucleating on the boundary between the i-phase and β-Ti in rapidly quenched TiFe alloys.[123]

For good reasons, therefore, the structural relations between quasicrytsals and related crystalline phases have traditionally been used to predict local atomic decorations of the i-phase: (i) the diffraction patterns of the i-phase and crystalline phases are often quite similar;[4,124] (ii) the best i-phase compositions, those being most stable and having fewest defects, are often found near the compositions of related crystalline phases;[117] and (iii) the i-phase and crystalline phases are often found forming together with a strong orientational relation (Fig. 11).[120-123] Phases that are isomorphic with the α(AlMnSi) and the Bergman (AlMgZn) crystalline phases are most frequently used for structural modeling of aluminum based quasicrystals; similar

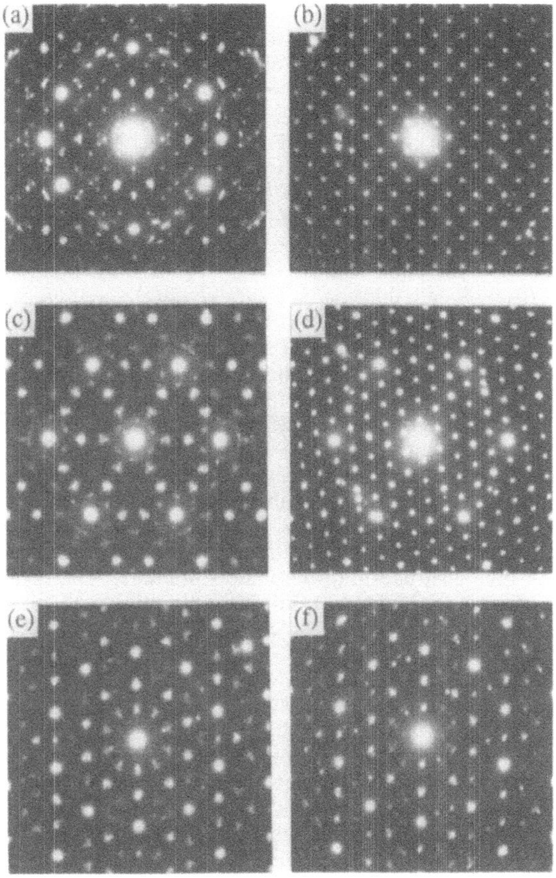

Figure 11. SAD patterns of an icosahedral—bcc phase mixture, Ti-V-Si, along prominent zone axes of the i-phase: (a) along the twofold icosahedral direction, coincident with (b) the [001] zone axis; (c) along the threefold icosahedral direction, coincident with (d) the [111] zone axis; (e) along the fivefold icosahedral direction, nearly coincident with (f) the [530] cubic zone axis. [From X. Zhang and K. F. Kelton, Phil. Mag. Lett. **63**, 39 (1991).]

phases have recently been reported near the i-phase-forming compositions of the titanium quasicrystals. As demonstrated by Elser and Henley,[118,125] these crystalline structures can be obtained by a 1/1 projection from the same 6D hypercubic lattice used to construct the i-phase.

These arguments for structural similarity were strengthened by Henley,[125] who argued that all icosahedral alloys can be classified by the ratio of the quasilattice constant, a_R, determined from the diffraction patterns, to the typical interatomic spacings, d, determined from crystalline alloys with the same composition. If $a_R/d \approx 2.0$, the i-phase belongs to the (Al,Zn)Mg class; if $a_R/d \approx 1.65$, it belongs to the (AlMnSi) class. Yang[126] pointed out that this was equivalent to proposing that if the i-phase compositions were close to the Frank-Kasper phases, containing only tetrahedra, they should belong to the (Al,Zn)Mg class, while those close in composition to crystalline phases, containing octahedra and tetrahedra, belong to the AlMnSi class. By this scheme, most quasicrystals belong to the (AlMnSi) class, including i(PdUSi), i(AlCuV), i(TiMnSi), and i(AlCuFe); i(AlLiCu) belongs to the (Al,Zn)Mg class. Recently Kalugin[127] has introduced a new scheme that differs with some of these classifications.

Based on these similarities, the icosahedral atomic structure has been determined by decorating a network of randomly packed icosahedral clusters[17,128-132] or projecting from a chemically decorated 6D hypercubic lattice.[133-137] The first method relies on the observation that the crystalline structures are dominated by icosahedrally symmetric clusters, such as the Mackay icosahedra, which are taken to be the same clusters packed differently for the i-phase. The cluster decorations are determined from scattering studies of the appropriate crystalline approximants. With the proper constraints, this does a good job in reproducing the diffraction patterns.[131,132] The projection method also relies on knowing the chemical decoration of the crystal approximant. Assuming that α(AlMnSi) is a 1/1 projection, Cahn, Gratias and Mozer determined the atomic decorations of the hypercubic lattice from a Patterson map constructed from scattering studies.[133,134] The atomic structure of i(AlMnSi) was then obtained by taking the appropriate slice through the hypercubic lattice. While the early work produced unphysically short bond lengths, several ad hoc schemes have been proposed to prevent this.[135,136] One severe criticism remains; the diffuse scattering present in the diffraction patterns is ignored in the construction of the Patterson maps. The resulting unphysical local atomic configurations would be a natural result if the structure were not a strict Penrose tiling, but a random packing.[138]

3.2. Similarity to Metallic Glasses

Since there is considerable evidence that an increased icosahedral short-range order occurs in the undercooled liquid and glass, and since the i-phase appears to have an extended icosahedral order, it is natural to inquire into the possible similarities between the icosahedral and metallic glass phases. It is possible that i-phases quenched from the liquid might form because of a proclivity for particularly strong icosahedral order within the liquid. It may then be possible to search for new i-phases by examining the diffraction spectra of liquid metals. Since the i-phase forms a link to the crystalline phase, such studies might also shed additional light on the local structure within metallic glasses.

Dubois and Janot[139] make the additional observation that while polymorphic crystallization of the i-phase (crystallization with no compositional change), might possibly elucidate the relationships between the quasicrystalline structures and the crystalline states, all such transformations require a considerable change in composition. These include reactions such as i($Al_{86}Mn_{14}$) to orthorhombic Al_6Mn,[45] i($Al_{73}Mn_{21}Si_6$) to hexagonal β($Al_{69}Mn_{23}Si_8$) or cubic α($Al_{71}Mn_{17}Si_{12}$),[120] or i($Al_{5.7}Cu_{1.1}Li_{3.2}$) to R($Al_{5.6}Cu_{1.2}Li_{3.2}$).[140] They therefore argue that these constitute narrow phases that do not overlap, likely related, but for which improper account of the compositional differences have been taken. Supporting the notion of similarities between the glass and quasicrystal, there are several examples of polymorphic devitrification of metallic glasses to quasicrystals. These transformations are discussed in detail below; a detailed study of the devitrification of amorphous AlCuV does indeed point to a strong local structural similarity. In this section, we present structural evidence for a similarity in these structures.

EXAFS measurements at the Mn K edge made on the icosahedral and orthorhombic phases in an Al_6Mn alloy have been reported by several authors.[141-144] Boyce *et al.*[143] pointed out that the Mn near-neighbor distribution in icosahedral Al-Mn is significantly broader than that in orthorhomic Al_6Mn, suggesting a superposition of a number of inequivalent Mn sites. Subsequent EXAFS measurements on icosahedral $Al_{1-x}Mn_x$ ($12 \leq x \leq 20$) and $Al_{74}Mn_{20}Si_6$, amorphous $Al_{85}Mn_{15}$, and crystalline Al_6Mn demonstrated that the amorphous and i-phases have very similar near-neighbor Al-Mn distributions, both of which are broader and more asymmetric than that for orthorhombic Al_6Mn.[144]

Warren *et al.*[145] measured the NMR spin-echo spectrum of ^{27}Al and ^{55}Mn for icosahedral $Al_{86}Mn_{14}$ and $Al_{74}Mn_{20}Si_6$ and for the amorphous and equilibrium crystalline forms of $Al_{86}Mn_{14}$. They found that the distribution of local electric-field-gradient magnitudes at the Al sites in icosahedral $Al_{86}Mn_{14}$ is only slightly narrower than in the amorphous phase, while there is no detectable difference in the distribution of local environments for the Mn sites in the two phases. This is further evidence that the amorphous and i-phases have a very similar local environment. Other NMR studies of Al-Mn quasicrystals[146,147] have led to this same conclusion: The atoms sit in a relatively broad distribution of sites of relatively low symmetry.

This is supported by Mössbauer-effect measurements on the i-phase in $Al_{86}(^{57}Fe_{0.02}Mn_{0.98})_{14}$, which indicate the existence of more than two inequivalent Mn sites.[148] The Mössbauer spectrum of the i-phase can be fitted at least as well with a continuous distribution of sites of the form suggested by an amorphous analog, as with a discrete set. Furthermore, Eibschutz *et al.* have shown that the distributions of the quadrupole splitting in the icosahedral and amorphous phases are comparable, in agreement with Mössbauer spectroscopy of amorphous and icosahedral Al-Fe.[149] Again, this indicates that the local structures of these two phases are quite similar.

Low-temperature heat capacity measurements on amorphous and icosahedral Al-Mn also support a similarity in short-range order. Low temperature heat capacity data are usually fit to an equation of the form $C_p = \gamma T + \beta T^3$, where γ is the electronic contribution and β is the lattice contribution. Berger *et al.*[150] found that both the electronic and lattice contributions to the low-temperature heat capacity in amorphous $Al_{85}Mn_{15}$ are similar to that for icosahedral $Al_{86}Mn_{14}$, suggesting a similar short-range order in the two phases.

A similarity in the atomic structures of the liquid and i-phase for Al-Mn has also been inferred from detailed crystallization studies.[45] We briefly review these results here. The i-phase in as-quenched $Al_{86}Mn_{14}$ occurs as dendritic nodules that are 1-10 μm in extent and separated by a layer of a solid solution α-Al containing 3-6 at. % Mn.[24,45] TEM investigations[45,49] agree that crystalline Al_6Mn nucleates at the phase boundary between the i-phase and α-Al. The transformation rate varies from nodule to nodule.[151,152] McAlister *et al.*[151] asserted that this was evidence for two types of nucleation sites. However, we believe that this is due to a low nucleation frequency,[45] since probability considerations show that nuclei will only develop in some nodules in a finite time. TEM studies by Köster and Schuhmacher[153] in $Al_{89}Mn_{11}$ demonstrated that the Al_6Mn forms a shell around the periphery of the i-phase nodule and grows by a peritectoid reaction into the i-phase. This morphology has not been reported for any other alloy composition.

Chen *et al.*[44] were the first to study the kinetics of crystallization of the i-phase in $Al_{86}Mn_{14}$ alloys. Based on their isothermal differential scanning calorimetry (DSC) data, they concluded that the crystallization of the i-phase to orthorhombic Al_6Mn is a diffusion-controlled transformation with a constant nucleation rate. Similar results were obtained by others.[45,151] Fitting isothermal and nonisothermal DSC data and isothermal changes in the electrical resistivity, and using the results from TEM observations of the evolving microstructure, we obtained the first estimate of the nucleation rate and growth velocity for this transformation.[45] Based on these data, we were able to estimate a lower bound of 0.03 J/m^2 on the interfacial energy between the i-phase and orthorhombic Al_6Mn. This is similar to the crystal — liquid interfacial energies for pure Al (0.09 J/m^2) and pure Mn (0.21 J/m^2).[154,155] Furthermore, it is much smaller than a typical orientationally averaged interfacial energy between unlike crystalline phases.[156] This suggests that the i-phase can accommodate the crystalline boundary in a similar fashion to the liquid, and indicates a similar atomic structure in the i-phase and the liquid.

Evidence for a similar short-range order in the amorphous and i-phases is not restricted to the Al-Mn system. Matsubara *et al.*[157] studied the atomic structures in amorphous and icosahedral $Al_{75}Cu_{15}V_{10}$ with anomalous-x-ray-scattering. Their radial distribution function indicates the presence of icosahedral clusters in the as-quenched amorphous sample and in other annealed amorphous samples. Koflat *et al.*[158] determined the total and differential atomic pair-correlation functions of amorphous and icosahedral $Pd_{58.8}U_{20.6}Si_{20.6}$ using differential anomalous-x-ray-scattering. They found that the pair distribution functions in the amorphous and i-phases are comparable up to the second-nearest neighbors. Similar results were obtained by Fuchs *et al.*[159] using x-ray and neutron diffraction. Levine *et al.*[160] found comparable plasmon widths in the amorphous and i-phases for both $Pd_{58.8}U_{20.6}Si_{20.6}$ and $Al_{75}Cu_{15}V_{10}$. In addition, von Löhneysen *et al.*[161] and Wosnitza *et al.*[162] found that the vibrational contribution to the specific heat is almost identical for amorphous and icosahedral $Pd_{58.8}U_{20.6}Si_{20.6}$, while there is only a small difference in the large electronic contributions in the two phases. All of these results suggest a similar short-range order in the amorphous and i-phases.

As already discussed, the structural relationships between polytetrahedrally packed crystalline phases, undercooled liquids, and glasses can be understood in terms of an unfrustrated icosahedral lattice on the surface, S^3, of a four dimensional sphere. Peaks in the structure factor are predicted at positions determined by the symmetries of the curved-space icosahedral crystal. These correspond to the peak positions observed in DRP models. Figure 12 shows the structure factors from vapor-deposited amorphous cobalt,[163] a computer-cooled Lennard-Jones glass,[164] the bimetallic glass $Mg_{70}Zn_{30}$,[165] and the metal-metalloid glass $Fe_{80}B_{20}$;[166] the qualitative features are similar. As illustrated for amorphous cobalt, the prominent amorphous peaks can be indexed by the integers n = 12,20,24,..., which also index the reciprocal lattice vectors of the curved-space icosahedral crystal;[76] the peak positions are related by $q_{20}/q_{12} \approx 1.7$ and $q_{24}/q_{12} \approx 2.0$.[77,78] This indicates the importance of the short-range icosahedral order in the amorphous structures. The peaks are presumably broadened due to the interruption of the icosahedral order by a tangled network of wedge disclination lines.[73] Additional broadening is present due to compositional fluctuations and density excess fluctuations over the best packed configuration arising from the rapid cooling rates of the quenching process. Annealing the as-quenched glass is known to densify the structure and sharpen the peaks by a process of structural relaxation.

Sachdev and Nelson[167] pointed out that the diffraction patterns of metallic glasses and icosahedral alloys are also similar. This can be seen by comparing Fig. 13, showing a schematic of the two-fold diffraction pattern from a simple icosahedral alloy, to the amorphous structure factors in Fig. 12. As first discussed by Sachdev and Nelson,[168] and Elser,[169] the peaks of the icosahedral diffraction pattern can be indexed conveniently in terms of the symmetry of an icosahedron, choosing a basis of vectors of length q_o pointing to the vertices of an icosahedron. Four members of this set lie in the two-fold symmetry plane, the four large spots close to the origin. The next most intense spots, formed by linear combinations of these fundamental vectors, q_A, q_B, and q_C, and their lengths are also indicated. Assuming that the first peak in the amorphous structure (Fig. 12) is broad enough to contain the q_o and q_A peaks, the n=20 and n=24 peaks then correspond to q_B and q_C respectively.

These observations strongly support the notion that the structures of the glass, quasicrystal, and certain crystal phases are similar. Reviewing the arguments presented previously, as a liquid is undercooled, a network of disclination lines is generated due to geometrical frustration. As the undercooling is increased, if it is kinetically possible, these disclinations crystallize into a lattice, giving rise to the complex intermetallic phases observed. If the cooling is too rapid, they entangle and form a glass. Within this view, an intermediate cooling rate may result in the system ordering the disclination lines as to preserve the long-range icosahedral orientational order, but not able to reach a crystalline order, producing a quasicrystal. Landau[170-172] and density functional[167] calculations have shown that the degree of metastability of the i-phase is determined by the balance of the stabilizing effect that the most prominent Bragg peaks are in approximate registry with the peaks in the structure factor of the amorphous phase, against the destabilizing penalty associated with the long-wavelength manifestation of the frustration associated with preserving the short-range icosahedral order.

This scenario for i-phase formation and structural similarities between these phases is supported by the measurements of local structural order, such as the interfacial energy between the glass and the i-phase and nucleation measurements from the liquid and glass. These are discussed in detail below.

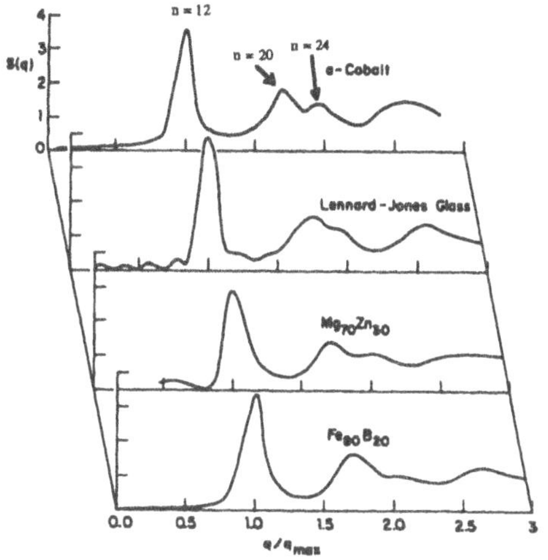

Figure 12. Structure factors of vapor-deposited amorphous cobalt, a computer-cooled Lennard-Jones glass, the bimetallic glass $Mg_{70}Zn_{30}$, and the metal-metalloid glass $Fe_{80}B_{20}$. The structure factors have been rescaled to make their primary peaks coincide. [From S. Sachdev and D. R. Nelson, Phys. Rev. B **32**, 4592 (1985).]

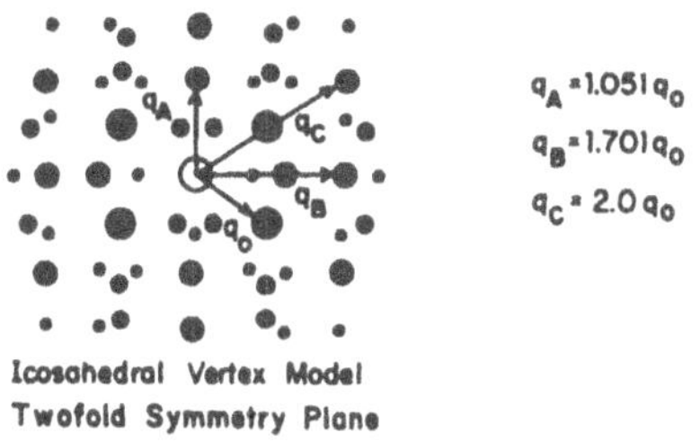

Figure 13. Diffraction pattern for the vertex icosahedral crystal. [From S. Sachdev and D. R. Nelson, Phys. Rev. B **32**, 4592 (1985).]

4. HOMOGENEOUS NUCLEATION THEORY

The observed metastability of many undercooled liquids can be understood by assuming that the phase transformation is initiated by large amplitude fluctuations of the appropriate order parameter, leading to the appearance of small regions of the stable crystalline phase. This process is called nucleation; the rate at which regions of the new phase appear is the nucleation rate.

The bulk of evidence supports the view that quasicrytals form from the liquid by a first order, nucleation-induced phase transformation. In this case, quantitative phase transformation studies can uncover valuable structural information. Since a knowledge of classical nucleation theory is necessary for the interpretation of these data, we review the salient points of that theory here. Nucleation in undercooled liquids frequently occurs at heterogeneous sites (*e.g.* container walls, foreign particles, and free surfaces), but these data are difficult to interpret theoretically. However, by carefully controlling the conditions of nucleation and growth, the effects of these heterogeneities can be limited in some cases, and quantitative interpretation becomes feasible; such data are discussed in the following section. We will review only the homogeneous nucleation theory here. Relevant issues dealing with heterogeneous nucleation will be discussed within the context of some data analysis in the following section.

4.1. Steady State Homogeneous Nucleation

Fundamental to the theory of nucleation is the concept of a barrier to nucleation. It is this barrier that leads to the observed metastability of undercooled liquids. Nucleation and the related process of growth are thermally activated processes. Near equilibrium, the barrier to nucleation is large and the probability of occurrence for a significant number of fluctuations is infinitesimal. For large departures from equilibrium, yet still within the metastability region, the barrier decreases to a few k_BT. Nucleation is typically described by the classical theory, a phenomenological theory in which actual clusters of atoms or molecules in the configuration of the transformation product arise spontaneously from fluctuations within the initial phase. Assuming spherical clusters of the same composition as the initial phase, and ignoring stress effects and cluster translation and rotation, the probability of obtaining a cluster of the new phase containing n atoms, P_n, is

$$P_n \propto \exp\left[-\frac{\Delta G_n}{k_B T}\right], \tag{4}$$

where ΔG_n is the reversible work of cluster formation and k_B is Boltzman's constant.

Following Gibbs, assuming a sharp interface between the cluster and the initial phase,

$$\Delta G_n = n\Delta G' + (36\pi)^{1/3} v^{2/3} n^{2/3} \sigma, \tag{5}$$

where $\Delta G'$ is the Gibbs free energy per molecule of the new phase less that of the initial phase, v is the molecular volume, and σ is the interfacial energy per unit area. Below the equilibrium transition temperature (*i.e.* the melting point if we are talking about nucleation in an undercooled liquid), $\Delta G'$ is negative. Since σ is always positive, ΔG_n increases for small cluster sizes due to the large surface to volume contributions, but decreases for large clusters where the volume free energy is dominant. A maximum in ΔG_n as a function of n is then obtained, corresponding to a critical cluster size,

$$n^* = \frac{32\pi}{3v} \frac{\sigma^3}{|\Delta G_v|^3}, \tag{6}$$

where ΔG_v is the Gibbs free energy per unit volume. Clusters smaller than n* will tend to shrink, while larger clusters will on average grow.

Within the kinetic model for classical nucleation theory, clusters evolve in size by a series of bimolecular reactions:

$$E_{n-1} + E_1 \underset{k_n^-}{\overset{k_{n-1}^+}{\rightleftharpoons}} E_n \tag{7a}$$

$$E_n + E_1 \underset{k_{n+1}^-}{\overset{k_n^+}{\rightleftharpoons}} E_{n+1}\,, \tag{7b}$$

where E_n represents a cluster of n molecules and E_1 a single molecule, k_n^+ is the rate of monomer addition to a cluster of size n and k_n^- is the rate of loss. The time-dependent cluster density $N_{n,t}$ is then determined by solving a system of coupled differential equations of the form,

$$\frac{dN_{nt}}{dt} = N_{n-1,t}k_{n-1}^+ - [N_{n,t}k_n^- + N_{n,t}k_n^+] + N_{n+1,t}k_{n+1}^- \,. \tag{8}$$

The nucleation rate is a time-dependent flux of clusters past a given cluster size n,

$$I_{n,t} = N_{n,t}k_n^+ - N_{n+1,t}k_{n+1}^- \,. \tag{9}$$

In general then, it is a function of the cluster size at which it is measured.

Generally, a steady-state cluster distribution is assumed. As first shown by Becker and Döring,[173] the steady-state nucleation rate is independent of the cluster size at which it is measured and has the form[155]

$$I^s = \frac{24Dn^{*2/3}N_A}{\lambda^2}\left[\frac{|\Delta G'|}{6\pi k_B T n^*}\right]^{1/2} \exp\left[-\frac{\Delta G_{n^*}}{k_B T}\right] \tag{10}$$

per mole, where D is the diffusion coefficient in the initial phase, N_A is Avogadro's number and λ is the atomic jump distance. Since the critical size, n^*, and ΔG_{n^*} decrease with undercooling, Eq. (10) predicts a corresponding sharp rise in the steady-state nucleation rate. It drops with further undercooling since the atomic mobility (reflected in the diffusion coefficient) drops sharply with decreasing temperature.

Eq. (10) has the form:

$$I^s = A^* \exp\left[-\frac{W^*}{k_B T}\right], \tag{11}$$

where W^* is the nucleation barrier. The nucleation rate is therefore proportional to the thermodynamic probability of having a fluctuation leading to the formation of a critical cluster and a dynamical factor that describes the rate at which that cluster traverses the barrier region, A^*.

Assuming that the free energy difference per unit volume, ΔG_v is proportional to the undercooling, $\Delta T = T - T_m$,

$$\Delta G_v = \frac{\Delta H_f \, \Delta T}{T_m}\,, \tag{12}$$

where ΔH_f is the enthalpy of fusion per unit volume and T_m is the equilibrium melting temperature. The steady-state nucleation rate can then be written as

$$I^s = A^* \exp\left[-\frac{16\pi T_m^2}{3k_B\Delta H_f^2}\frac{\sigma^3}{T\Delta T^2}\right]. \tag{13}$$

The amount of undercooling possible is therefore strongly dependent on the interfacial energy between the initial and final phases, giving a sensitive method for estimating that quantity. Available undercooling data for pure metals were compiled recently.[155] As originally pointed out by Turnbull,[154] the gram-atomic interfacial energy is linearly related to the measured enthalpies of fusion, with a slope of approximately 0.43. Quantitative fits to nucleation data taken as a function of temperature in Hg and several silicate glasses suggest that the interfacial energy increases with temperature. Turnbull[174] and Spaepen *et al.*[175,176] have argued that this must be due to a sizable negative entropy near the interface arising from ordering in the liquid, the 'negentropy' effect.

4.2. Time Dependent Homogeneous Nucleation and Glass Formation

The number of nuclei produced as a function of time, N_v, is given by

$$N_v = \int_0^t I_{n,t}\, dt \tag{14}$$

For a steady state nucleation rate, $I_{n,t} \approx I^s$; N_v will therefore be linearly related to time. Nucleation studies in some glasses, however, show an initially low nucleation rate, only approaching the steady state rate for long annealing times (Fig. 14). This transient behavior is common; it has been observed in metallic glasses, glass-ceramics, enamels, and many silicate glasses. A recent study of the transient nucleation rate as a function of preannealing treatments, has provided a quantitative check on the validity of the kinetic model for cluster evolution.[177]

The origin of such time dependent nucleation is readily understood within the classical theory. Since glasses are formed by cooling the melt rapidly, there is generally insufficient time for the required atomic rearrangements to maintain the steady state cluster distribution. The number density of clusters increases with departure from equilibrium (undercooling for liquids and glasses) so the resulting glass will have a lower density of clusters near the critical size than the appropriate steady state value. With annealing, the evolution to the steady state distribution will occur following Eq. (8), resulting in an increasing nucleation rate. From a computer simulation of the cluster evolution, Kelton *et al.*[178] showed that the time-dependent nucleation rate at the critical size under isothermal annealing is best described by an expression due to Kashchiev:[179]

$$I_{n^*,t} = I^s\left[1 + 2\sum_{m=1}^{\infty}(-1)^m \exp\left(-\frac{m^2 t}{\tau_K}\right)\right], \tag{15}$$

where the transient time is given by

$$\tau_K = \frac{24k_B T n^*}{\pi^2 k^+_{n^*}\, |\Delta G'|}. \tag{16}$$

Since most glasses contain a population of quenched-in crystalline nuclei, and are not truly amorphous, Uhlmann[180] first suggested an operational criterion for glass formation that the crystalline volume fraction at the end of the quench should be less than some arbitrary value, typically chosen to be 10^{-6}. Since the distribution increases with undercooling, increasing the thermodynamic driving force for nucleation, while the kinetics become more sluggish, inhibiting the system from maintaining the steady state distribution, the actual nucleation rate during the quench is likely to be much less than the steady state value. By directly simulating the evolution of the cluster distribution under the nonisothermal conditions of a quench, Kelton and Greer[181] demonstrated that the nucleation rate decreases significantly with increasing quench rate (Fig. 15a). The effect is even more dramatic in typical metallic glass forming alloys (Fig. 15b) It is likely that only because of this effect can some metal alloys be quenched into the amorphous state.[181] As we will discuss in the following section, a proper account of these transient effects must be made in analyzing devitrification from the metallic glass to the i-phase in Al-Cu-V alloys.

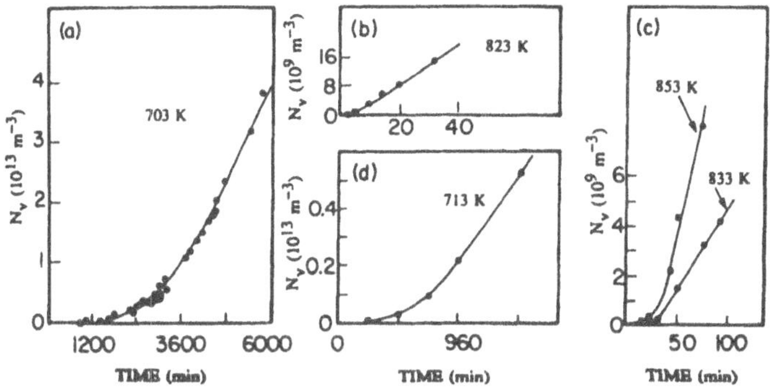

Figure 14. The number of nuclei measured as a function of time for several silicate glasses: (a) $Li_2O{\cdot}2SiO_2$ (b) $Na_2O{\cdot}BaO{\cdot}SiO_2$ (c) $Na_2O{\cdot}CaO{\cdot}SiO_2$ (d) $Li_2O{\cdot}2SiO_2$, demonstrating the universality of time-dependent effects. [From I. Gutzow, D. Kashchiev, and I. Avramov, J. Non-Cryst. Solids **73**, 477 (1985).]

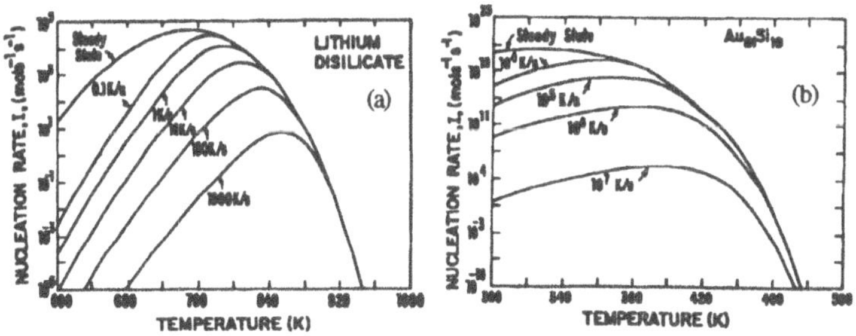

Figure 15. The homogeneous nucleation rate calculated as a function of temperature and quench rate for (a) lithium disilicate and (b) $Au_{81}Si_{19}$, a metallic glass former. Notice the significant depression of the nucleation rate below its steady state value with increasing quench rate. [From K. F. Kelton and A. L. Greer, J. Non-Cryst. Solids **79**, 295 (1986).]

5. KINETICS OF THE AMORPHOUS TO ICOSAHEDRAL PHASE TRANSITION

The existence of significant icosahedral order in undercooled liquids should imply a relatively small interfacial energy between the i-phase and the liquid, $\sigma_{i\text{-}l}$, compared to the interfacial energy between the liquid and any competing crystalline phase, $\sigma_{c\text{-}l}$. This, in turn, implies that the nucleation frequency of the i-phase should be much greater than that of competing crystalline phases. It is very hard to quantify these arguments because the nucleation rate of any phase from an undercooled liquid is extremely difficult to measure directly. However, the discovery of metallic glasses that devitrify to the i-phase provides the possibility to determine the nucleation rate of the i-phase from the glass. It is possible to study the mode and kinetics of devitrification of a metallic glass by controlled annealing. By a careful analysis of these kinetic data, we have been able to estimate, for the first time, the interfacial energy between the i-phase and the amorphous phase.

5.1. Nucleation of the Icosahedral Phase from an Undercooled Liquid

There is abundant experimental evidence that the nucleation rate of the i-phase in undercooled liquids is quite large. Using an electrohydyrodynamic (EHD) atomization technique, Bendersky and Ridder[182] produced very small liquid droplets of $Al_{86}Mn_{14}$ to study the nucleation behavior of the i-phase. They found that for the smallest droplets (diameter < 30 nm), which are subjected to the fastest quench rate in this process, the structure could be interpreted as 'microquasicrystalline' with an average icosahedral grain size of ≈ 2 nm, indicating that the nucleation rate is large. Further evidence for a large nucleation rate was provided by detailed investigations of sputtered thin films of i-phase-forming alloys. Robertson *et al.*[183] studied the structure of $Al_{72}Mn_{22}Si_6$ films sputtered on NaCl at 45, 150, and 230 °C. They found a typical metallic-glass structure factor for the 45 °C films and an icosahedral structure (plus fcc Al) for the 230 °C films. However, they demonstrated that a combined grain-size and phason strain broadening applied to the 230 °C structure factor produces a structure factor that is essentially identical to that of the 45 °C film. The best fit was obtained with a grain size of ≈ 2.5 nm. Robertson *et al.* were led to the conclusion that "either the amorphous is not really amorphous but rather 'microquasicrystalline,'... or these ordered metallic glasses may be thought of as highly defective quasicrystals in which the orientational order is lost through the random orientation of local units." The fundamental distinction is that a microquasicrystalline sample should transform to a demonstrably icosahedral structure by a grain-growth process, while a metallic glass should transform to an i-phase by a nucleation and growth process.

A relatively simple experiment to distinguish between a normal grain-growth process and a nucleation and growth process, based on the characteristic isothermal calorimetric signal associated with the transformation from a sample with an amorphous diffraction pattern to one with sharp diffraction rings, has been described by Chen and Spaepen.[184,185] For either process, the calorimeter measures an exothermic signal, *i.e.* an enthalpy released by the sample upon transformation. For nucleation and growth this enthalpy is the difference between that of the amorphous and i-phases, while for grain-growth it corresponds to the reduction of the total grain boundary enthalpy in the system. Chen and Spaepen point out that the characteristic shape of the isothermal calorimetric signal is fundamentally different for these two processes: For nucleation and growth the differential power signal has a peak at non-zero time, while for a normal grain-growth process the signal decreases monotonically with time. Using this test, with x-ray diffraction and electron microscope studies, they demonstrated that sputtered films of $Al_{82.6}Mn_{17.4}$ and $Al_{82\text{-}83}Fe_{17\text{-}18}$ transform to the i-phase by a grain-growth process. Therefore, although x-ray diffraction data seem to indicate an amorphous structure for the sputtered films, they are actually microquasicrystalline. From the measured change in enthalpy, Chen *et al.* estimated an average interfacial enthalpy of ~0.14 J/m^2, which is about half of the average grain boundary energy in crystalline Al.[156] The DSC data suggest that the microquasicrystalline structure persists even for samples sputtered onto substrates at -100 °C.

We are thus led to the conclusion that the barrier against nucleation of the i-phase is very small, *i.e.* the nucleation rate is extremely large.

5.2. Qualitative Studies of the Amorphous to Icosahedral Phase Transition

There have also been several studies of the amorphous to icosahedral phase transition in Al-TM alloys.[29-31,186] However, excepting the work by Chen and Spaepen, the mechanism of the transformation has not been investigated in detail because the time scale of the transformation could not be controlled and/or it was obscured by the presence of other phases. In fact, Chen and Spaepen's results suggest that the 'amorphous' phases in Al-TM alloys are not truly amorphous, although this could depend on the preparation technique. These alloys are therefore not a good choice for a detailed study of the amorphous to icosahedral phase transition. Fortunately, other systems exist.

There are three candidate systems for this study: Pd-U-Si,[32] Al-Mn-Si,[33] and Al-Cu-V.[108,187,188] For all three alloys, it is possible to choose the composition such that the rapidly quenched metallic glass transforms polymorphically to the i-phase, *i.e.* the glass and the i-phase have the same atomic composition. As we will discuss below, a detailed kinetic analysis is still quite involved, but without the polymorphic constraint it is essentially impossible to extract anything meaningful.

Drehman *et al.*[189] showed that although the metallic glass can be formed over a broad compositional range in Pd-U-Si, the i-phase is formed over a range of only ~1 at.% U and Si. They found that for $Pd_{58.8}U_{20.6}Si_{20.6}$, the metallic glass transforms to a single phase icosahedral sample. Several authors[190-192] independently reached the conclusion that this is a nucleation and growth transformation. Shen *et al.* used x-ray diffraction to study the isothermal transformation kinetics at various temperatures. Isothermal transformation data are usually analyzed using the Johnson-Mehl-Avrami (JMA)[193,194] equation;

$$x(t) = 1 - \exp\left[-(kt)^n\right] , \tag{17}$$

where x is the volume fraction of the transformed phase, t is the annealing time, k is an effective rate constant, and n is a constant that indicates the mode of the transformation (the Avrami exponent). This equation can be written as

$$\ln\left[\ln\left(\frac{1}{1-x}\right)\right] = n \ln(k) + n \ln(t) . \tag{18}$$

An "Avrami plot" of $\ln[-\ln(1-x)]$ versus $\ln(t)$ should then yield a straight line with slope n and intercept $n \ln(k)$. Shen *et al.*[190] obtained Avrami exponents in the range 1 to 1.7, with a systematic increase as the annealing temperature is increased. These values are anomalously low; for a polymorphic transformation in three dimensions the Avrami exponent should be greater than or equal to 3.[195] Shen *et al.*[190] proposed that this anomaly arises from simultaneous coarsening of the fine-grained i-phase during the transformation. However, Drehman *et al.*[191] found no evidence for coarsening in the temperature range of the transformation.

Conflicting results have also been obtained for the activation energy of the transformation, usually determined by nonisothermal DSC measurements. For these measurements, the DSC peak position depends on the temperature scanning rate. Several mathematical schemes for analyzing these peak shifts have been proposed, all of which are based on the JMA theory and which make various approximations to yield a tractable analytic expression. The most frequently used is the Kissinger method,[196] which relates the heating rate R, the effective activation energy of the transformation Q, and the temperature of the maximum of the peak T_p, as

$$\ln\left(\frac{R}{T_p^2}\right) = -\frac{Q}{k_B T_p} + C\,, \qquad (19)$$

where C is a constant. Using this method, activation energies of 4.69 eV,[191] 6.5 eV,[192] and 7 eV,[197] have been measured for the amorphous to icosahedral phase transition in $Pd_{58.8}U_{20.6}Si_{20.6}$ alloys.

Just as for Pd-U-Si, the i-phase in Al-Mn-Si and Al-Cu-V alloys forms over a much narrower compositional range than does the metallic glass. The ideal composition for the i-phase in these alloys is $Al_{55}Mn_{20}Si_{25}$[33] and $Al_{75}Cu_{15}V_{10}$.[108] Tsai *et al.*[198] have investigated the mode and kinetics of the amorphous to icosahedral phase transition in these alloys and found that at these compositions it is a polymorphic nucleation and growth transformation to a single-phase icosahedral sample. From their isothermal DSC measurements they obtained Avrami exponents in the range of 1.7—2.0 for $Al_{55}Mn_{20}Si_{25}$ and 1.7—1.9 for $Al_{75}Cu_{15}V_{10}$. As discussed previously, these Avrami exponents are anomalously low, although they are consistent with the results from $Pd_{58.8}U_{20.6}Si_{20.6}$.

5.3. Detailed Study of the Amorphous to Icosahedral Phase Transformation in Al-Cu-V Alloys

We have recently published a detailed study of the amorphous to i-phase transformation kinetics for $Al_{75}Cu_{15}V_{10}$ over a wide temperature range from 603 to 725 K, made by measuring the rate of heat evolution with DSC and changes in the electrical resistance.[199] Nonisothermal transformation kinetics were measured by DSC for heating rates of 0.7—80 K/min. The transformation mechanism and the developing microstructure were observed directly by TEM. From these observations, we were able to propose and test a kinetic model for this transformation. We showed that the anomalous Avrami exponents arise from an inhomogeneous distribution of quenched-in i-phase nuclei. From fits to the kinetic model we were able to obtain the first estimate of the interfacial energy between the amorphous phase and the i-phase in any alloy system. Since this study provides valuable additional structural information, and since the new techniques developed there are valid for polymorphic devitrification studies of metallic glasses, we shall briefly review the data and the analysis used to obtain this estimate for $\sigma_{a\text{-}i}$.

Transformation kinetics were measured for samples quenched at two different wheel surface velocities, 58 m/s and 80 m/s, referred to as slow- and fast-quenched samples, respectively. X-ray and TEM diffraction studies confirmed that both samples were amorphous. Partially transformed samples showed spherical nodules of the i-phase growing into the amorphous matrix, indicating an isotropic growth velocity, and a well-defined boundary between the grains and the amorphous matrix, demonstrating that the i-phase grows by atomic attachment at the boundary (Fig. 16). EDS measurements on individual i-phase nodules place the composition at $Al_{75}Cu_{15}V_{10}$, the same composition as the glass. In agreement with Tsai *et al.*[198] therefore, we found that this amorphous to i-phase transformation is polymorphic (partitionless) and proceeds via nucleation and interface-limited growth.

In contrast with Tsai *et al.*,[198] however, we found that the i-phase nodules are not homogeneously distributed in partially transformed samples. Figures 16a and 16b show bright field images of a partially transformed ribbon from the side next to the wheel (wheel side) and side away from the wheel (free side), respectively. The free side of the ribbon is almost completely transformed, while the wheel side is less than 50% transformed. This is due to a larger number of quenched-in i-phase nuclei on the free side of the ribbon because of the slower quench rate there. It is important to note the large distribution of nodule sizes on the wheel side of the ribbon (Fig. 16a). This is evidence that nucleation events occur during the annealing treatments. The implication is that the large nodules formed at an earlier time, since all nodules grow with the same growth velocity.

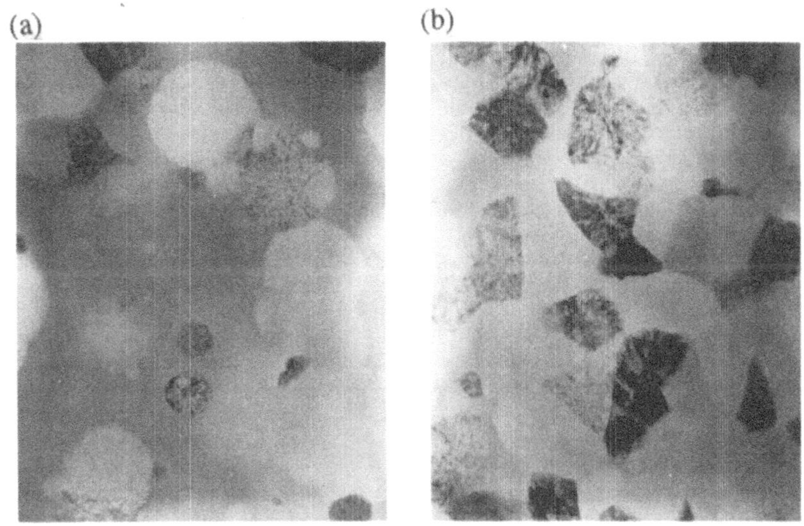

Figure 16. Bright field TEM images of a partially transformed $Al_{75}Cu_{15}V_{10}$ ribbon from (a) the side next to the wheel (wheel side) and (b) the side away from the wheel (free side). Due to an inhomogeneous distribution of quenched-in i-phase nuclei, the free side of the ribbon is almost completely transformed, while the wheel side of the ribbon is less than 50% transformed. [From J. C. Holzer and K. F. Kelton, Acta Metall. Mater. **39**, 1883 (1991).]

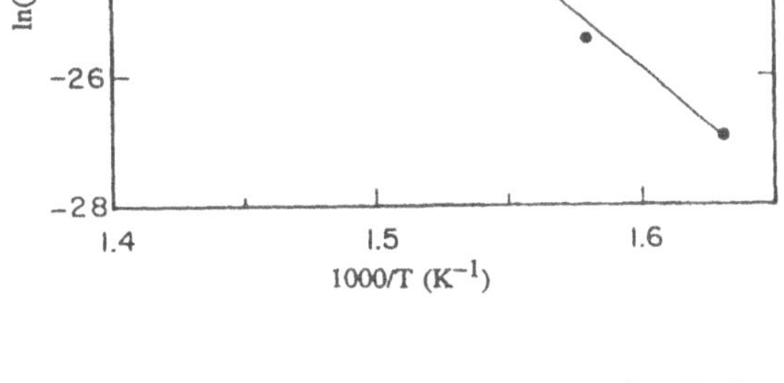

Figure 17. Arrhenius plot of the $Al_{75}Cu_{15}V_{10}$ i-phase growth velocity determined from the maximum diameter of i-phase nodules measured by TEM. [From J. C. Holzer and K. F. Kelton, Acta Metall. Mater. **39**, 1883 (1991).]

The growth velocity of the i-phase was measured directly by electron microscopy observations on partially transformed samples. The growth velocities, calculated by assuming that the largest nodule results from the growth of a quenched-in nucleus, are plotted in Fig. 17. A linear least squares fit to our data yields $u_o = 1.5 \times 10^{13\pm0.5}$ m/s and $Q_u = (3.03 \pm 0.13)$ eV, where

$$u = u_o \exp\left[\frac{-Q_u}{k_B T}\right]. \tag{20}$$

These results agree with those of Tsai *et al.*[198] to within 10% over the temperature range of there measurements.

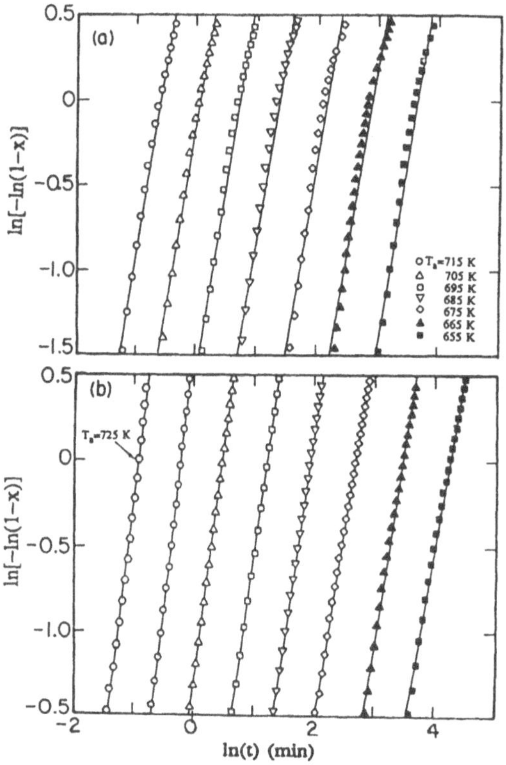

Figure 18. Avrami plots constructed for $0.2 \leq x \leq 0.8$ from the DSC isothermal measurements of the heat evolution for (a) slow-quenched and (b) fast-quenched samples. Relatively straight lines are obtained, but the slopes are "anomalous" (see text). [From J. C. Holzer and K. F. Kelton, Acta Metall. Mater. **39**, 1883 (1991).]

Figures 18a and 18b show Avrami plots for the slow- and fast-quenched samples, respectively, assuming that the volume fraction transformed at time t scales linearly with the fraction of the total heat released. The average slope is n = 2.25 ± 0.15 for the slow-quenched samples and n = 2.6 ± 0.3 for the fast-quenched samples. There is no systematic variation of slope with temperature for the slow-quenched samples, but for the fast-quenched samples, the slope increases from 2.1 at 655 K to 3.0 at 725 K. A similar systematic variation of slope with temperature was observed by Shen *et al.*[191] in the amorphous to i-phase transformation of Pd-U-Si alloys.

As discussed previously, for a polymorphic transformation in three dimensions, Avrami exponents between 3 and 4 should be obtained, in contradiction with the DSC isothermal results. However, the standard JMA analysis, used to interpret the devitrification data of others, assumes that quenched-in nuclei, if present, are distributed homogeneously throughout the sample. It was noted previously that the quenched-in nuclei are actually distributed inhomogeneously. Any attempt to interpret isothermal transformation data according to the standard JMA theory is therefore fundamentally in error. The Avrami exponents obtained previously are only 'anomalous' because the theory itself is invalid for these samples.

The nonuniform distribution of nuclei through the thickness of the ribbon demonstrates, not surprisingly, that there must exist a temperature gradient during the quench. Modeling the sample as an infinite slab of thickness Δ, and maintaining the temperature of the sample next to the wheel at 300 K, with no heat flow from the free side of the ribbon, the two-dimensional density of quenched-in nuclei was predicted as a function of z, the distance from the Cu substrate. This density, $n_o(z)$, was used to obtain a modified form of the JMA equation by dividing the sample into parallel planes perpendicular to z. Assuming that both the quenched-in nuclei and those generated during the anneal are distributed randomly in each plane y, the actual area fraction transformed in each plane, a(y,t), is obtained from the extended area fraction transformed, $a^e(y,t)$, in the usual way;[195]

$$a(y,t) = 1 - \exp[-a^e(y,t)] \,. \tag{21}$$

The volume fraction transformed, x(t), is then obtained by integrating over the thickness of the sample,

$$x(t) = \frac{1}{\Delta}\int_0^{\Delta} a(y,t)\, dy \,. \tag{22}$$

The solution for the extended area fraction transformed in plane y at time t, $a^e(y,t)$, is

$$a^e(y,t) = a_o^e(y,t) + a_n^e(y,t) \,, \tag{23}$$

$$a_o^e(y,t) = \frac{1}{\Delta}\int_{\max[0,y-r_o(t)]}^{\min[\Delta,y+r_o(t)]} \pi n_o(z)\left[r_o^2(t) - (z-y)^2\right] dz \,, \tag{24}$$

$$a_n^e(y,t) = \int_0^t \pi I(\tau)\left[r_n^2(\tau,t)z - \frac{(z-y)^3}{3}\right]\Bigg|_{z=\max[0,y-r_n(\tau,t)]}^{z=\min[\Delta,y+r_n(\tau,t)]} d\tau \,, \tag{25}$$

$$r_o(t) = \int_0^t u(t_1)\, dt_1 \,, \tag{26}$$

$$r_n(\tau,t) = \int_{\tau}^{t} u(t_1)\, dt_1 \,, \tag{27}$$

where

$a_o^e(y,t)$ = extended area fraction transformed in plane y at time t due to growth of quenched-in nuclei

$a_n^e(y,t)$ = extended area fraction transformed in plane y at time t due to growth of new nuclei

$r_o(t)$ = radius of all quenched-in nuclei at time t

$r_n(\tau,t)$ = radius at time t of nuclei that formed at time τ

$I(\tau)$ = number of nuclei that form per unit time per unit volume at time τ (the nucleation rate)

$u(t)$ = isotropic growth velocity at time t

$\min[\Delta, y+r_o(t)]$ = smaller of Δ and $y+r_o(t)$

$\max[0, y-r_o(t)]$ = larger of 0 and $y-r_o(t)$

Equations (21)-(27) represent a closed-form solution for the volume fraction transformed, x, at any time, given the time dependence of the nucleation rate, I(t), the growth velocity, u(t), and the density of quenched-in nuclei, $n_o(z)$.

It is normally assumed that the volume fraction transformed scales linearly with the change in resistivity. However, since the samples transform more rapidly on the free side of the ribbon, a more realistic model was also developed to analyze the resistivity data. Since the electric field lines are parallel to the surfaces of the infinite slab (the sample), the sample can be approximated by a series of parallel resistors, corresponding to the different planes used in the modified JMA equation. Using an effective-medium theory[200] to obtain the resistivity of each plane,

$$2\rho(y,t) = \rho_i(3f - 2) + \rho_a(1 - 3f) + \{[\rho_i(3f - 2) + \rho_a(1 - 3f)]^2 + 8\rho_i\rho_a\}^{1/2} \,, \tag{28}$$

where $f = a(y,t)$, ρ_i is the resistivity of the i-phase, and ρ_a is the resistivity of the amorphous phase. The equivalent sample resistivity at time t is obtained by taking the parallel combination of these resistors;

$$\frac{1}{\rho(t)} = \frac{1}{\Delta}\int_0^t \frac{dy}{\rho(y,t)} \,. \tag{29}$$

All isothermal transformation data were fit well by these models (Figs. 19 and 20), allowing an estimate of the nucleation rate to be obtained as a function of temperature (Fig. 21).

As demonstrated by Fig 21, the nucleation rate has an apparent Arrhenius temperature dependence over the range measured. A linear least squared fit to these data yields $I_o = 1.5 \times 10^{40\pm0.5}$ nuclei/m^{3}s and $Q_I = 3.3 \pm 0.1$ eV, where

$$I = I_o \exp\left(\frac{-Q_I}{k_B T}\right) . \tag{30}$$

Since the heating rate for the DSC scans is constant , the time dependence of the nucleation rate and growth velocity are readily determined from their temperature dependence, given in Eqs. (30) and (20), respectively. As mentioned previously, Eqs. (21)-(27) can then be used to model nonisothermal transformation data, by approximating the nonisothermal transformation by a series of short isothermal anneals of length inversely proportional to the scan rate. Assuming that the rate of heat evolution at each temperature is proportional to the rate of change of the volume fraction transformed, the DSC traces were fit by adjusting only the preterm in the nucleation rate, I_0 (Fig. 22). The general features of the nonisothermal data are predicted, including the high temperature shoulder on the slow-quenched samples (Fig. 22a). The preterms obtained from these fits were all within one standard deviation of the values obtained from the isothermal fits.

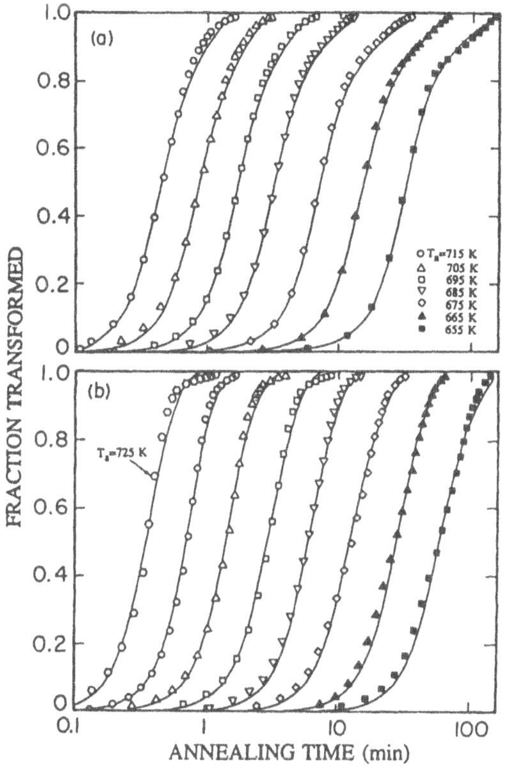

Figure 19. Volume fraction of i-phase as a function of annealing time at all isothermal DSC annealing temperatures for the (a) slow-quenched and (b) fast-quenched samples. [From J. C. Holzer and K. F. Kelton, Acta Metall. Mater. **39**, 1883 (1991).]

The good agreement between the measured and simulated isothermal and nonisothermal transformation data and the TEM observations of an inhomogeneous distribution of quenched-in nuclei demonstrate the validity of this model. Assuming classical nucleation theory it is then possible to obtain an estimate for the interfacial energy between the i-phase and the undercooled liquid. Estimating the atomic mobility at the interface from the growth velocity, Eq. (10) can be used to obtain

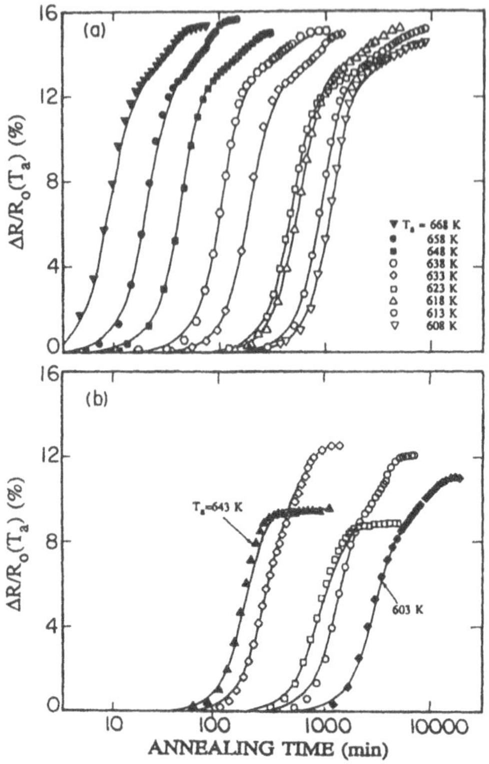

Figure 20. Percentage change in resistance of rapidly quenched $Al_{75}Cu_{15}V_{10}$ ribbons at all annealing temperatures to the (a) slow-quenched and (b) fast-quenched samples. [From J. C. Holzer and K. F. Kelton, Acta Metall. Mater. **39**, 1883 (1991).]

$$\xi = \ln\left[\frac{I^s\Delta G}{uT}\right] = K - \frac{16\pi}{3k_B}\left(\frac{W}{\rho_i}\right)^2 \frac{\sigma_{a\text{-}i}^3}{T\Delta G^2}, \qquad (31)$$

where ΔG is the Gibbs free energy of the glass less that of the i-phase, K is a constant, W is the gram-atomic weight of the alloy, ρ_i is the density of the i-phase, and $\sigma_{a\text{-}i}$ is the interfacial energy between the glass and the i-phase. The interfacial energy can therefore be calculated from the slope of a plot of ξ vs. $1/T\Delta G^2$. Using the measured growth velocities and the nucleation rates obtained from the fits to the isothermal transformations, and calculating ΔG from measurements of the specific heats at constant pressure for the amorphous, liquid, and icosahedral phases,[199] produces a positive slope, implying a negative interfacial energy. This is clearly an unphysical result; the introduction of an interface must increase the energy of the system. Furthermore, TEM shows that the grain morphology is approximately spherical, a geometry that minimizes the surface to volume ratio. If the interfacial energy were actually negative then the i-phase would grow in such a way that the amount of interface is maximized.

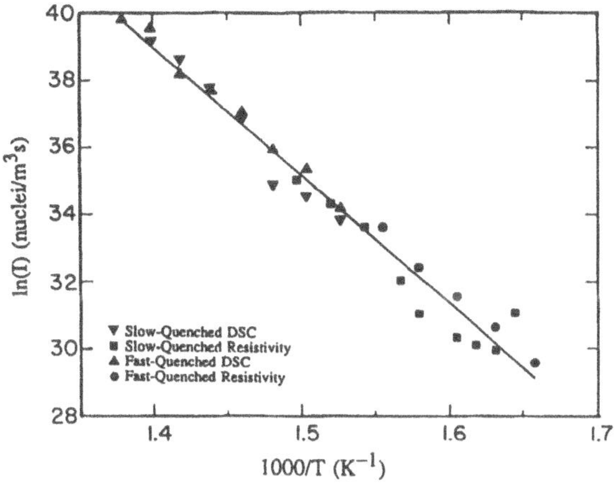

Figure 21. Arrhenius plot of the $Al_{75}Cu_{15}V_{10}$ i-phase nucleation rate determined form fitting amorphous to i-phase isothermal transformation data. [From J. C. Holzer and K. F. Kelton, Acta Metall. Mater. **39**, 1883 (1991).]

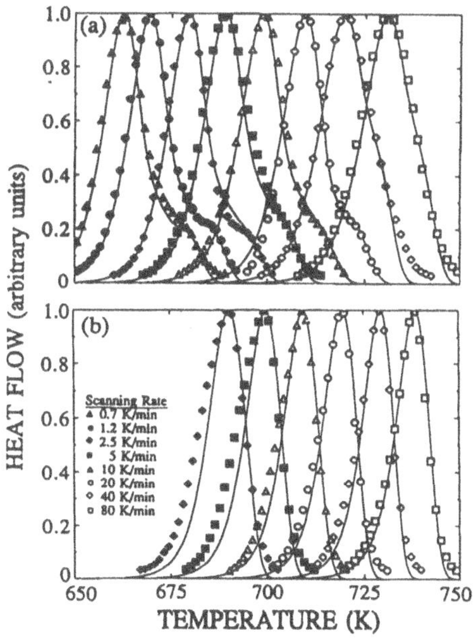

Figure 22. DSC traces, normalized to the peak heights,of the amorphous to i-phase transformation in $Al_{75}Cu_{15}V_{10}$ at various scan rates for the (a) slow-quenched and (b) fast-quenched samples. [From J. C. Holzer and K. F. Kelton, Acta Metall. Mater. **39**, 1883 (1991).]

One likely explanation for the failure of the above analysis is that the nucleation rate is not the steady state value. An analysis of Eqs. (20), (30), and (31) shows that the positive slope from Eq. (31) arises because the nucleation rates determined from the isothermal fits yield an unphysically high activation energy for a steady state rate. As illustrated in Fig. 15, this is expected if transient nucleation is important and the nucleation rates are measured below the peak temperature in the nucleation rate.[181] Due to the rapid quench required to prepare the sample, the initial nucleation rate at any temperature can be significantly lower than the steady state nucleation rate. Since the nucleation rates obtained from isothermal transformation fits are probably an average nucleation rate over the course of the transformation, and since the difference between the initial nucleation rate and the steady state nucleation rate increases with decreasing temperature, an effectively larger activation energy for this nucleation rate results. If true, the nucleation rates depicted in Fig. 21 and represented by Eq. (30) are not steady state nucleation rates and cannot be used in an analysis according to Eq, (31) to estimate the interfacial energy, $\sigma_{a\text{-}i}$.

However, it is still possible to obtain upper- and lower-bound estimates for the interfacial energy. $\sigma_{a\text{-}i}$ determines the temperature of the peak in the steady state nucleation rate, all other parameters being equal. From Fig 21, the peak in the nucleation rate must occur at a temperature above the highest temperature isothermal transformation data point, 725 K, setting an upper bound on the interfacial energy of $\sigma_{a\text{-}i} \leq 0.015$ J/m^2. It is important to note that this upper bound on σ does not depend on any assumption of transient nucleation. A lower-bound estimate for the interfacial energy can be obtained by considering the relation of the transient time, τ, to the time scale of the transformation at each temperature. For the details of this estimate the reader is referred to Ref. 199 and the references therein. By fitting isothermal transformation curves with a time-dependent nucleation rate, a lower-bound estimate was obtained for the interfacial energy, $\sigma_{a\text{-}i} \geq 0.002$ J/m^2. This is a lower bound on $\sigma_{a\text{-}i}$ since if $\sigma_{a\text{-}i}$, and hence τ, is small enough, then the time to complete transformation is too large unless I^s is unphysically large.

The interfacial energy between the i-phase and the glass thus lies in the range 0.002 J/m^2 $\leq \sigma_{a\text{-}i} \leq 0.015$ J/m^2, which is much smaller than for example the interfacial energy between crystalline and liquid Al (the major alloy component) at the equilibrium melting temperature, $\sigma_{c\text{-}l}^{Al} = 0.108$ J/m^2.[155] Such a small interfacial energy demonstrates that the local structure at the interface must be very similar between the i-phase and the glass (or undercooled liquid).

6. DISCUSSION OF LOW INTERFACIAL ENERGY

This estimate of the interfacial energy between the amorphous and icosahedral phase is perhaps the most convincing evidence to date that the local structure in the two phases is very similar. Only for the polymeric alkanes does one observe such a low liquid — crystal interfacial energy.[201] For these systems it is known that the local atomic structure is identical in the liquid and crystal.

It would be interesting to test this estimate for the interfacial energy by constructing a model for the interface. Spaepen[175] has described a structural model for the solid — liquid interface in a monatomic system constructed between a close-packed crystal plane and a DRP of hard spheres (the liquid). The interface was constructed by depositing hard spheres onto the crystalline plane, layer by layer, according to the following construction rules: (i) form tetrahedral holes preferentially, (ii) disallow all octahedral holes, and (iii) maximize the density. As we have discussed, tetrahedral holes are typical for the liquid, while some octahedral holes are necessary for the crystal. The first liquid layer of Spaepen's model has no two-dimensional symmetry parallel to the interface, while it is completely localized in the perpendicular direction. The second liquid layer has some degree of disorder in the perpendicular direction, though not as much as a true DRP of hard spheres. Nevertheless, Spaepen assumed that the properties of the interface could be understood with a description of just the first two interfacial layers. This assumption is supported by recent molecular dynamics simulations[202] and density functional calculations,[203-205] which indicate that the interfacial boundary is quite narrow for large clusters.

Consistent with Turnbull's 'negentropy' suggestion,[174] the interfacial free energy in this model results primarily from the configurational entropy loss in the liquid as it becomes more localized in one dimension near the crystal plane. Spaepen and Meyer[176] estimated the surface tension from this model and showed that it is in reasonable agreement with experimental values. Elser[206] recently obtained the exact solution for the configurational entropy of this model, 0.077241 per interface atom, which corresponds to an interfacial free energy at the melting point of 0.85 times the heat of fusion per atom in the crystal plane. Density functional calculations give a value of 0.87 times the heat of fusion per atom in the crystal plane,[205] while Thompson[207] obtained a slightly smaller estimate of 0.71 times the heat of fusion per atom in the crystal plane for the interfacial tension between a (110) body-centered cubic plane and the liquid.

An analogous model for the icosahedral — liquid interface would be difficult to construct because the exact atomic structure of the i-phase is not known. We can speculate, however, that the interface must look something like that of a high order crystal approximant such as α(AlMnSi). Of course, undercooling experiments on the approximant phase or devitrification studies of a glass to the approximant phase, such as were discussed for i(AlCuV), would be the only way to verify that the approximant — liquid interfacial free energy is indeed small. Spaepen's construction rules could be used to model the structure of the liquid layers above a pseudo five- or threefold plane of the approximant phase, or perform a molecular dynamics simulation for the liquid over the fixed crystalline plane in order to study the structure of the interface.

These models for the liquid — solid interface and the corresponding calculations of the interfacial energy strictly apply only to a flat interface at the equilibrium melting point, whereas the interfacial tension that must be used in calculating the nucleation rate (Eq. (13)) is that for a curved interface at the nucleation temperature. To date, the temperature dependence of the interfacial energy for a curved interface has not been calculated. However, if the interfacial negentropy is independent of temperature, then the interfacial energy of a flat interface should increase with temperature, *i.e.* it should have a positive temperature coefficient. This is consistent with the smaller interfacial energy (0.62 times the heat of fusion per atom in the crystal) determined by nucleation rate measurements for mercury at 2/3 T_m.[80]

It is interesting to also express the estimate for the amorphous — icosahedral interfacial energy as a fraction of the heat of fusion per atom in the icosahedral plane. Although no atomic model exists for the i-phase in $Al_{75}Cu_{15}V_{10}$, we can still estimate the atomic density for an average icosahedral plane from the measured density of the i-phase, $\rho_i = 2.79$ g/cm^3. The heat of fusion per atom in an average icosahedral plane, ΔU, can be expressed as

$$\Delta U = \frac{\Delta H_f}{N_A^{1/3}} \left(\frac{\rho_i}{W}\right)^{2/3} , \tag{32}$$

where ΔH_f is the enthalpy of fusion of the i-phase and W is the gram-atomic weight of $Al_{75}Cu_{15}V_{10}$. From the estimated enthalpy of fusion of the i-phase, $\Delta H_f \approx 2160$ J/g-at.,[199] and the estimate for the amorphous — icosahedral interfacial energy, $0.002\ \mathrm{J/m^2} \leq \sigma_{a\text{-}i} \leq 0.015\ \mathrm{J/m^2}$, we find that the constant K in $\sigma_{a\text{-}i} = K\Delta U$ must lie in the range $0.04 \leq K \leq 0.3$, which is again much smaller than the value for crystalline phases.

In the kinetic analysis used to estimate the amorphous — icosahedral interfacial energy we ignored any temperature dependence of $\sigma_{a\text{-}i}$. Strictly speaking therefore, this estimate is an average value over the temperature range 608—715 K, corresponding to a reduced temperature range of ~ 0.71—0.84 T_m. It should be noted, however, that nothing in the fitting of the kinetic data suggests that there must be any temperature dependence in the interfacial energy, indicating that the temperature coefficient must be nearly zero.

It might be possible to obtain a direct measurement of the icosahedral — liquid interfacial energy by performing undercooling experiments on this alloy. In an emulsion of the alloy in a suitable carrier fluid it should be possible to isolate impurities and hopefully achieve some degree of undercooling. If the i-phase is the phase that nucleates in this undercooled melt, then it would be possible to estimate the icosahedral — liquid interfacial tension. It would be very interesting to compare this value to the amorphous — icosahedral interfacial tension. The equilibrium i-phase alloys (*e.g.* AlCuFe) would also be candidates for an undercooling experiment, since one might expect that the equilibrium phase is the one that will nucleate first in the undercooled liquid.

Another interesting aspect of such a low interfacial energy is its implication for faceting. We mentioned previously that faceted quasicrystals have been observed in several alloy systems. In conventional crystals equilibrium faceting can be understood in terms of the anisotropy in the crystal — liquid interfacial energy, *i.e.* crystalline planes with a lower interfacial energy are favored. It would be somewhat surprising, however, if the anisotropy in such a small interfacial energy could give rise to the observed faceting. This suggests that the faceting observed in quasicrystals is due to preferred growth directions (kinetics) rather than interfacial energy (thermodynamics).

7. SUMMARY AND FUTURE PROSPECTS

In summary, there is abundant evidence for similar polytetrahedral order in glasses, undercooled liquids, complex crystalline phases, and the i-phase. The local tendency for short-range polytetrahedral order in glasses, undercooled liquids, and the complex crystalline phases is extended to long-range polytetrahedral order in the i-phase. The previous evidence for the close structural similarity between the glass and the i-phase is strongly supported by our estimate of the extremely low interfacial energy between the icosahedral and amorphous phases. As we mentioned previously, such a low value has been found only in systems for which it is known that the local atomic structure is identical, the polymeric alkanes.

Several possible experiments are suggested by these results. First, an effort should be made to verify that a small icosahedral — amorphous interfacial energy is indeed a general property. The kinetics of the amorphous to icosahedral phase transformation should be investigated in other alloy systems. $Pd_{58.8}U_{20.6}Si_{20.6}$ is a prime candidate for such a study. Second, as an extended investigation of the effects of strong local icosahedral order on nucleation and growth, a detailed study like that made of the amorphous to i-phase transformation in Al-Cu-V should be repeated on a glass that devitrifies polymorphically to a crystal approximant. Some Al-Mn-Si metallic glasses, for example, transform to α(AlMnSi); it must be verified, however, that this is polymorphic and results in a single phase. Third, a model for the solid — liquid interface should be constructed for the i-phase or a crystal approximant. This would provide a theoretical assessment of the interfacial structure and identify whether the interfacial energy is primarily enthalpic or entropic in origin. Molecular dynamics and/or Monte-Carlo calculations of the interface between the liquid and an i-phase or crystal approximant would also be interesting, determining, for example, whether the interfacial region extends further into the liquid than for the simple crystalline phases previously studied, where there is a much larger difference in local structure. Finally, undercooling experiments on liquids that nucleate quasicrystals could provide additional, and more direct, estimates of the icosahedral — liquid interfacial energy.

ACKNOWLEDGEMENTS

This work was partially supported by the National Science Foundation under Grant Number DMR-8903081. One of the authors (J.C.H.) gratefully acknowledges the cooperation of W. L. Johnson and the financial support of the Department of Energy (Grant Number DEFG0386ER45242).

REFERENCES

1. Plato, *Timaeus*, in *The Collected Dialogues of Plato*, (trans. B. Jowett; ed. E. Hamiltion and H. Cairns), 54c 1180ff; 1973, Princeton, Princeton University Press.
2. C. Kittel, *Introduction to Solid State Physics*, Chapt. 1; 1976, New York, John Wiley & Sons.
3. B. K. Vainshtein, *Modern Crystallography I*, Chapt. 2; 1981, Berlin, Springer-Verlag.
4. L. E. Levine, J. C. Holzer, P. C. Gibbons, and K. F. Kelton, Phil. Mag. B, in press.
5. F. C. Frank, Proc. Royal Soc. London **215A**, 43 (1952).
6. P. J. Steinhardt, D. R. Nelson, and M Ronchetti, Phys. Rev. Lett. **47**, 1297 (1981); Phys. Rev. B **28**, 784 (1983).
7. J. D. Honeycutt and H. C. Anderson, J. Phys. Chem. **91**, 4950 (1987).
8. T. L. Beck and R. S. Berry, J. Chem. Phys. **88**, 3910 (1988).
9. H. Jonsson and H. C. Anderson, Phys. Rev. Lett. **60**, 2295 (1988).
10. D. Shechtman, I. Blech, D. Gratias, and J. W. Cahn, Phys. Rev. Lett. **53**, 1951 (1984).
11. K. F. Kelton, *Quasicrystals: Structure and Stability*, International Materials Reviews, in press.
12. L. Pauling, Nature (London) **317**, 512 (1985); Phys. Rev. Lett. **58**, 365 (1987); Proc. Natl. Acad. Sci. **85**, 8376 (1988).
13. M. J. Carr, J. Appl Phys. **59**, 1063 (1986).
14. T. R. Anantharaman, Current Sci. **57**, 578 (1988); in *Quasicrystals and Incommensurate Structures in Condensed Matter*, (ed. M. J. Yacaman, D. Romeau, V. Castano, and A. Gomez), 199, 1989, Singapore, World Scientific.
15. R. D. Field and H. L. Fraser, Mater. Sci. and Eng. **68**, L17 (1985).
16. K. S. Vecchio and D. B. Williams, Metall. Trans. **19A**, 2875 (1988).
17. D. Shechtman and I. Blech, Metall. Trans. **16A**, 1005 (1985).
18. D. Shechtam, D. Gratias, and J. W. Cahn, Compt. Rend. Acad. Sci. II **300**, 909 (1985).
19. K. Hiraga, M. Watanabe, A. Inoue, and T. Masumoto, Sci. Rep. Inst. Tohoku Univ. **A-32**, 309 (1985).
20. L. Bursill and J. Lin, Nature **316**, 50 (1985).
21. R. Portier, D. Shechtman, D. Gratias, and J. W. Cahn, J. Mic. Spectro. Elec. **10**, 107 (1985).
22. K. M. Knowles, A. L. Greer, W. O. Saxton, and W. M. Stobbs, Phil. Mag. B **52**, L31 (1985).
23. R. Gronsky, K. M. Krishnan, and L. Tanner, in Proc. EMSA Meeting **43**, 34 (1985).
24. K. F. Kelton and T. W. Wu, Appl. Phys. Lett. **46**, 1059 (1985).
25. L. Bendersky, Phys. Rev. Lett. **55**, 1461 (1985).
26. P. A. Bancel, P. A. Heiney, P. W. Stephens, A. I. Goldman, and P. M. Horn, Phys. Rev. Lett. **54**, 2422 (1985).
27. A. J. Melmed and R. Klein, Phys. Rev. Lett. **56**, 1478 (1986).
28. H. B. Elswijk, P. M. Bronsveld, and J. Th. M. De Hosson, Phys. Rev. B **37**, 4261 (1988).
29. D. A. Lilienfeld, M. Nastasi, H. H. Johnson, D. G. Ast, and J. W. Mayer, Phys. Rev. Lett. **55**, 1587 (1985).
30. D. A. Lilienfeld, M. Nastasi, H. H. Johnson, D. G. Ast, and J. W. Mayer, J. Mater. Res. **1**, 237 (1986).
31. D. A. Lilienfeld, M. Nastasi, H. H. Johnson, D. G. Ast, and J. W. Mayer, Phys. Rev. B **34**, 2985 (1986).
32. S. J. Poon, A. J. Drehman, and K. R. Lawless, Phys. Rev. Lett. **55**, 2324 (1985).
33. A. Inoue, Y. Bizen, and T. Masumoto, Metall. Trans. **19A**, 383 (1988).

34. A. Csanady, P. B. Barna, J. Mayer, and K. Urban, Scripta Metall. **21**, 1535 (1987).
35. Y. Saito, K. Mihama, and H. S. Chen, Phys. Rev. B **35**, 4085 (1987).
36. J. A. Sekhar and T. Rajasekharan, Nature **320**, 153 (1986).
37. B. Grushko and G. R. Stafford, Metall. Trans. **20A**, 1351 (1989).
38. B. Grushko and G. R. Stafford, Scripta Metall. **23**, 1043 (1989).
39. S. R. Nishitani, H. Kawaura, K. F. Kobayashi, and P. H. Shingu, J. Cryst. Growth **76**, 209 (1986).
40. W. A. Cassada, G. J. Shiflet, and S. J. Poon, Phys. Rev. Lett. **56**, 2276 (1986).
41. D. M. Follstaedt and J. A. Knapp, Phys. Rev. Lett. **56**, 1827 (1986).
42. J. Eckert, L. Schultz, and K. Urban, Appl. Phys. Lett. **55**, 117 (1989).
43. J. D. Budai and M. J. Aziz, Phys. Rev. B **33**, 2876 (1986).
44. H. S. Chen, C. H. Chen, A. Inoue, and J. T. Krause, Phys. Rev. B **32**, 1940 (1985).
45. K. F. Kelton and J. C. Holzer, Phys. Rev. B **37**, 3940 (1988); Mat. Sci. and Eng. **99**, 389 (1988).
46. K. Sadananda, A. K. Singh, and M. A. Imam, Phil. Mag. Lett. **58**, 25 (1988).
47. S. Baranidharan, E. S. R. Gopal, G. Parthasarathy, J. A. Sekhar, Scripta Metall. **21**, 1623 (1987).
48. J. Eckert, L. Schultz, and K. Urban, Europhys. Lett. **13**, 349 (1990).
49. K. Urban, N. Moser, and H. Kronmüller, Phys. Stat. Sol. (a) **91**, 411 (1985).
50. J. Mayer, K. Urban, and J. Fidler, Phys. Stat. Sol. (a) **99**, 467 (1987).
51. L. J. Huang and B. X. Liu, Appl. Phys. Lett. **57**, 1401 (1990).
52. M. D. Ball and D. J. Lloyd, Scripta Met. **19**, 1065 (1985).
53. W. Ohashi and F. Spaepen, Nature **330**, 555 (1987).
54. A. P. Tsai, A. Inoue, and T. Masumoto, Jpn. J. Appl. Phys. **26**, L1505 (1987).
55. C. A. Guryan, A. I. Goldman, P. W. Stephens, K. Hirogo, A. P. Tsai, A. Inoue, and T. Masumoto, Phys. Rev. Lett. **62**, 2409 (1989).
56. A. P. Tsai, A. Inoue, and T. Masumoto, Jpn. J. Appl. Phys. **27**, L1587 (1988).
57. A. P. Tsai, A. Inoue, Y. Yokoyama, and T. Masumoto, Mater. Trans. Jpn. Inst. Metals **31**, 98 (1990).
58. P. A. Bancel, Phys. Rev. Lett. **63**, 2741 (1989).
59. M. Audier and P. Guyot, in *Proceedings of the Adriatico Research Conference on Quasicrystals*, (M. V. Jaric and S. Lundqvist, eds.), World Scientific, Singapore (1989).
60. Z. Zhang, N. C. Li, and K. Urban, J. Mater. Res. **6**, 366 (1991).
61. C. Dong, J. M. Dubois, M. de Boissieu, M. Boudard, and C. Janot, J. Mater. Res. **6**, 2637 (1991).
62. S. Ranganathan and K. Chattopadhyay, Key Engineering Materials **13-15**, 229 (1987).
63. K. H. Kuo, J. Non-Cryst. Solids **117/118**, 756 (1990).
64. *Introduction to Quasicrystals*, in *Aperiodicity and Order*, vol. 1 (M. V. Jaric, ed.) Academic Press, New York, 1988.
65. *Extended Icosahedral Structures*, in *Aperiodicity and Order*, vol. 3 (M. V. Jaric, ed.) Academic Press, New York, 1988.
66. D. R. Nelson and F. Spaepen, in *Solid State Physics*, (H. Ehrenreich and D. Turnbull, eds.) vol. 42, p. 1, Academic Press, N.Y., 1989.
67. J. D. Bernal, Proc. Royal. Soc., London **A280**, 299 (1964).
68. J. D. Bernal, Nature **188**, 910 (1960).
69. H. S. M. Coxeter, *Regular Polytopes*, Dover, New York, 1975.
70. H. S. M. Coxeter, *Introduction to Geometry*, Wiley, New York, 1969.
71. M. Kleman and J. F. Sadoc, J. Phys. (Paris) Lett. **40**, L569 (1979).
72. J. F. Sadoc, J. Phys. (Paris) Colloq. **41**, C8-326 (1980).
73. D. R. Nelson, Phys. Rev. Lett. **50**, 982 (1983);Phys. Rev. B **28**, 5515 (1983).
74. J. Sethna Phys. Rev. Lett. **51**, 2198 (1983).
75. J. Sethna, Phys. Rev. B **31**, 6278 (1985).
76. D. R. Nelson and M. Widom, Nucl. Phys. B **240**, 113 (1984).
77. S. Sachdev and D. R. Nelson, Phys. Rev. Lett. **53**, 1947 (1984).

78. S. Sachdev and D. R. Nelson, Phys. Rev. B **32**, 1480 (1985).
79. D. B. Fahrenheit, Phil. Trans Roy. Soc. **39**, 78 (1724).
80. D. Turnbull, J. Chem. Phys. **20**, 411 (1952).
81. L. D. Landau and E. M. Lifshitz, *Quantum Mechanics*, Pergamon, Oxford, 1965, p. 106.
82. M. V. Jaric, Nucl. Phys. B **265**, 647 (1986).
83. R. D. Mountain and A. C. Brown, J. Chem. Phys. **80**, 2730 (1984).
84. W. B. Pearson, *The Crystal Chemistry and Physics of Metals and Alloys*, 1972, New York, John Wiley & Sons.
85. M. Cooper and K. Robinson, Acta Cryst. **20**, 614 (1966).
86. A. L. Mackay, Acta Cryst. **15**, 916 (1962).
87. F. C. Frank and J. S. Kasper, Acta Cryst. **11**, 184 (1958); **12**, 483 (1959).
88. A. K. Sinha, in *Progress in Materials Science*, Vol. 15, 75-185, Pergamon, New York, 1972.
89. C. B. Shoemaker and D. P. Shoemaker, Acta Cryst. B. **28**, 2957 (1972).
90. M. H. Cohen and D. Turnbull, J. Chem. Phys. **31**, 1164 (1959).
91. W. Klement, R. H. Willens, and P. Duwez, Nature **187**, 869 (1960).
92. D. Turnbull and M. H. Cohen, J. Chem. Phys. **34**, 120 (1961); M. H. Cohen and D. Turnbull, Nature **189**, 131 (1961).
93. G. S. Cargill III, J. Appl. Phys. **41**, 12 and 2248 (1970).
94. G. S. Cargill III, in *Solid State Physics*, (F. Seitz and D. Turnbull, eds.), vol. 30, p. 227, Academic Press, N. Y., 1975.
95. J. L. Finney, Proc. Roy. Soc. London **319A**, 497 (1970).
96. G. D. Scott, Nature **188**, 908 (1960); Nature **194**, 956 (1962).
97. T. Ichikawa, Phys. Stat. Sol. **19**, 707 (1973).
98. D. E. Polk, Scripta Metall. **4**, 117 (1970); Acta Metall. **20**, 485 (1972).
99. P. H. Gaskell, in *Glassy Metals II, Topics in Applied Physics*, **53** (Springer, Berlin, 1983) pp. 5-49.
100. J. F. Sadoc and J. Dixminer, Mat. Sci. Eng. **23**, 187 (1976).
101. Y. Waseda, Prog. Mater. Sci. **26**, 1 (1981).
102. T. M. Hayes, J. W. Allen, J. Tauc, B. C. Giessen and J. J. Hauser, Phys. Rev. Lett. **40**, 1282 (1978).
103. H. J. Frost, Acta Metall. **30**, 889 (1982).
104. S. Rundqvist, Ark. Kemi **20**, 67 (1962); Acta Chem. Scand. **16**, 995 (1962).
105. D. S. Boudreaus and J. M. Gregor, J. Appl. Phys. **48**, 152 (1977); J. Appl. Phys. **48**, 5057 (1977).
106. K. S. Vecchio and D. B. Williams, Phil. Mag. B **57**, 535 (1988).
107. F. Spaepen, Li-Chyong Chen and W. Ohashi, Bull. Am. Phys. Soc. **33**, 604 (1988).
108. A. P. Tsai, A. Inoue, and T. Masumoto, Jpn. J. Appl. Phys. **26**, L1994 (1987).
109. A. R. Kortan, F. A. Thiel, H. S. Chen, A. P. Tsai, A. Inoue and T. Masumoto, Phys. Rev. B. **40**, 9398 (1989).
110. A. Blatter, M. Von Allmen, and N. Baltzer, J. Appl. Phys. **62**, 276 (1987).
111. Eun-Kee Jeong, J. C. Holzer, A. E. Carlsson, M. S. Conradi, P. A. Fedders, and K. F. Kelton, Phys. Rev. B **41**, 1695 (1990).
112. P. Sainfort and B. Dubost, J. Physique Coll. **47**, 321 (1986).
113. M. Audier, P. Sainfort, and B. Dubost, Phil. Mag. B **54**, L105 (1986).
114. Y. Ma, E. A. Stern, and F. W. Gayle, Phys. Rev. Lett. **58**, 1956 (1987).
115. P. Sainfort and P. Guyot, Scripta Metall. **21**, 1517 (1987).
116. R. S. Schaefer, L. A. Bendersky, D. Shechtman, W. S. Boettinger, and F. S. Biancaniello, Metall. Trans. **17A**, 2117 (1986).
117. C. H. Chen and H. S. Chen, Phys. Rev. B **33**, 2814 (1986).
118. V. Elser and C. L. Henley, Phys. Rev. Lett. **55**, 2883 (1985).
119. P. Guyot and M. Audicr, Phil. Mag. B **52**, L15 (1985).
120. K. K. Fung and Y. Q. Zhou, Phil. Mag. B **54**, L27 (1986).
121. D. C. Koskenmaki, H. S. Chen, and K. V. Rao, Phys. Rev. B **33**, 5328 (1986).

122. X. Zhang and K. F. Kelton, Phil. Mag. Lett. **63**, 39 (1991).
123. C. Dong, K. Chattopadhyay, and K. H. Kuo, Scripta Metall. **21**, 1307 (1987).
124. M. Audier and P. Guyot in *Extended Icosahedral Structures*, in *Aperiodicity and Order*, vol. 3 (M. V. Jaric, ed.), pp. 1-36, Academic Press, New York, 1988.
125. C. L. Henley, Phil. Mag. B. **53**, L59 (1988).
126. Q. B. Yang, Phil. Mag. Lett. **57**, 171 (1988).
127. P. A. Kalugin, preprint.
128. P. W. Stephens and A. I. Goldman, Phys. Rev. Lett. **56**, 1168 (1986);**57**, 2331 (1986).
129. V. Elser in *Proceedings of the XVth International Colloquium on Group Theoretical Methods in Physics* (R. Gilmore and D. H. Feng, eds.), World Scientific, Singapore (1987).
130. P. W. Stephens in *Extended Icosahedral Structures*, in *Aperiodicity and Order*, vol. 3 (M. V. Jaric, ed.), pp. 37-104 Academic Press, New York, 1988.
131. J. L. Robertson, J. Non-Cryst. Solids **106**, 225 (1988).
132. J. L. Robertson and S. C. Moss, Phys. Rev. Lett. **66**, 353 (1991).
133. D. Gratias, J. W. Cahn and B. Mozer, Phys. Rev. B **38**, 1643 (1988).
134. J. W. Cahn, D. Gratias and B. Mozer, J. Phys. France **49**, 1225 (1988).
135. M. V. Jaric and S. Narashimhan, Phase Transitions **16/17**, 351 (1989).
136. Z. Olami and S. Alexander, Phys. Rev. B **37**, 3973 (1988); C. Oguey and M. Duneau, Europhys. Lett. **7**, 49 (1989).
137. M. Duneau and C. Oguey, J. Physique **50**, 135 (1989).
138. C. L. Henley, in *Quasicrystals*, (T. Fujiwari and T. Ogawa, eds.), Springer, Berlin (1990).
139. J. M. Dubois and C. Janot, Phil. Mag. B **61**, 649 (1990).
140. B. Dubost, C. Collinet and I. Ansara in *Quasicrystalline Materials*, (C. Janot and J. M. Dubois, eds.) World Scientific, Singapore, p. 39 (1988).
141. E. A. Stern, Y. Ma, and E. E. Bouldin, Phys. Rev. Lett. **55**, 2172 (1985).
142. A. Sadoc, A. M. Flank, P. Lagarde, P. Sainfort, and R. Bellissent, J. Physique **47**, 105 (1986).
143. J. B. Boyce, J. C. Mikkelsen, Jr., F. Bridges, and T. Egami, Phys. Rev. B **33**, 7314 (1986).
144. J. B. Boyce, F. B. Bridges, and J. J. Hauser, J. Physique **47**, C8-1029 (1986).
145. W. W. Warren, Jr., H. S. Chen, and J. J. Hauser, Phys. Rev. B **32**, 7614 (1985).
146. M. Rubinstein, G. H. Stauss, T. E. Phillips, K. Moorjani, and L. H. Bennett, J. Mater. Res. **1**, 243 (1986).
147. K. R. Carduner, B. H. Suits, J. A. DiVerdi, M. D. Murphy, and D. White, J. Mater. Res. **2**, 431 (1987).
148. M. Eibschutz, H. S. Chen, and J. J. Hauser, Phys. Rev. Lett. **56**, 169 (1986).
149. P. J. Schurer, B. Koopmans, F. van der Woude, and P. Bronsveld, Solid State Commun. **59**, 619 (1986).
150. C. Berger, J. C. Lasjaunias, and C. Paulsen, Solid State Commun. **65**, 441 (1988).
151. A. J. McAlister, L. A. Bendersky, R. J. Schaefer, and F. S. Biancaniello, Scripta Metall. **21**, 103 (1987).
152. K. Yu-Zhang, M. Harmelin, A. Quivy, Y Calvayrac, J. Bigot, and R. Portier, Mat. Sci. and Eng. **99**, 385 (1988).
153. U. Köster and B. Schuhmacher, Mat. Sci. and Eng. **99**, 417 (1988).
154. D. Turnbull, J. Appl. Phys. **21**, 1022 (1950).
155. K. F. Kelton, *Crystal Nucleation in Liquids and Glasses*, in *Solid State Physics: Advances in Research and Applications*, Vol. 45, (H. Ehrenreich and D Turnbull, eds.), Academic Press, New York, 1991.
156. H. Gleiter and B. Chalmers, in *Progress in Materials Science*, Vol. 16, pp. 13-39, Pergamon Press, Oxford, 1972.
157. E. Matsubara, Y. Waseda, A. P. Tsai, A. Inoue, and T. Masumoto, J. Mater. Sci. **25**, 2507 (1990).

158. D. D. Koflat, S. Nanao, T. Egami, K. M. Wong, and S. J. Poon, Phys. Rev. Lett. **57**, 114 (1986).
159. R. Fuchs, S. B. Jost, H. Rudin, H. J. Güntherodt, and P. Fischer, Z. Phys. B - Condensed Matter **68**, 309 (1987).
160. L. E. Levine, P. C. Gibbons, and K. F. Kelton, Phys. Rev. B **40**, 9338 (1989).
161. H. von Löhneysen, J. Wosnitza, R. van den Berg, E. Compans, and S. J. Poon, Jpn. J. Appl. Phys. **26**, 887 (1987).
162. J. Wosnitza, R. van den Berg, H. v. Löhneysen, and S. J. Poon, Z. Phys. B - Condensed Matter **70**, 31 (1988).
163. P. K. Leung and J. C. Wright, Phil. Mag. **30**, 185 (1974).
164. K. Kimure and F. Yonezawa in *Topological Disorder in Condensed Matter*, (F. Yonezawa and T. Ninomiya, eds.) Springer, Berlin (1983).
165. H. Rubin, S. Jost and H. J. Güntherodt, J. Non-Cryst. Solids **61/62**, 291 (1984).
166. A. Defrain, L. Bosio, R. Cortes and P. G. Da Costa, J. Non-Cryst. Solids **61/62**, 439 (1984).
167. S. Sachdev and D. R. Nelson, Phys. Rev. B. **32**, 4592 (1985).
168. D. R. Nelson and S. Sachdev, Phys. Rev. B **32**, 689 (1985).
169. V. Elser, Phys. Rev. B **32**, 4892 (1985).
170. P. Bak, Phys. Rev. Lett. **54**, 1517 (1985).
171. N. D. Mermin and S. M. Troian, Phys. Rev. Lett. **54**, 1524 (1985).
172. O. Biham, D. Mukamel and S. Shtrikman, in *Introduction to Quasicrystals*, in *Aperiodicity and Order*, vol. 1 (M. V. Jaric ed.) Academic Press, New York, (1988).
173. R. Becker and W. Döring, Ann. Phys. **24**, 719 (1935).
174. D. Turnbull, in *Physics of Non-Crystalline Solids*, (J. A. Prins, ed.), North-Holland, Amsterdam, 1964, p. 41.
175. F. Spaepen, Acta Metall. **23**, 729 (1975).
176. F. Spaepen and R. B. Meyer, Scripta Metall. **10**, 257 (1976).
177. K. F. Kelton and A. L. Greer, Phys. Rev. B **38**, 10089 (1988).
178. K. F. Kelton, A. L. Greer, and C. V. Thompson, J. Chem. Phys. **79**, 6261 (1983).
179. D. Kashchiev, Surf. Sci. **14**, 209 (1969).
180. D. R. Uhlmann, in *Materials Science Research*, Vol. 4, Plenum, New York (1969), p. 172; J. Non-Cryst. Solids **7**, 337 (1972).
181. K. F. Kelton and A. L. Greer, J. Non-Cryst. Solids **79**, 295 (1986).
182. L. A. Bendersky and S. D. Ridder, J. Mater. Res. **1**, 405 (1986).
183. J. L. Robertson, S. C. Moss, and K. G. Kreider, Phys. Rev. Lett. **60**, 2062 (1988).
184. L. C. Chen and F. Spaepen, Nature **336**, 366 (1988).
185. L. C. Chen, F. Spaepen, J. L. Robertson, S. C. Moss, and K. Hiraga, J. Mater. Res. **5**, 1871 (1990).
186. J. A. Knapp and D. M. Follstaedt, Phys. Rev. Lett. **55**, 1591 (1985).
187. S. Gargon, P. Sainfort, G. Regazzoni, and J. M. Dubois, Scripta Metall. **21**, 1493 (1987).
188. J. M. Dubois, M. de Boissieu, A. Pianelli, J. Pannetier, and C. Janot, Scripta Metall. **23**, 1069 (1989).
189. A. J. Drehman, S. J. Poon, and K. R. Lawless, Mater. Res. Soc. Proc. **58**, 249 (1986).
190. Y. Shen, S. J. Poon, and G. J. Shiflet, Phys. Rev. B **34**, 3516 (1986).
191. A. J. Drehman, A. R. Pelton, and M. A. Noack, J. Mater. Res. **1**, 741 (1987).
192. J. C. Holzer and K. F. Kelton, unpublished results.
193. W. A. Johnson and R. F. Mehl, Trans. Am. Inst. Min. Engrs. **135**, 416 (1939).
194. M. Avrami, J. Chem. Phys. **7**, 1103 (1939).
195. J. W. Christian, *The Theory of Transformations in Metals and Alloys*, 2nd edition, pp. 528-542. Pergamon Press, Oxford (1975).
196. H. E. Kissinger, J. Res. Natn. Bur. Stand. **57**, 217 (1956).
197. P. Grütter, H. Bretscher, G. Indlekofer, H. Jenny, R. Lapka, P. Oelhafan, R. Weisendaager, T. Zigg, and H. J. Güntherodt, Mat. Sci. and Eng. **99**, 357 (1988).

198. A. P. Tsai, A. Inoue, Y. Bizen, and T. Masumoto, Acta Metall. **37**, 1443 (1989).
199. J. C. Holzer and K. F. Kelton, Acta Metall. Mater. **39**, 1883 (1991).
200. A. Davidson and M. Tinkham, Phys. Rev. B **13**, 3261 (1976).
201. D. Turnbull and F. Spaepen, J. Polymer Sci. **63**, 237 (1978).
202. J. H. Sikkenk, J. O. Indekeu, J. M. van Leeuwen, and E. O. Vossmack, Phys. Rev. Lett. **59**, 98 (1987).
203. C. Ebner and W. F. Saam, Phys. Rev. Lett. **38**, 1486 (1977).
204. W. A. Curtin, Phys. Rev. Lett. **59**, 1228 (1987).
205. W. E. McMullen and D. W. Oxtoby, J. Chem. Phys. **88**, 1967 (1988).
206. V. Elser, J. Phys. A **17**, 1509 (1984).
207. C. V. Thompson, Ph.D. Thesis, Harvard University (1981).

Crystal-Quasicrystal Transitions
M.J. Yacamán and M. Torres (Editors)

STRUCTURAL ASPECTS OF ALLOYS WITH QUASICRYSTALLINE PHASES

R. Perez, J. A. Juarez-Islas, J. Reyes Gasga and M. Jose Yacaman

Laboratorio de Cuernavaca, Instituto de Fisica UNAM, P. O. Box 139-B

62191 Cuernavaca, Mor. MEXICO.

Abstract

In this investigation we report some structural experimental results obtained in quasicrystalline phases for different type of alloys. Quaternary alloys based on Al-Cu-Co-Fe were obtained using a gravity chill casting technique. Electron diffraction patterns in addition to X-ray diffraction patterns show two kind of quasicrystalline phases. The icosahedral and decagonal phases coexist in this type of alloys with binary and ternary compounds of intermetallic nature. Defects are commonly found in these quasicrystalline phases. The defects display image contrast with similar characteristics to stacking faults and dislocations in crystalline materials. The image contrast characteritics suggest two different type of planar faults. Ternary and binary alloys of Al-Mn-Si and Al-Mn have also been obtained. In this case, two different alloys preparation methods have been employed, the gravity chill casting technique and also the gun technique based on shock waves. The intermetallic and also the quasicrystalline phases have been study using transmission electron microscopy and X-ray diffraction techniques. The decagonal phase in alloys of Al-Cu-Co-Si experienced transformations to crystalline structures when it was exposed to the electron beam in a transmission electron microscope (TEM). A study based on electron diffraction patterns and HREM images of this type of transformation is presented.

1. INTRODUCTION

There have been in the past several experimental techniques for the obtention of quasicrystalline phases in different alloy systems (1-6). Most of these techniques have been based on rapid solidification from the liquid melt. The melt spinning technique is one of the most widely used methods for the preparation of quasicrystals (1). However, other techniques like the splat quenching from the liquid melt (2), electron beam remelting (3), sputter or vapor deposition (4), solid state reactions (5), heat treatment of amorphous phases (6) and mechanical alloying (7) have also been employed for the obtention of quasicrystals.

Conventional casting techniques have recently been used as an alternative route for the normal production of decagonal and icosahedral phases (8,9). In this investigation we report the coexistance of icosahedral and decagonal phases obtained after casting an Al-Cu-Co-Fe alloy into a wedge-shaped copper mould. Both kind of quasicrystalline phases show planar faults and dislocation type of defects. These kind of defects have already been reported in the literature (10). However, their physical nature is not completely understood at the present although some comments on their geometrical nature have already been reported(10). Large attention has been paid in the past to the quasicrystalline phases obtained in alloys of Al-Mn-Si, however, not much attention has been devoted to the relationships between the intermetallic phases and the metastable quasicrystalline compounds (11-13). We present in this communication a study of the relation between intermetallic phases such as Al_6Mn and Al_4Mn with metastable structures such as the icosahedral and decagonal phases. Two different methods of alloy preparation have been used. The gravity chill casting technique and also the gun technique based on shock waves. Another aspect of interest is related with the transformations of the quasicrystalline phases to crystalline structures induced by the electron beam in a transmission electron microscope (TEM). In the past, very few investigations on this area have been carried out. Mainly in reports related with the Al-Mn alloy (3,14). We have studied in our case the transitions induced by the electron beam in quaternary alloys of Al-Cu-Co-Si with quasicrystalline phases. The transition has been followed in-situ in the microscope using diffraction patterns and also HREM images.

2. QUASICRYSTALLINE PHASES IN Al-Cu-Co-Fe ALLOYS.

An $Al_{65}Cu_{22}Co_{6.5}Fe_{6.5}$ (in wt%) alloy was prepared by induction melting of the high purity elements (99.99 %) in an alumina crucible under argon atmosphere. The liquid melt was cast at 100K superheat above the liquidus temperature into a wedge-shaped copper mould of dimensions reported elsewhere (15). Areas of the wedge-shaped ingot in contact (at the wedge edge) with the copper mould and in regions which were longitudinally and centrally in the plane normal to the diverging wedge were observed using optical and scanning electron microscopes. Samples of certain regions were removed for X-ray diffraction measurements using a Siemmens instrument with filtered Cu$k\alpha$ radiation. In addition, electron diffraction patterns from the quasicrystalline phases were obtained with a JEOL 4000EX microscope operated at 200 KV. The first observations carried out in the wedge-shaped ingot were taken from areas which were located in the narrow part of the wedge edge as is shown in Fig. 1. Observations with the optical microscope were taken from these areas. They show a characteristic dendritic growing which is illustrated in fig. 2. The dentrites display shapes which resemble five-fold, eight fold and ten-fold symmetrical objects. This is more pronounced in Fig. 3, where scanning electron micrographs of these kind of dentrites are illustrated. These type of images (in particular Fig. 3A and Fig. 2) show a central nucleus from which arms are growing given rise to morphological symmetries of approximately ten-fold, eigth-fold and five-fold types. As the growing of these arms continue, secondary arms (tree like) are developed. These secondary arms stop growing at the grain boundaries. Electron diffraction patterns obtained from these areas have clear indications of the presence of the two kind of quasicrystalline phases. This is illustrated in

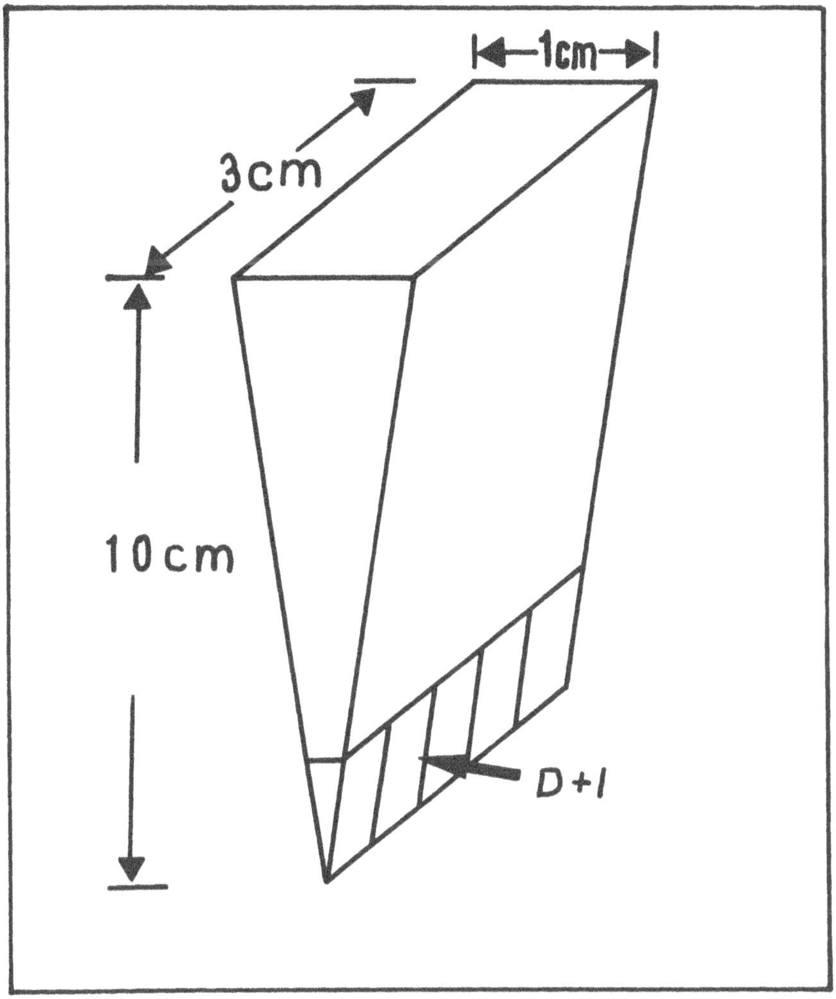

Fig. 1. Schematic drawing of the structure zone location in a chill-cast wedge of an Al-Cu-Co-Fe alloy. The region where the decagonal and icosahedral phases were obtained are clearly indicated in this figure as a D + I region (D = decagonal, I = Icosahedral).

Fig. 2. Optical micrograph of the microstructure observed in the chill cast ingot. The area corresponds to those regions in contact with the copper mould and in the thinnest part of the wedge sample. The arrows show the characteristic type of dendrites.

Fig. 3. SEM images of the characteristic dendrites obtained in the narrow areas of the wedge ingot. A) this dendrite has an apparent ten-fold morphology. B) This dendrite has an apparent five-fold morphology. C) This array has an apparent eight-fold symmetrical axis.

Fig. 4 and Fig. 5. The main zone axis diffraction patterns of the icosahedral phase are shown in Fig. 4 and the decagonal diffraction patterns are shown in Fig. 5. Both sets of patterns were obtained from quasicrytalline grains in the same observed specimen. Specimens taken from these type of regions have also been used for X-ray diffractometry. In order to obtain the decagonal and icosahedral corresponding peaks in the spectrum, "interplanar" distances from the icosahedral and decagonal electron diffraction patterns have been previously obtained. These distances in real space have been obtained using the traditional approach of the measurements of the main distances in reciprocal space directly from the electron diffraction micrographs and subsequently considering the camera length of the microscope in use. For the binary and ternary compounds the identification has been based on the interplanar spacings and the hkl values reported in the literature (16). Decagonal and icosahedral peaks were mainly obtained in the angular range of $38^{\circ} < 2\theta < 46^{\circ}$, although, there is an icosahedral peak in an angular value of 73.29°. Fig. 6 shows the X-ray spectrum in the angular range of $25^{\circ} < 2\theta < 45^{\circ}$. Two peaks were detected for the decagonal phase at angular values of 38.56° and 42.64° respectively. On the other hand, three peaks were obtained for the icosahedral phase at angular values of 43.32, 44.38 and 44.9° respectively. The other peaks in the spectrum correspond to different crystalline phases. Thus, for example, peak 1 is related with Al_9Co_2, peaks 2 and 5 to Al_2Cu and finaly peak 3 and 4 to Al_7Cu_2Fe. The icosahedral and decagonal phases were only obtained in the thinnest areas close to the wedge edge of the sample. The dendrites obtained in these regions show a nucleus with an apparent pentagonal shape. This is probably related with an icosahedral phase which acts as a nucleation site for the growing of the decagonal phases giving rise to dendrites with morphological shapes which resemble ten-fold, eight-fold and five-fold symmetrical objects.

3. QUASICRYSTALLINE DEFECTS IN ICOSAHEDRAL AND DECAGONAL PHASES

Indications of the presence of defects in the icosahedral and also in the decagonal phase have already been reported (10).Thus, for example, dislocations, planar faults and antiphase boundaries have already been discussed. Although, the nature of these kind of defects is not at the present completely understood, there have been in the literature some reports on their geometrical nature (10). The icosahedral and decagonal phases which are obtained in quaternary alloys of Al-Cu-Co-Fe have also a high density of planar and dislocation-type of defects. The planar defects display image contrast characteristics which are similar to stacking faults in crystals. This is illustrated in Fig. 7, where a bright field (BF) and a dark field (DF) images obtained from the icosahedral phase are displayed. Strong beam conditions are used and the orientation is closed to a five-fold axis. This figure shows a symmetric BF and an asymmetric DF images. The same behavior is displayed by stacking faults in crystals. In crystalline structures, the responsible mechanism for this image contrast characteristic is related with the strong scattering of the electrons due to the periodic array of atoms. However, in the quasicrystalline phases there is no periodic arrangement of atomic species, but the scattering effect is similar to the crystalline case. Planar faults are also obtained in the decagonal phase as is shown in Fig. 8. The quasicrystal is oriented closed to a two-fold axis. The image contrast obtained from these faults is similar to stacking faults where the fault plane is closed perpendicular to the crystalline surface. Another aspect of

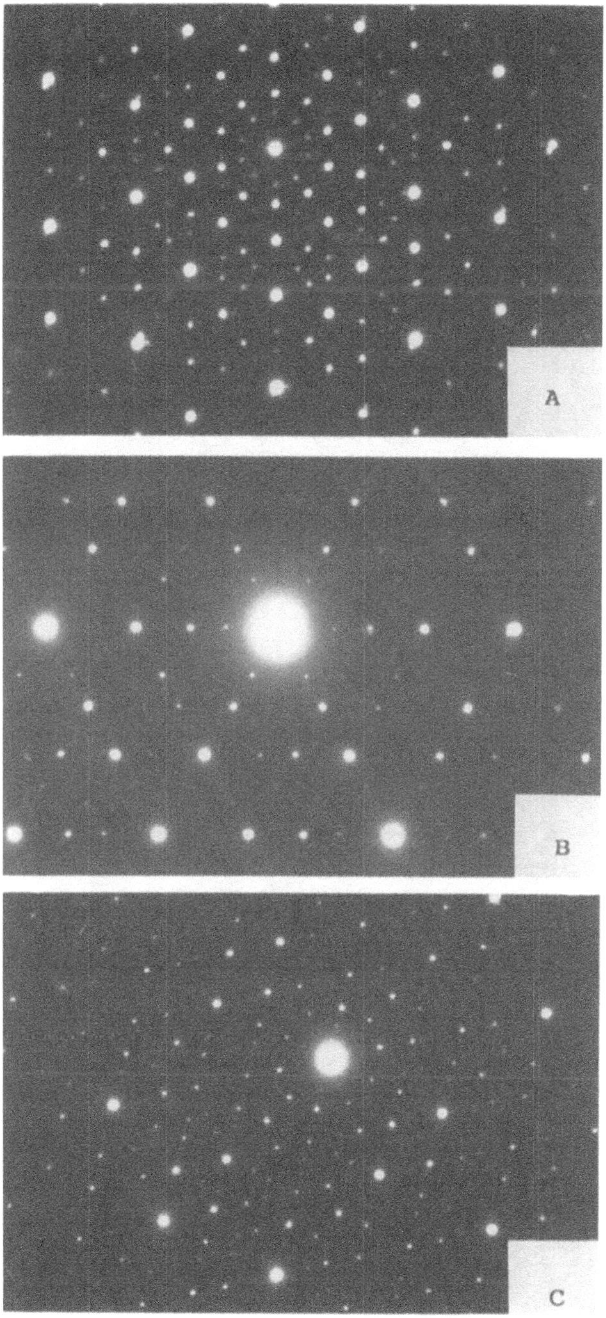

Fig. 4. Diffraction patterns from the icosahedral phase along the five-fold (A), three-fold (B) and two-fold (C) zone axis.

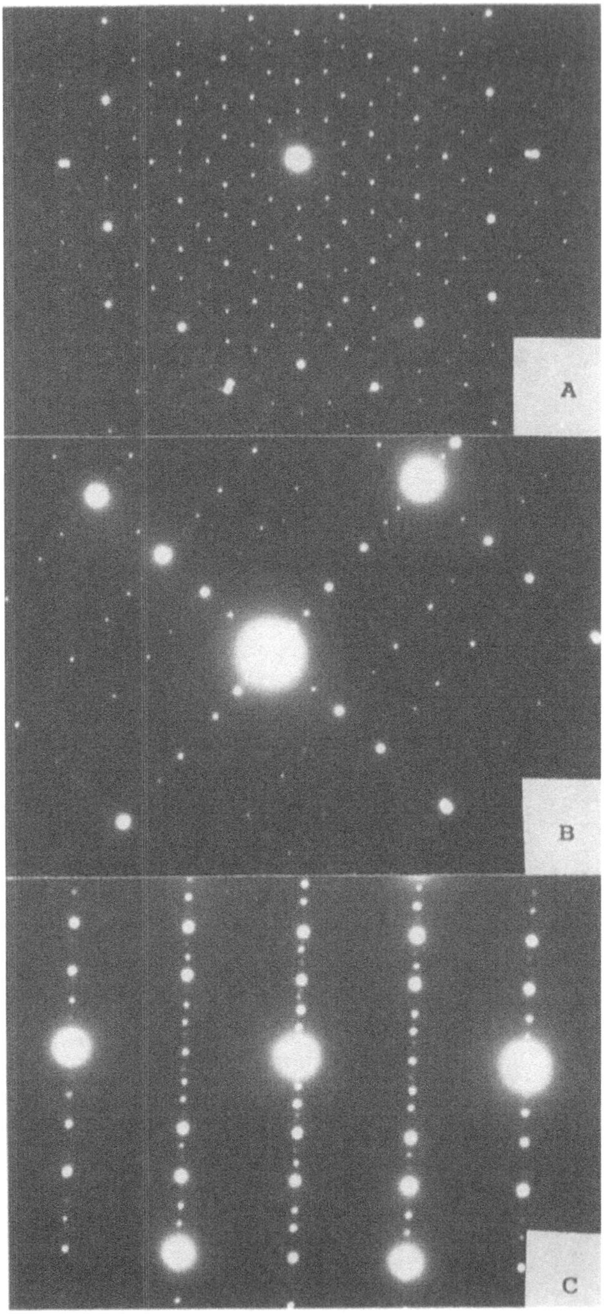

Fig. 5. Diffraction pattern from the decagonal phase along the ten-fold (A) and two-fold (B,C) zone axis.

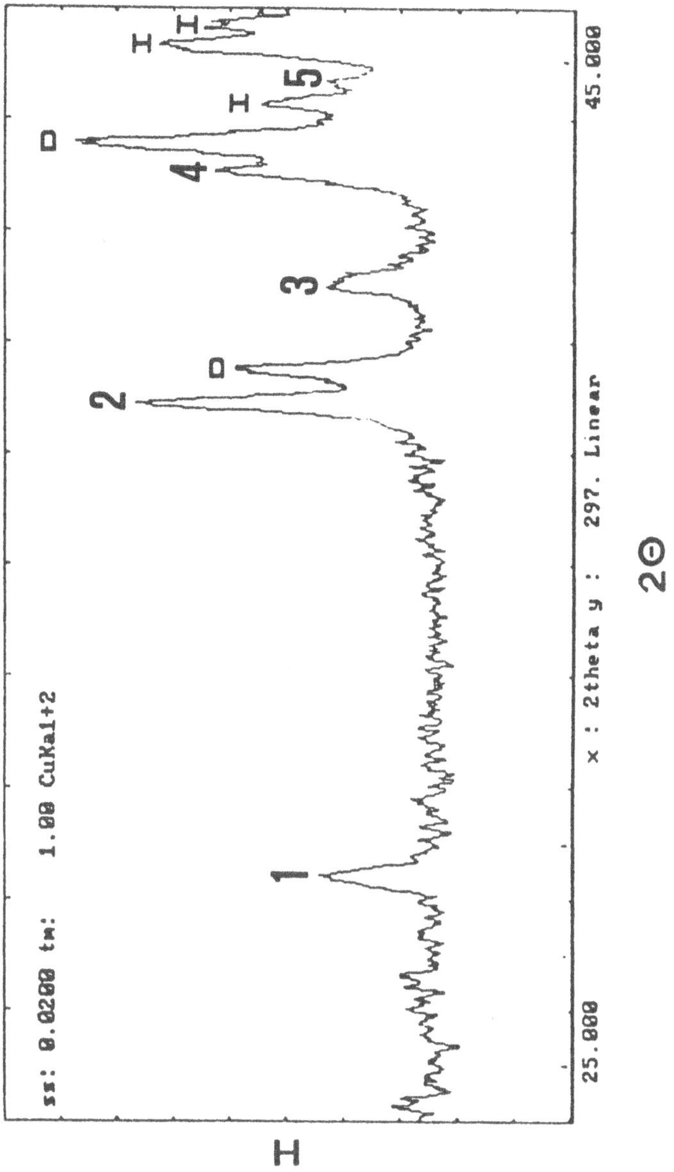

Fig. 6. X-ray diffraction pattern of the phases observed in the samples obtained from the narrow region of the wedge. In this figure only the angular range of $25^o < 2\theta < 45^o$ in the spectrum is shown.

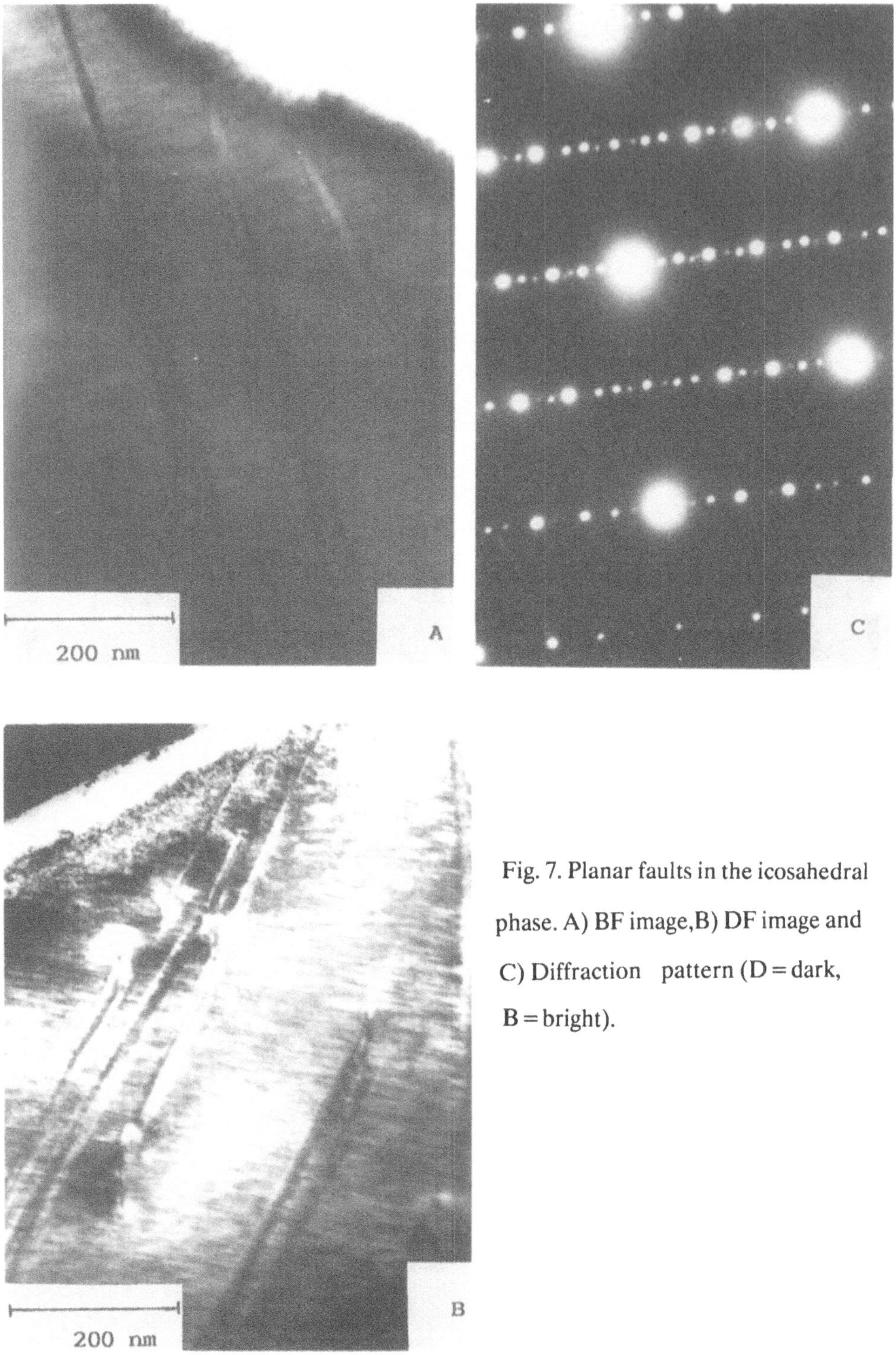

Fig. 7. Planar faults in the icosahedral phase. A) BF image,B) DF image and C) Diffraction pattern (D = dark, B = bright).

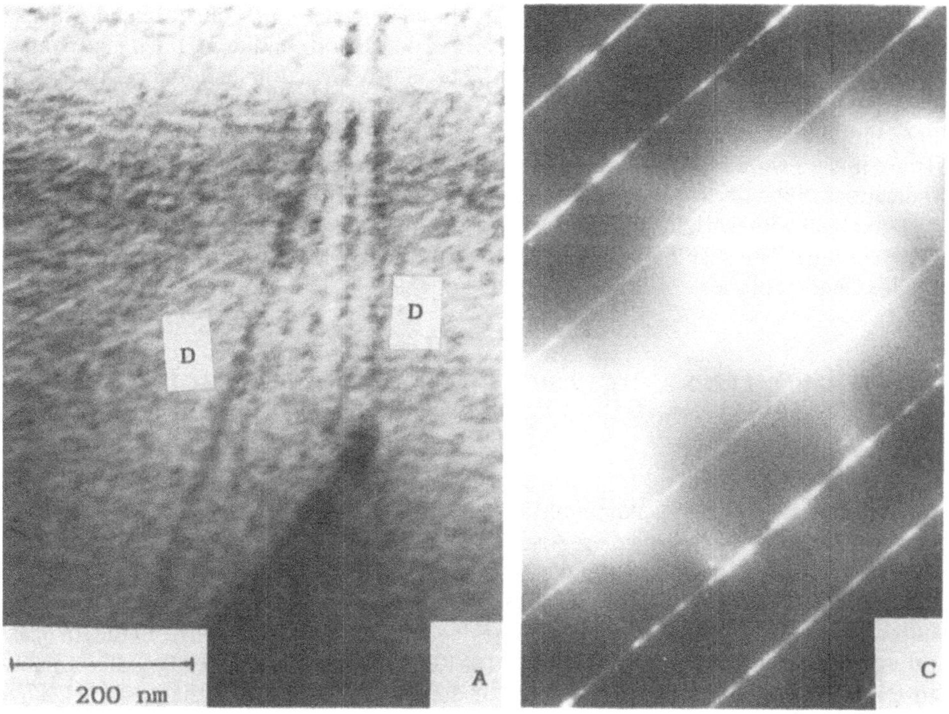

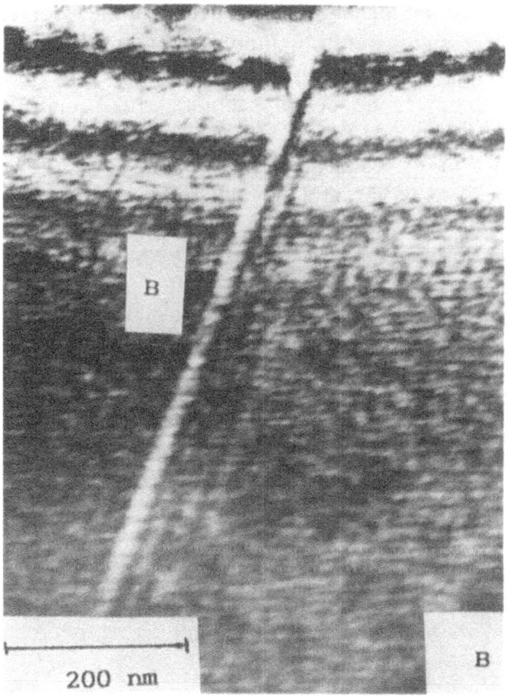

Fig. 8. Planar faults in the decagonal phase. A) BF image,B) DF image and C) Diffraction pattern.

interest is related with the nature of the planar faults in icosahedral phases. In crystalline structures, there are mainly two different type of stacking faults intrinsic and extrinsic. When both faults are imaged under similar strong diffraction conditions, the image contrast displayed with one type of fault is exactly the reverse of the corresponding contrast of the other fault. This is due to the difference in sign of the phase angle associated with the fault (17). Fig. 9 shows BF and DF images of planar faults obtained under strong bean conditions. These images were taken from an icosahedral grain. The planar faults in this figure show indications of the reverse in the overall image contrast. Thus, for example, the BF case has a planar fault with a bright-bright edge image contrast and the other fault with a dark-dark edge contrast. This result strongly suggest the presence of two different types of faults in the icosahedral phase.

4. INTERMETALLICS AND QUASICRYSTALLINE PHASES IN Al-Mn AND Al-Mn-Si ALLOYS.

Alloys of Al-Mn containing 10, 15, 28 and 32 wt% Mn were made by induction melting of the high purity (99.99%) aluminum and electrolitic (99.7%) manganese under argon atmosphere in alumina crucibles. Also, an alloy of Al-32wt%Mn-5wt%Si was made by induction melting of two master alloys of Al-50wt%Mn and Al-30wt%Si in addition to aluminium of high purity again in an alumina crucible under argon atmosphere. Two different techniques have been used for the final preparation of the alloys. In the first case, the Al-10 and 15wt%Mn alloys were casted at 100 K superheat above the liquidus temperature into a wedge-shaped copper mould of dimensions reported in the past (15). Thermocouples introduced at the side of the mould indicated cooling rates in the ranges of approximately 20, 50 and 150 K/s for the upper, middle and lower portions of the wedge as is indicated in Fig. 10. Specimens from these ingots were obtained and regions of characteristic morphology were located by optical microscopy and subsequently analyzed by X-ray diffractometry and scanning electron microscopy. In the second technique, small quantities (~100mg) of alloys with compositions of Al-28wt%Mn,Al-32wt%Mn and Al-32wt%Mn-5wt%Si have been melted using the induction furnace. The melted drop is subsequently fragmented with a shock wave (produced with argon gas) and the tiny dropplets (1μm) with velocities of approximately a few hundred meters were hit on a copper substrate. The flake-like products are already transparent to the electron beam in a TEM. Observations have, therefore, been obtained with the TEM equipped with and EDX attachment. For the ingots obtained by gravity chill casting, the regions used in this study are clearly indicated in Fig. 10. Five different phases have been identified by microanalysis and X-ray diffractometry. Primary α-Al solid solution and interdendritic eutectic was observed at the narrow end of the ingot with composition Al-10wt%Mn (Fig. 11). Al_6Mn and cellular α-Al solid solution was observed in ingots of Al-10wt%Mn (Fig. 11). Identification of this intermetallic was carried out by X-ray diffractometry and SEM microanalysis (Table 1). Al_4Mn and cellular α -Al solid solution was observed in ingots of Al-10wt%and 15wt%Mn. Fig. 12 shows the X-ray diffraction pattern obtained from specimens which contain this intermetallic. Decagonal phase in α- Al solid solution was observed in ingots of Al-10wt%Mn as nodules which showed a central nucleus with dendrites growing radially outwards as is

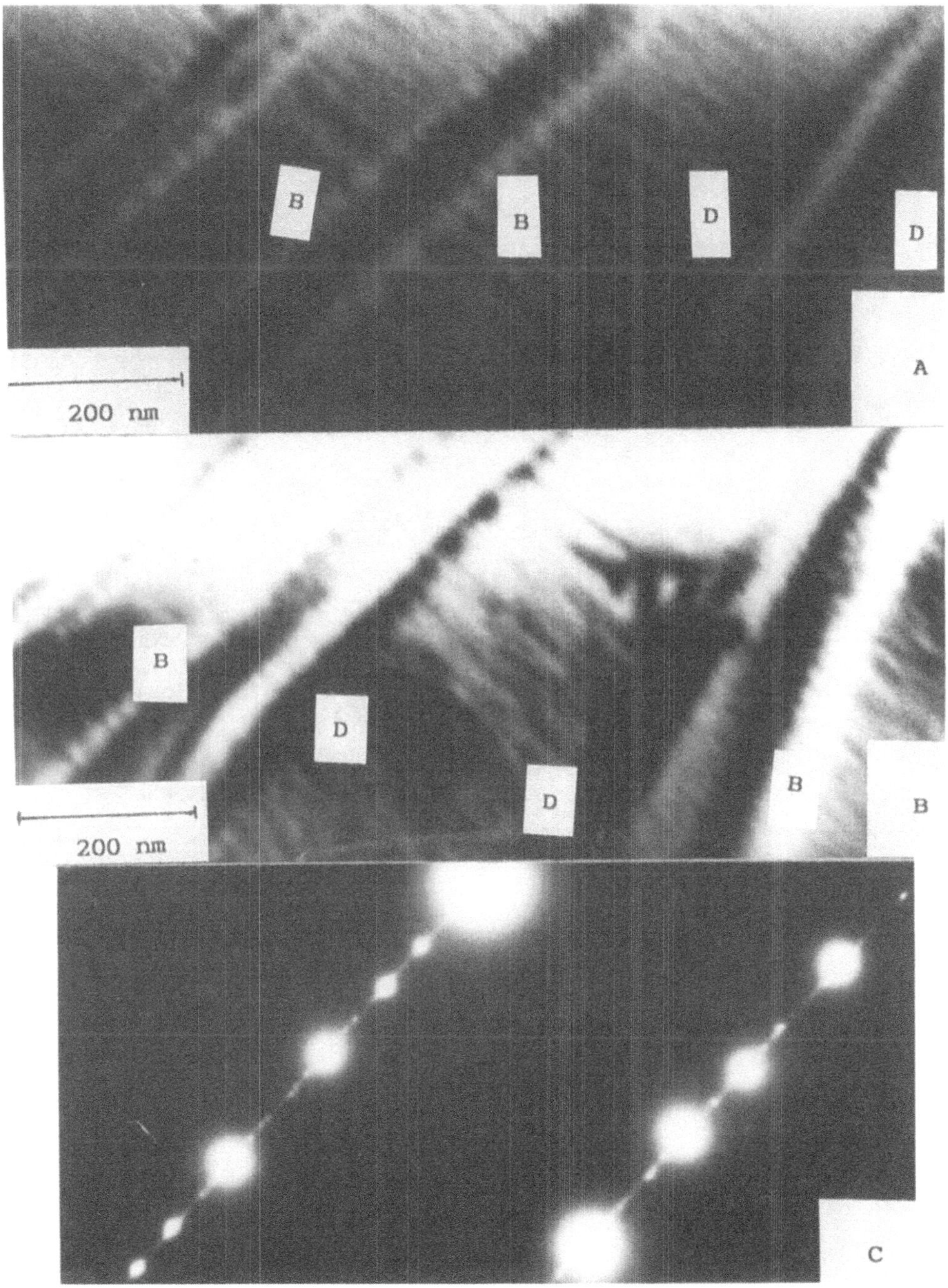

Fig. 9. planar faults in the icosahedral phase. A) BF image,B) DF image and C) Diffraction pattern. Image contrast reversed in some of the planar faults (D = dark, B = bright).

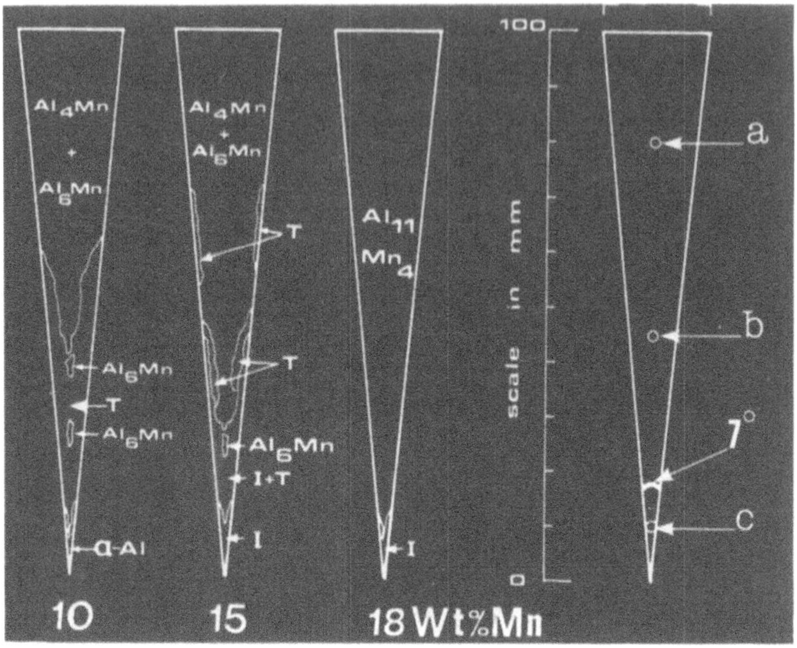

Fig. 10. Structure zone location in chill cast wedges of Al-10and 15wt%Mn alloys (D = decagonal, I = icosahedral).

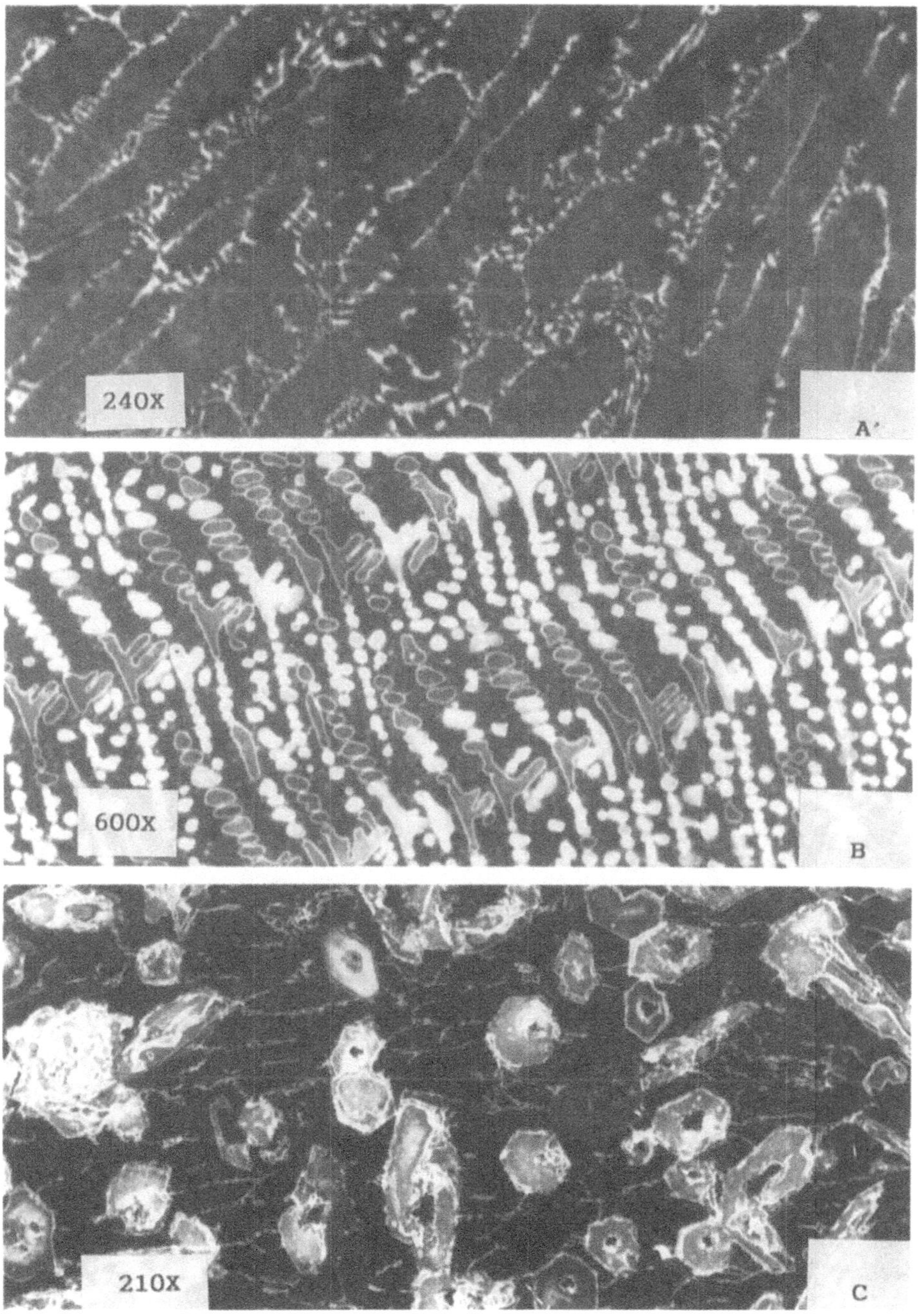

Fig. 11. Microstructure observed in chill cast ingots. A) α-Al solid solution and interdendritic eutectic in Al-10wt%Mn. B) Al6Mn cellular α -Al solid solution in Al-10wt%Mn. C) Al4Mn and cellular α-Al solid solution in Al-15wt%Mn.

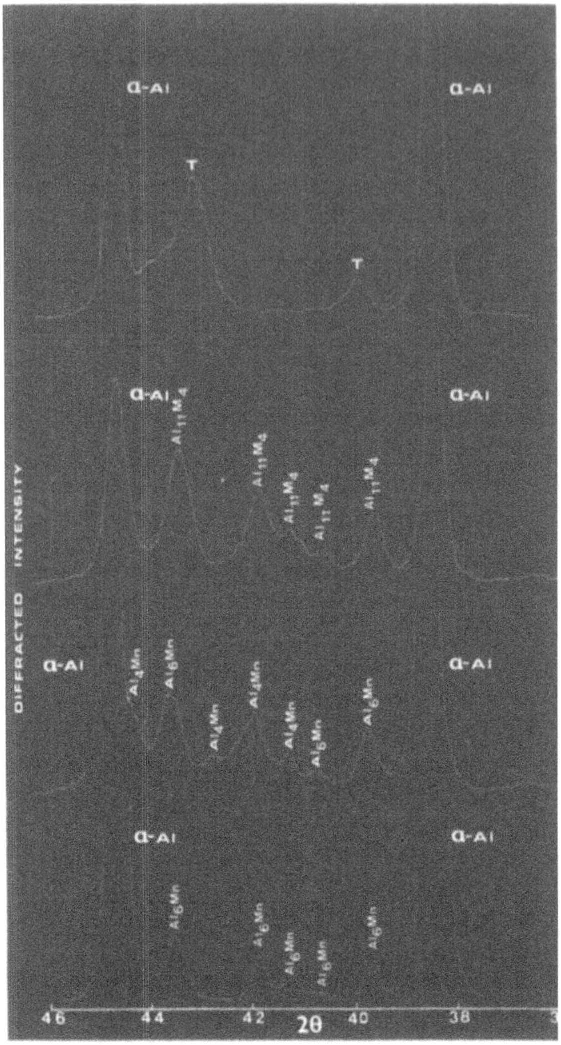

Fig. 12. X-ray diffraction pattern obtained with CuK α radiation in the range of $38^{\circ} < 2\theta < 46^{\circ}$.

shown in Fig. 13. The icosahedral phase shows the morphology of an isolated equiaxed quasicrytal in an α-Al matrix. They grow radially at the narrow end of the wedge in the Al-15wt%Mn ingot (Fig. 14A). Fig. 14B shows the X-ray diffraction pattern of the region containing the icosahedral phase which also shows the presence of the decagonal phase. The icosahedral phase was detected in regions where the cooling rate was approximately 150 K/s, however, the decagonal phase was detected up to cooling velocities of approximately 50 K/s. Electron transparent areas from the flake-like splats of Al-28wt%Mn, Al-32wt%Mn and Al-38wt%Mn-5wt%Si alloy obtained by using the shock wave technique were chosen to carry out observation of the icosahedral phase. As is shown in Fig. 15, the Al-28wt%Mn alloy consisted almost entirely of icosahedral phase. Grains of this phase have a central nucleus with elongated branches which end at the grain boundaries. The icosahedral phase obtained in the Al-32wt%Mn alloy in shown in fig.15B. The grain revealed very elongated dendrites from the central core or sometimes fine particle agregates which appear to meet at the central nucleus of the grain. The icosahedral phase obtained in Al-30WT%Mn-5wt%Si alloy was formed of fine particles aggregates (Fig.15C). The grain boundaries in this case were nearly rectilinear. In summary, at cooling rates approximately 20 K/S, Al_4Mn is the initial phase to be formed in Al-10 and 15wt%Mn. The decagonal phase is found at cooling rates of approximately 50K/s in alloys consisting of Al-10 and 15wt%Mn, furthermore, the icosahedral phase was only obtained at cooling rates of 150K/s. The icosahedral phase obtained in rapidly solidified alloys of Al-28wt%Mn, Al-32wt%Mn and Al-38wt%Mn-5wt%Si showed the presence of second phase particles. These particles were identified as orthorhombic Al_6Mn.

5. STRUCTURE TRANSFORMATION OF THE DECAGONAL PHASE IN Al-Cu-Co-Si ALLOY INDUCED BY THE ELECTRON BEAM IN A TEM.

The $Al_{62}Cu_{20}Co_{15}Si_3$ (wt%) alloy was prepared by a melting and growth process carried out in a double spherical mirror furnace(18). During this process the alloy took the shape of a sphere with one or two centers of nucleation where many decaprisms arranged in a radial manner grow. These small (approx. 1 or 2mm long) decaprisms were separated from the matrix, grounded and supported on Cu grids for their electron microscope analysis. The quasicrystal-crystal transition was observed in-situ in a high resolution electron microscope (JEOL4000EX) with an attached video system. The electron microscope has been operated at 400KV. Under this condition the electron beam energy is able to induce phase transitions in the decagonal phase of the Al-Cu-Co-Si alloy in just a few (4 to 5) minutes. Fig. 16 shows a sequence of the changes observed in the diffraction patterns along one of the two-fold axis. These changes are observed in-situ in the microscope. Fig.16A shows the decagonal phase spots and also those corresponding to the (111) zone axis of the superimposed bcc structure. Fig.16B-D show the diffraction patterns after different electron radiation times. These figures clearly illustrate the systematic disappearance of the diffraction spots which correspond to the decagonal phase. In Fig.16D, practically all the spots correspond to the crystalline phase and two different geometrical patterns can be recognized. The bcc(111) zone axis diffraction pattern is superimposed on a rectangular pattern. The rectangular pattern has dimensions which correspond to 0.2 nm X 0.12 nm in real space Fig. 17, on the

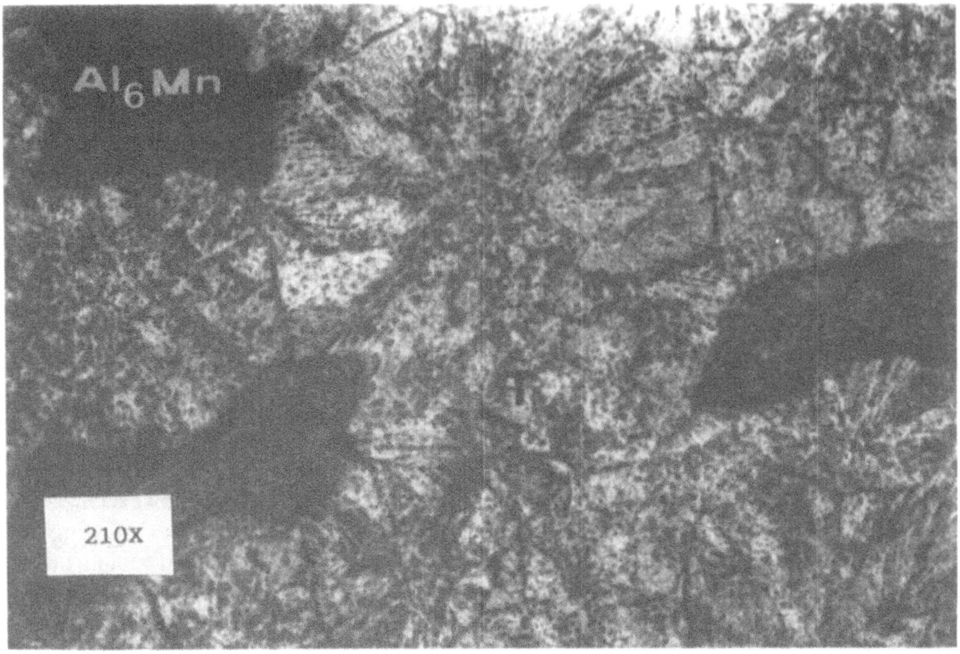

Fig. 13. Decagonal phase with some dendrites of Al6Mn.

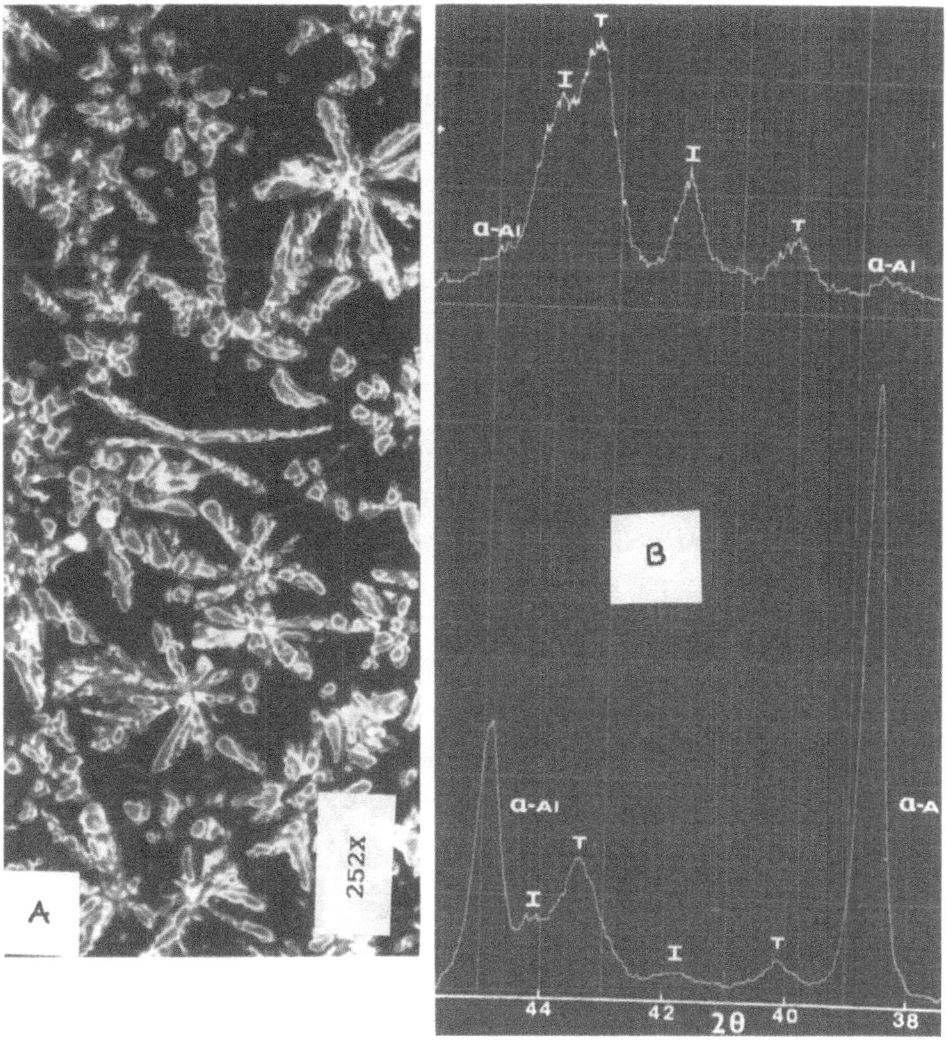

Fig. 14. A) Icosahedral phase obtained in a chill cast wedge of Al-15wt%Mn. B) X-ray diffraction pattern showing decagonal and icosahedral peaks.

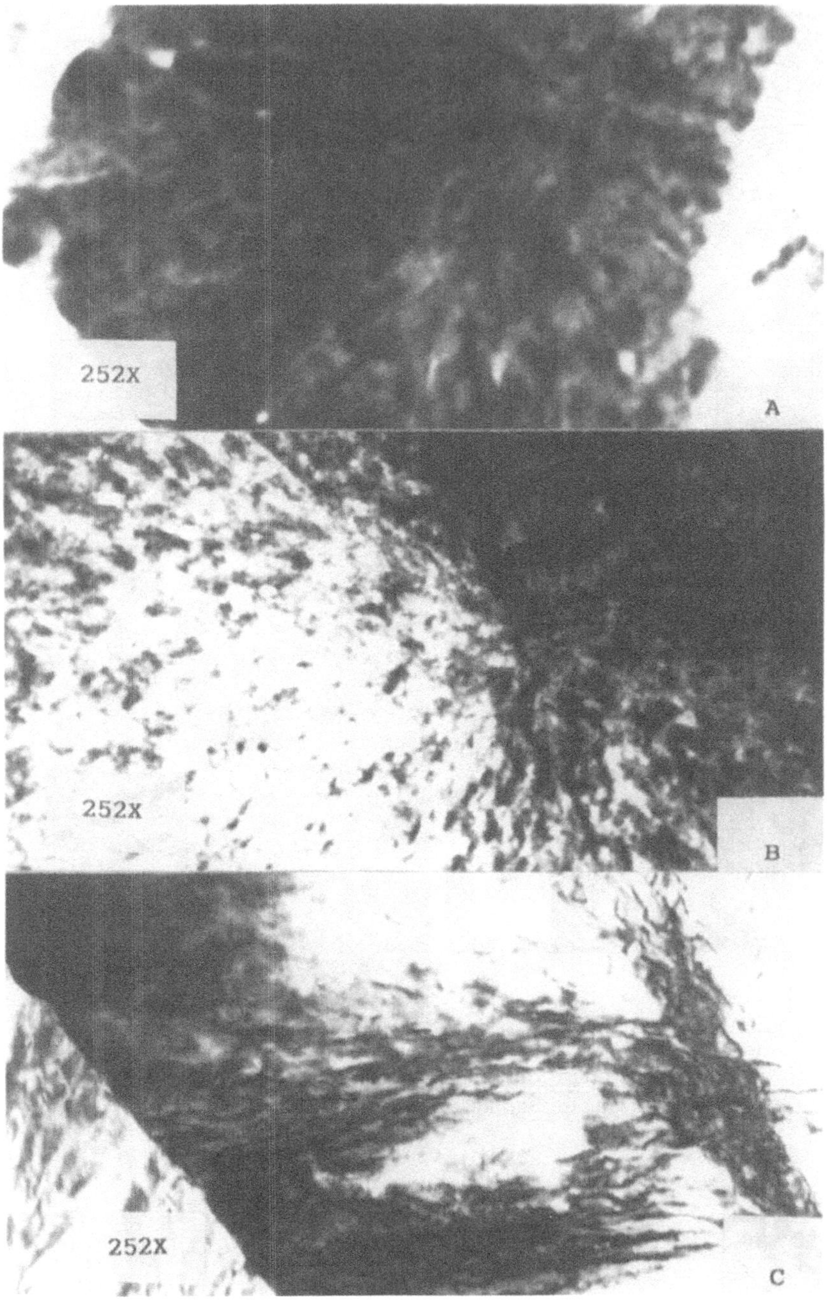

Fig. 15. Icosahedral phase in A) Al-28wt%Mn, B) Al-32wt%MN and C) Al-38wt%Mn-5wt%Si, obtained by the shock wave gun technique

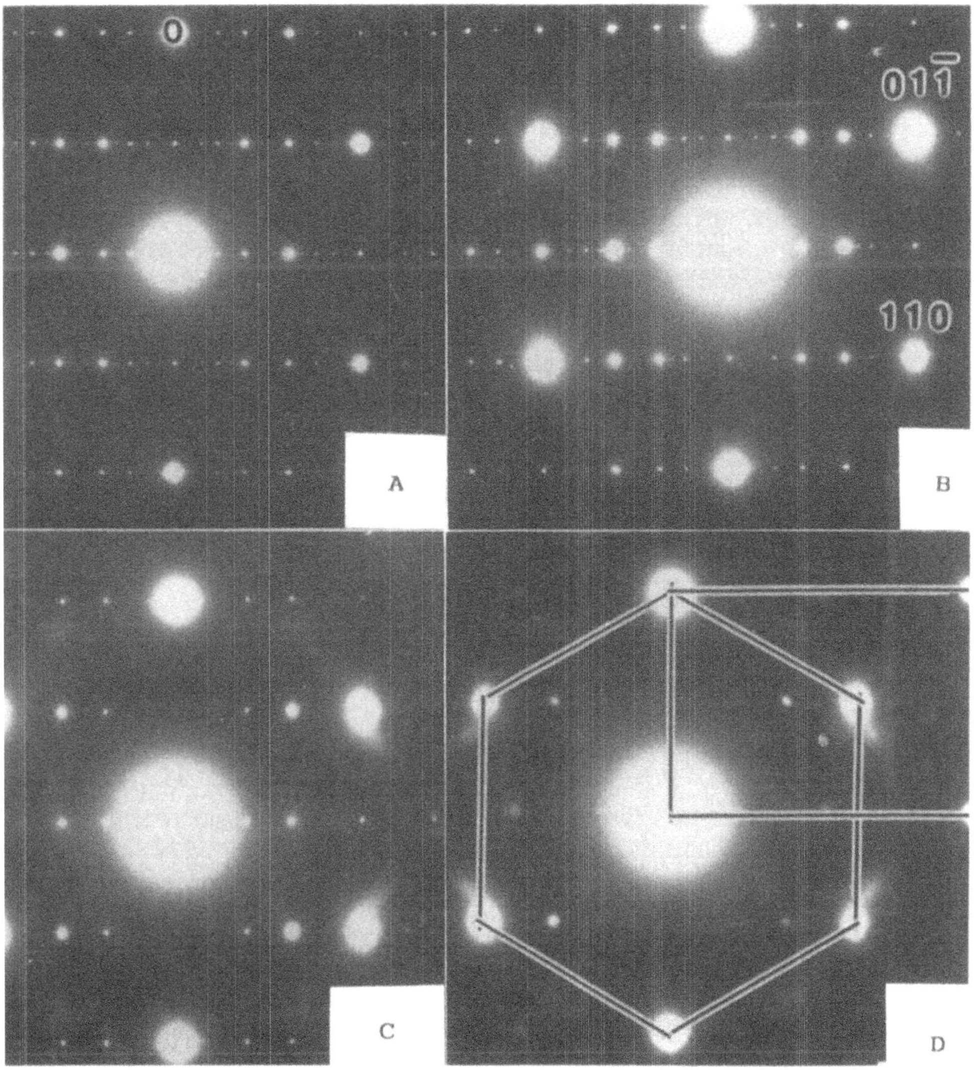

Fig. 16. Sequence of the diffraction pattern changes produced by the electron beam along one of the two-fold axis in the decagonal phase. In D) two patterns can clearly be observed.

other hand, shows the HREM images corresponding to this two-fold orientation after different radiation times. Fig. 17A shows the caracteristic 0.2 nm periodicity parallel to the two-fold axis and the quasiperiodicity perpendicular to this axis. Slowly the electron beam transforms this quasicrystalline image contrast in a crystalline hexagonal array. The final stage is observed in Fig. 17D, where small domains whith a hexagonal array can clearly be seen. The same kind of behavior is observed along the other two-fold axis. Fig. 18 shows the diffraction patterns sequence as a function of the electron radiation time. The final stage shows the systematic disappearance of the decagonal spots and the appearance of two different geometric patterns: the (100) bcc zone axis diffraction pattern and a superimposed rectangular pattern with dimensions in real space of 0.2 nm X 0.14 nm. These quasicrystalline orientation do not give, however, a clear indication on the involved deformation mechanism for the quasicrystalline-cristalline transition. Partial information can be obtained from the diffraction pattern along the ten-fold axis. This is ilustrated in Fig. 19, where the last Figure (Fig.19D) shows the clear disappearance of the ten-fold spots, leaving a pattern which strongly resembles the diffraction spots obtained from kinematical diffraction theory of twinned crystalline specimens (19). This diffraction pattern can be generated from a rectangular array of spots which correspond in real space with parameters of 0.19 nm X 0.13 nm. The HREM images which correspond to this orientation is transformed by the electron beam into small domains with rectangular image contrast feature (Fig.20). These domains keep twinning relationships between them as has been proved by image processing (Fig.21).

6. CONCLUSIONS

A simple gravity chill casting technique is used for the preparation of quasicrystalline phases in quaternary Al-Cu-Co-Fe alloys. Both quasicrystalline phases, decagonal and icosahedral are found to coexists in samples prepared whith this method. These quasicrystalline phases are more frequently found in sample areas at the edge of the wedge and in those regions which were in contact with the copper mould. Planar faults and dislocations are also present in these kind of phases. The planar faults display image contrast which is similar to stacking faults in crystalline materials. From the image contrast characteristics of the planar faults in the icosahedral phase, two different faults can possible be obtained. Results based on the image contrast properties suggest that planar faults in the decagonal phase have the fault plane perpendicular to a decagonal zone axis. The study of the relationships between the quasicrystals and the intermetallic phases in alloys of Al-Mn and Al-Mn-Si indicate that at rates of cooling of approximately 20 K/s during solidification, Al_4Mn is the initial phase to be formed in Al-10 and 15wt%Mn. The decagonal phase is formed at solidification velocities of approximately 50 K/s in alloys containing 10 and 15wt%Mn, however in regions of these ingots where the cooling rate was closed to 150 K/s, the icosahedral phase is found. The icosahedral phase formed in the alloy of 15wt%Mn grows in competition with the decagonal phase which is eventually replaced by the icosahedral phase under high solidification velocities.The icosahedral phase obtained in rapidly solidified alloys of Al-28wt%Mn, Al-32wt%Mn and Al-38wt%Mn-5wt%Si showed the presence of a second phase particles which seem to be related with orthorhombic Al_6Mn. The electron beam in a 400 KV TEM induces phase transformations in the decagonal phase of the Al-Cu-Co-Si alloy. The

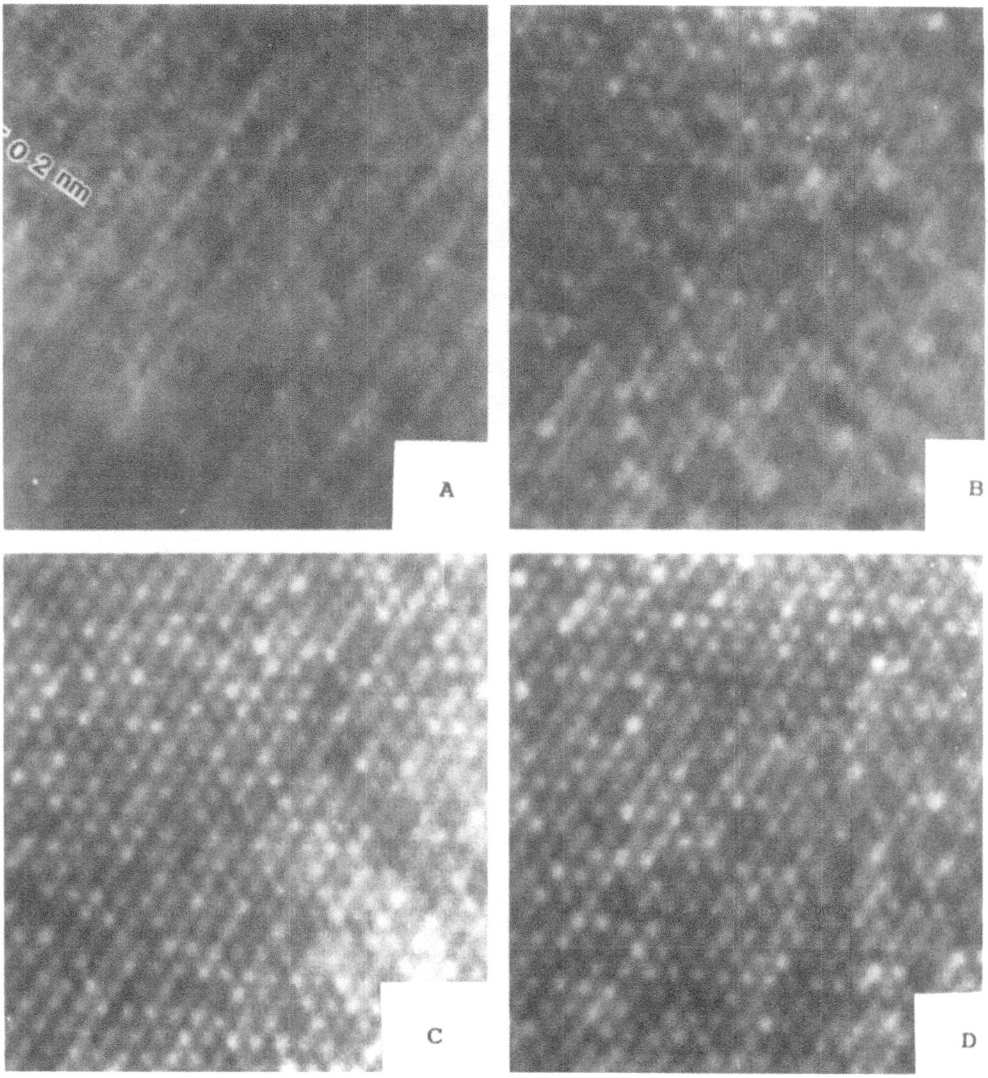

Fig. 17. HREM images taken along the two-fold axis corresponding to fig. 16. A transformation from the characteristic two-fold image contrast to a hexagonal array of contrast features can be seen from A) to D). Their approximate radiation times were: A)t = 0, B) t = 2 min. C) t = 4 min and D) t = 6 min.

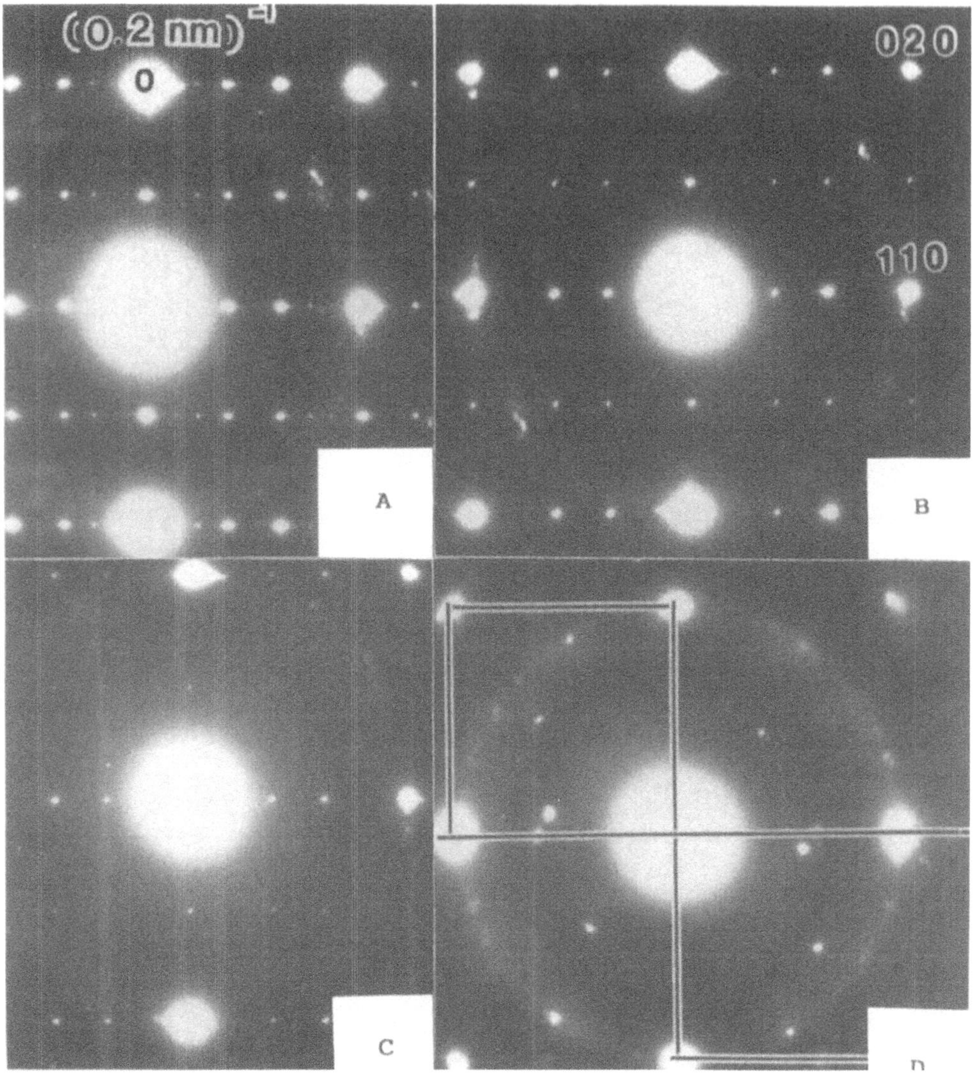

Fig. 18. Changes in the diffraction pattern along the other two fold axis of the decagonal phase. Two geometric patterns can also be observed at the end of the radiation time.

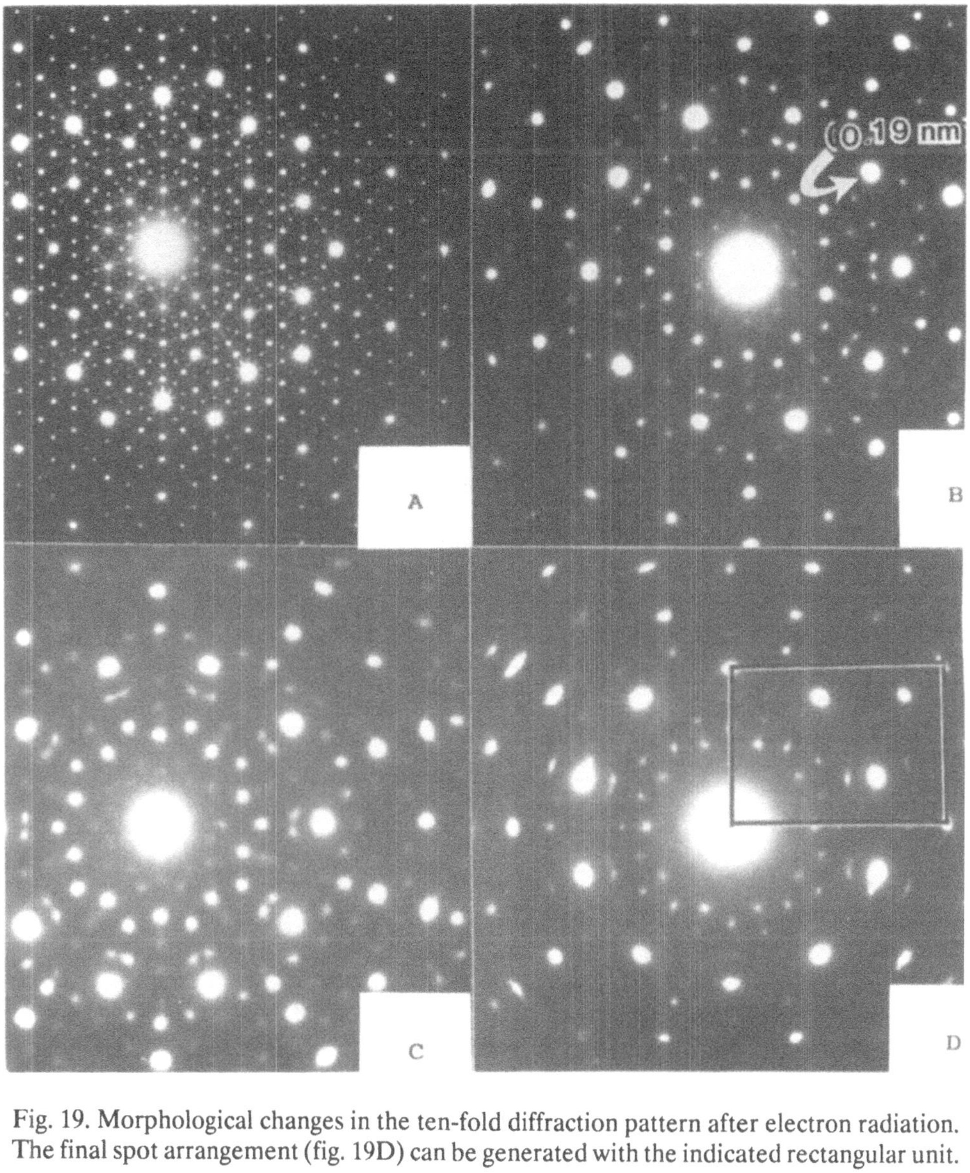

Fig. 19. Morphological changes in the ten-fold diffraction pattern after electron radiation. The final spot arrangement (fig. 19D) can be generated with the indicated rectangular unit.

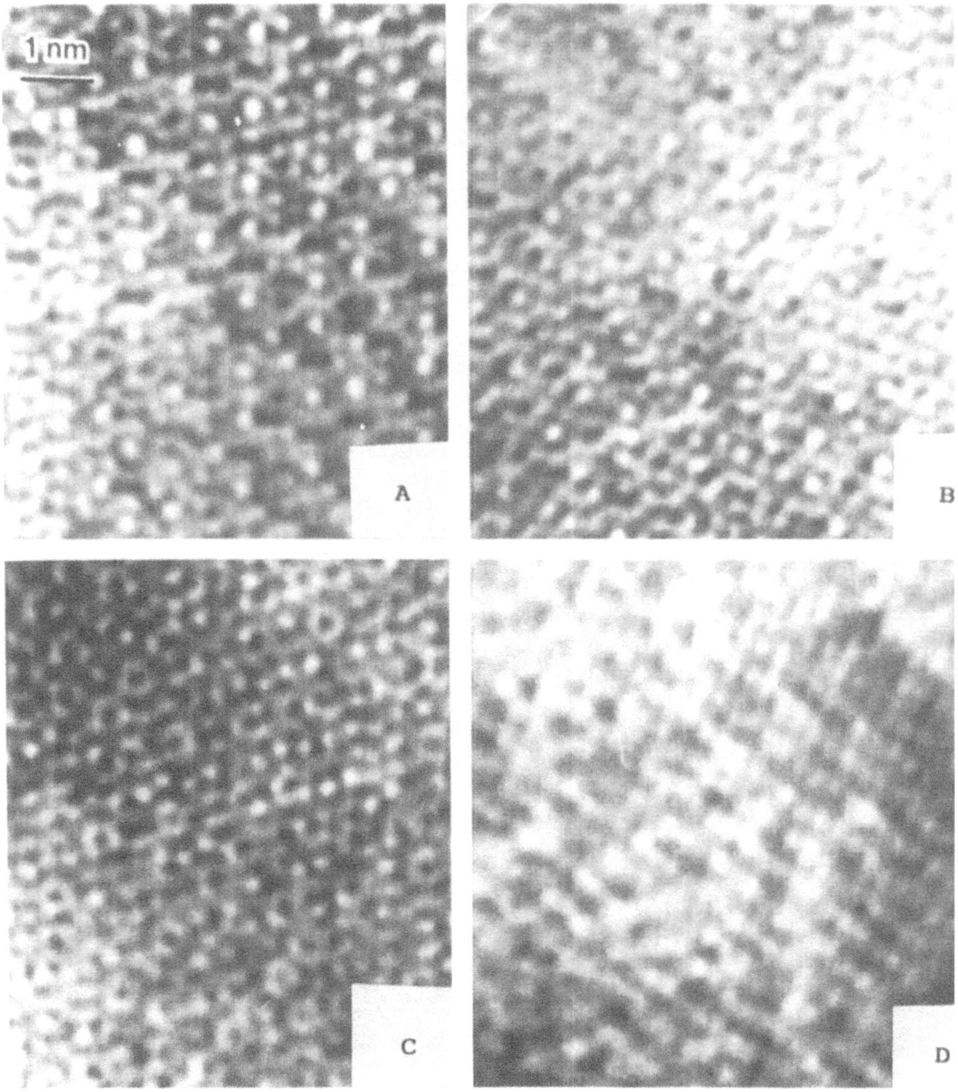

Fig. 20. HREM images along the ten-fold axis. The original image is slowly transformed to a rectangular array of image contrast features using the same electron radiation times of fig.16.

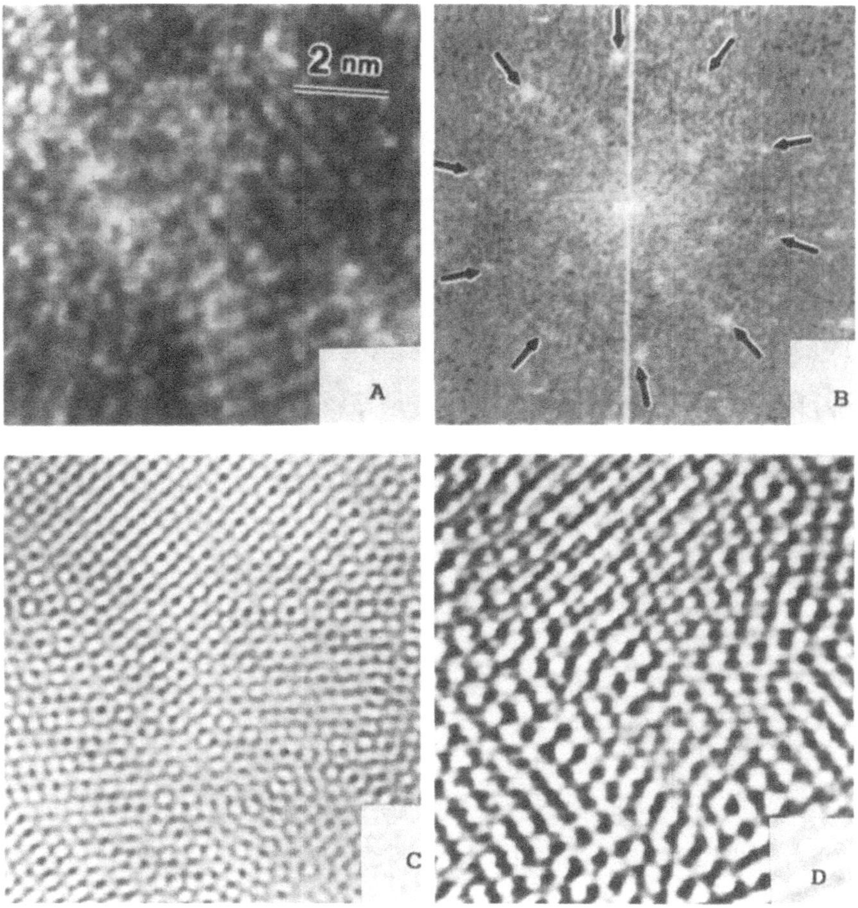

Fig. 21. Computer processing of the HREM image obtained at the end of the transformation (fig. 20 D). A) The original image.B) Its Fourier spectrum and C) Filtered image which results by setting apertures at each main spot in the spectrum.

deformation mechanism related with the quasicrystalline-crystalline transformation seems to be associated with a twinning phenomena. Experimental partial evidence is obtained by studying the evolution of a ten-fold diffraction pattern under electron radiation.

7. ACKNOWLEDGEMENTS

The authors are grateful to L. Rendon, S. Tehuacanero, C. Zorrilla, A. Gonzalez and J.L. Albarran for technical help. One of the authors (R.P.) is also grateful to CONACYT throught project No 0048E for financial support.

8. REFERENCES

1.- D. Shechtman, I. Blech, D. Gratias and J.M.Cahn. Phys. Rev.Lett. 53, 1951 (1984).

2.- R. Perez. Jour. of Mater. Science. 27, 5751 (1992).

3.- D. A. Lilienfeld, N. Nastasi, H. H. Johnson, D. G. Ast and J. W. Mayer. Phys.Rev.Lett. 55,1587 (1985).

4.- K. G. Kreider, F. S. Biancaniello and H. J. Kaufmann. Scripta Metall. 21, 657 (1987).

5.- D. M. Follstaedt and J. A. Knapp. Phys. Rev. Lett. 56, 1827 (1986).

6.- S. J. Poon, A. J. Drehman and K. R. Lawless. Phys. Rev. Lett.55, 2324 (1985).

7.- J. Eckert, L. Schultz and K. Urban, Acta Metall. et Mater. 39No 7, 1497 (1991).

8.- K. Hiraya, B. P. Zang, M. Hirabashi, A. Inoue and T.Masumoto. Jap. Jour. of Appl. Phys. 27 No 12, L2252 (1988).

9.- A. P. Tsai, A. Inoue and T. Masumoto, Mater. Trans. JIM 30 No2, 150 (1989).

10.- Z. Zhang and K. Urban, Phil. Mag. Lett. 60, No 3, 97 (1989).

11.- K. Chattopadhyay, S. Ranganathan, G. N. Subbunna and N. Thangaraj. Scripta Metall. 19, 767 (1985).

12.- D. Shechtman and I. A. Blech. Met. Trans. A. 16A, 1005 (1985).

13.- R. J. Schaefer, L. A. Berdensky, D. Shechtman, W. J. Boettinger and F. S. Biancaniello. Met. Trans. A. 17A ,216(1986).

14.- K. Urban, N. Moser and H. Kronsmuller, Phys, Stat. Sol. (a)91, 411 (1985).

15.- J. A. Juarez-Islas, D. Harrington and H. Jones, Jour. ofMater. Scien. 24, 2076 (1989).

16.- Selected Powder Diffraction for Metals and Alloys. Int.Center for Diffraction Data, JCPDS, Swarthmorse. 16 (1987).

17.- P. B. Hirsch at al. Electron Microscopy of thin Crystals,New York, R. E. Krieger Publ. Co. (1977).

18.- J. Reyes, A. Lara, H. Riveros and M. Jose-Yacaman, Mater. Scien. and Eng. In the press.

19.- C. N. Yang, M. Jose-Yacaman and K. Heinemann. Jour. of Cryst. Growth 47 No 2, 283 (1979).

Crystal-Quasicrystal Transitions
M.J. Yacamán and M. Torres (Editors)

Transitions between quasiperiodic and periodic phases by one-parameter Schur rotations

M. Baake, D. Joseph, and P. Kramer

Institut für Theoretische Physik, Universität Tübingen, Auf der Morgenstelle 14,
D-7400 Tübingen, Germany

Abstract

Observed transitions from the octagonal to a cubic phase and from the icosahedral to primitive or body-centered cubic phases have a natural mathematic description in terms of one-parameter rotations in higher-dimensional space. We give the relation between the reduction of point groups and this so-called Schur rotation and present tilings and diffraction patterns of δ-scatterers for a series of rotation angles. A Schur rotation also connects octagonal and dodecagonal phases and should prove useful for the description of alloys like $V_{15}Ni_{10}Si$ which exists in both phases with the same stoichiometry. We analyze values of the rotation angle which yield rational reductions and determine the lattices of the corresponding periodic phases. The Schur rotations starting from the icosahedral phases reach primitive and body-centered cubic lattices but admit, with one exception, only tetrahedral point symmetry.

1. Introduction

A variety of experiments on transitions between quasiperiodic and periodic phases suggest that part of the symmetry be preserved [1, 2, 3]. In the phenomenologic Landau theory of transitions between periodic phases, compare e.g. [4], it is assumed that a subgroup of the space group be preserved in the process. The phase transition is then controlled by order parameters derived from the subduction scheme. We shall show that a similar approach can be formulated for quasicrystal–crystal phase transitions, provided we use an appropriate lattice embedding which encompasses both structures. To do so, we make use of minimal lattice embeddings with maximal symmetry for the quasiperiodic structures [5]. Then, we consider the point group of a quasicrystal and its subgroups. In the examples given below, we determine maximal point groups shared by the corresponding quasicrystal and crystal structures. We show that the restriction to these maximal point subgroups leaves a one-parameter phase freedom, the phase being the so-called Schur rotation angle [6]. It has an equivalent description in terms of dimensionless quantities that are observable, e.g., in the diffraction pattern. Such quantities are natural candidates for (local) order parameters in the Landau theory, cf. [7]. Next, we analyze those values of this angle which admit a lattice $\Gamma_{||}$ in the physical space $\mathbb{E}_{||}$. In this fashion, a fixed N-dimensional lattice embedding for a quasicrystal structure determines a family of 2D or 3D crystal structures with definite point groups and lattice properties.

In Section 2, we implement and explain out theoretical scheme for the transition from octagonal quasicrystals to crystals with fourfold point symmetry and obtain a unified description of the rational approximants of the well-known Ammann-Beenker pattern as a by-product. In Section 3, we connect quasicrystalline and crystalline phases with four-, eight-, and twelvefold symmetry. Both examples are illustrated in terms of corresponding tilings and diffraction patterns. In Sections 4 and 5, the three 6D embeddings of icosahedral structures into hypercubic lattices of P-, F-, and I-type are considered under the restriction to the tetrahedral group, where again all rational approximants are obtained via the rotation scheme.

2. Four- and eightfold symmetric phases and the lattice $\mathbb{Z}^4$

Recently, Kuo [1] has reported on an interesting transition of a Cr-Ni-Si alloy from a quasiperiodic phase with octagonal symmetry (a so-called octagonal T-phase) to a periodic phase of β-Mn type through various intermediate phases with fourfold symmetry. It is the aim of this section to present a mathematical basis for the development of explicit structure models of this transition.

In the experiment cited, the octagonal T-phase was looked at along the axis of eightfold symmetry wherefore the electron diffraction image showed d_8 symmetry. Heating the probe, Kuo observed the transition to a periodic phase where d_8 is replaced by d_4. To outline our ideas, we will restrict ourselves to a description of the 2D plane perpendicular to the eightfold axis, which is justified by the structure of the T-phase. In the quasiperiodic case, the diffraction pattern is rather well described by a structure based on the well-known octagonal quasilattice [8, 9]. The minimal embedding into higher-dimensional space requires dimension 4 and the canonical choice is the hypercubic lattice $\mathbb{Z}^4$. The physical space $\mathbb{E}_{||}$ is determined as one of the two unique invariant subspaces w.r.t. d_8 which is a subgroup of $\Omega(4)$, the point group of the lattice $\mathbb{Z}^4$. Here, $d_8 = \langle\langle \mathbf{g}_8, \mathbf{s} \rangle\rangle$ with the representation

$$T(\mathbf{g}_8) := \begin{pmatrix} 0 & 0 & 0 & -1 \\ 1 & 0 & 0 & 0 \\ 0 & 1 & 0 & 0 \\ 0 & 0 & 1 & 0 \end{pmatrix}, \quad T(\mathbf{s}) := \begin{pmatrix} 1 & 0 & 0 & 0 \\ 0 & 0 & 0 & -1 \\ 0 & 0 & -1 & 0 \\ 0 & -1 & 0 & 0 \end{pmatrix}. \tag{1}$$

This reducible representation of d_8 splits into two inequivalent irreducible ones,

$$U_8\,T(\mathbf{g})\,U_8^{-1} = T^{red}(\mathbf{g}), \quad \mathbf{g} \in d_8, \tag{2}$$

$$T^{red}(\mathbf{g}) = \begin{pmatrix} c & -s & 0 & 0 \\ s & c & 0 & 0 \\ 0 & 0 & c' & -s' \\ 0 & 0 & s' & c' \end{pmatrix}, \text{ and } \quad T^{red}(\mathbf{s}) := \mathbb{1} \otimes \begin{pmatrix} 1 & 0 \\ 0 & -1 \end{pmatrix}, \tag{3}$$

where

$$\begin{aligned} c &:= \cos(\delta), & s &:= \sin(\delta), \\ c' &:= \cos(5\delta), & s' &:= \sin(5\delta), \end{aligned} \tag{4}$$

and $\delta := \frac{2\pi}{8}$. The choice of 5δ as rotation angle in $\mathbb{E}_\perp$ might look unusual but does not change the pattern and its structure and facilitates the reduction to the subgroup d_4 since $5 \equiv 1 \bmod 4$. The reduction matrix U_8 in Eq. 2 reads

$$U_8 := \sqrt{\frac{1}{2}} \begin{pmatrix} 1 & \sqrt{\frac{1}{2}} & 0 & -\sqrt{\frac{1}{2}} \\ 0 & \sqrt{\frac{1}{2}} & 1 & \sqrt{\frac{1}{2}} \\ 1 & -\sqrt{\frac{1}{2}} & 0 & \sqrt{\frac{1}{2}} \\ 0 & -\sqrt{\frac{1}{2}} & 1 & -\sqrt{\frac{1}{2}} \end{pmatrix}, \tag{5}$$

wherefrom one can extract the projection images $\pi_{||}(\mathbf{e}_i), \pi_\perp(\mathbf{e}_i)$ of the lattice basis in $\mathbb{E}_{||}$ and $\mathbb{E}_\perp$, respectively.

Now, if we restrict the representation of d_8 to the subgroup $d_4 = \langle\langle \mathbf{g}_4, \mathbf{s}\rangle\rangle$, $\mathbf{g}_4 = \mathbf{g}_8^2$, we find

$$T^{red}(\mathbf{g}_4) := \mathbb{1} \otimes \begin{pmatrix} \cos(\frac{\pi}{2}) & -\sin(\frac{\pi}{2}) \\ \sin(\frac{\pi}{2}) & \cos(\frac{\pi}{2}) \end{pmatrix}, \quad \text{and } T^{red}(\mathbf{s}) \text{ as above,} \tag{6}$$

i.e., the representations in $\mathbb{E}_{||}$ and $\mathbb{E}_\perp$ are identical. The consequences of this general situation were pointed out in the similar situation of the tetrahedral group [6]: According to Schur's lemma, the invariant subspaces $\mathbb{E}_{||}$ and $\mathbb{E}_\perp$ are no longer unique, we now have a nontrivial phase freedom,

$$[T^{red}(\mathbf{h}), R(\phi)] = 0, \quad \mathbf{h} \in d_4, \text{ with } \quad R(\phi) := \begin{pmatrix} \cos(\phi) & -\sin(\phi) \\ \sin(\phi) & \cos(\phi) \end{pmatrix} \otimes \mathbb{1}\,. \tag{7}$$

Therefore we have

$$T^{red}(\mathbf{h}) = U(\phi)\,T(\mathbf{h})\,U^{-1}(\phi), \quad \mathbf{h} \in d_4 \tag{8}$$

with

$$U(\phi) = R(\phi) \cdot U_8, \quad U(0) = U_8. \tag{9}$$

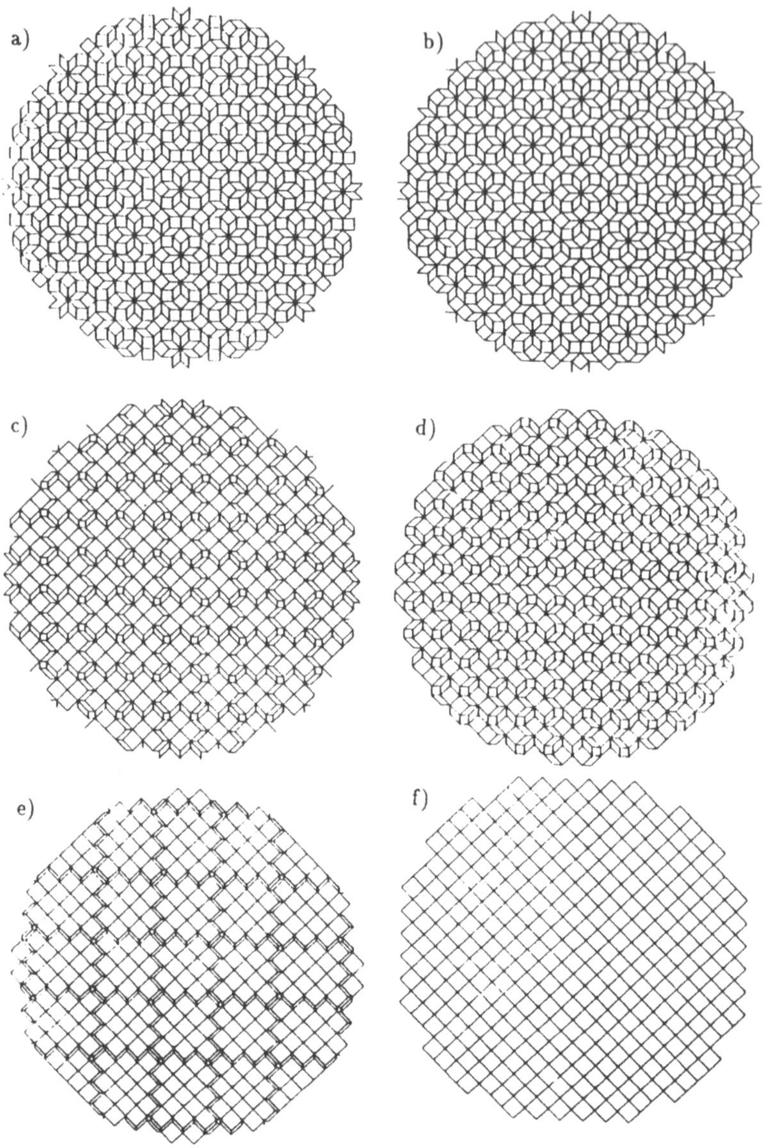

Fig. 1 a) The octagonal quasiperiodic pattern at $\phi = 0^o$, b) a quasiperiodic pattern with fourfold symmetry at $\phi = 2^o$, c) $\phi = 12^o$, d) $\phi = 20^o$, e) $\phi = 30^o$, f) a periodic pattern with fourfold symmetry at $\phi = 45^o$, all derived from $\mathbb{Z}^4$.

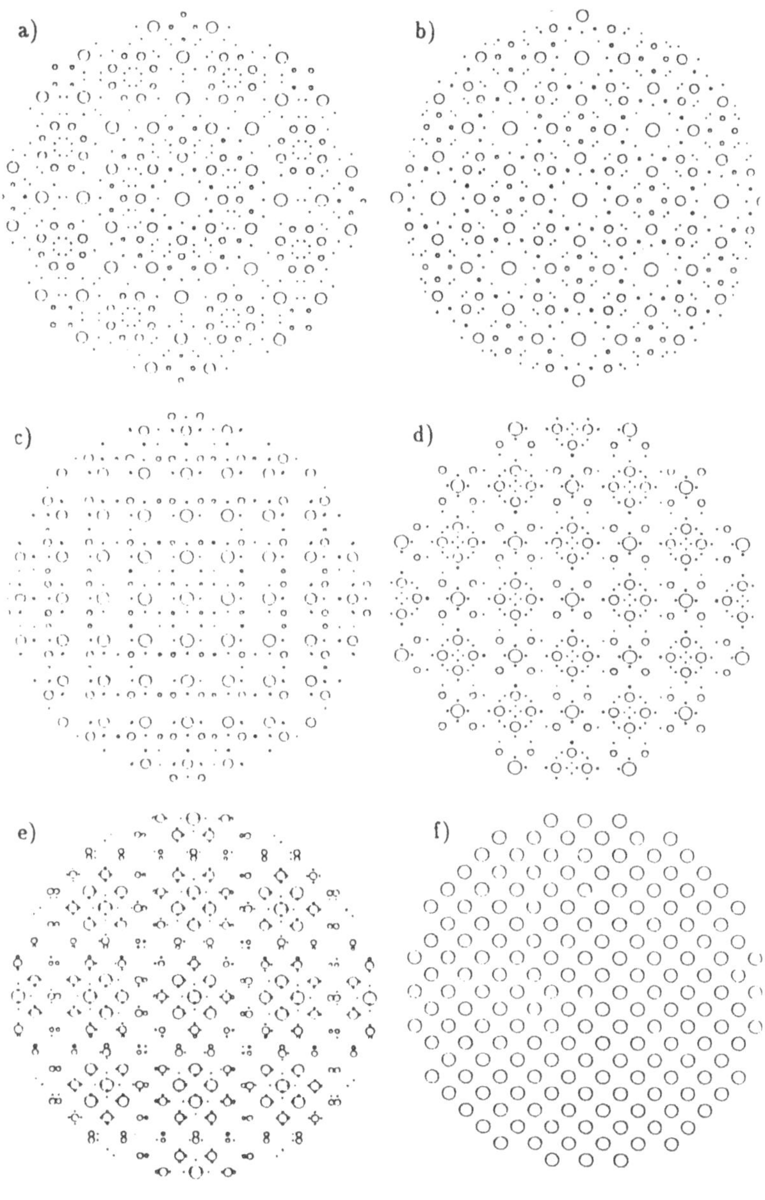

Fig. 2 a) The Fourier image of the octagonal quasiperiodic pattern at $\phi = 0°$, b) of a quasiperiodic pattern with fourfold symmetry at $\phi = 2°$, c) $\phi = 12°$, d) $\phi = 20°$, e) $\phi = 30°$, f) of a periodic pattern with fourfold symmetry at $\phi = 45°$, all derived from $\mathbb{Z}^4$.

This means that we obtain a whole one-parameter family of tilings with d_4 symmetry by rotating $\mathbb{E}_{\|}$ according to the Schur rotation $R(\phi)$ with the "Schur angle" ϕ. For a dense, but countable set of angles ϕ one finds periodic tilings, the generic case is the quasiperiodic one.

It is illustrative to write down the reduction matrix explicitly,

$$U(\phi) := \sqrt{\frac{1}{2}} \begin{pmatrix} \tilde{c}-\tilde{s} & \sqrt{\frac{1}{2}}(\tilde{c}+\tilde{s}) & 0 & -\sqrt{\frac{1}{2}}(\tilde{c}+\tilde{s}) \\ 0 & \sqrt{\frac{1}{2}}(\tilde{c}+\tilde{s}) & \tilde{c}-\tilde{s} & \sqrt{\frac{1}{2}}(\tilde{c}+\tilde{s}) \\ \tilde{c}+\tilde{s} & -\sqrt{\frac{1}{2}}(\tilde{c}-\tilde{s}) & 0 & \sqrt{\frac{1}{2}}(\tilde{c}-\tilde{s}) \\ 0 & -\sqrt{\frac{1}{2}}(\tilde{c}-\tilde{s}) & \tilde{c}+\tilde{s} & -\sqrt{\frac{1}{2}}(\tilde{c}-\tilde{s}) \end{pmatrix}, \tag{10}$$

$\tilde{c} = \cos(\phi), \tilde{s} = \sin(\phi)$, because one can see that $\pm\mathbf{e}_1$ and $\pm\mathbf{e}_3$ on the one hand and $\pm\mathbf{e}_2$ and $\pm\mathbf{e}_4$ on the other hand build, in the projection to $\mathbb{E}_{\|}$ or to $\mathbb{E}_{\perp}$, two regular 4-stars with a relative angle of 45° independent of ϕ. Therefore, the influence of ϕ is on the size of the 4-stars only, and one finds the relative scale to be

$$\eta(\phi) = \frac{|\,\pi_{\|}(\mathbf{e}_1)\,|}{|\,\pi_{\|}(\mathbf{e}_2)\,|} = \frac{|\,\pi_{\perp}(\mathbf{e}_2)\,|}{|\,\pi_{\perp}(\mathbf{e}_1)\,|} = \frac{|\cos(\phi)-\sin(\phi)|}{|\cos(\phi)+\sin(\phi)|} = |\tan(\phi-\pi/4)|\,. \tag{11}$$

The quantity η is just one simple possibility to give ϕ a geometric meaning in $\mathbb{E}_{\|}$ alone and contains nothing new in comparison with ϕ. It can be measured from the diffraction image if one has found a consistent indexing of the spots in the (generic) quasiperiodic case, where $\pi_{\|}(\mathbf{e}_1), \ldots, \pi_{\|}(\mathbf{e}_4)$ are linearly independent over the integers.

The reduction is rational if and only if $\tan\phi = (p - q\sqrt{2})/(p + q\sqrt{2})$ with coprime integers p and q. Then, we find square lattices $\Gamma_{\|}$ and $\Gamma_{\perp}$ in $\mathbb{E}_{\|}$ and $\mathbb{E}_{\perp}$, respectively, such that $\Gamma_{\|} \times \Gamma_{\perp}$ is a sublattice of $\mathbb{Z}^4$. Let us choose $\Gamma_{\|}$ and $\Gamma_{\perp}$ to be maximal w.r.t. this property. Then, both $\Gamma_{\|}$ and $\Gamma_{\perp}$ possess the same lattice constant which turns out to be $\sqrt{p^2+2q^2}$ for p odd and $\sqrt{\frac{1}{2}p^2+q^2}$ for p even. Consequently, the index of the translation subgroup compatible with $\mathbb{E}_{\|} \oplus \mathbb{E}_{\perp}$ is

$$I = [\mathbb{Z}^4 : \Gamma_{\|} \times \Gamma_{\perp}] = \left\{ \begin{array}{ll} (p^2+2q^2)^2, & \text{if } p \text{ odd} \\ \frac{1}{4}(p^2+2q^2)^2, & \text{if } p \text{ even} \end{array} \right\}. \tag{12}$$

From this formula one can try to pick out a suitable periodic phase with the right lattice constant relative to the octagonal quasiperiodic phase at $\phi = 0^o$, say. Here, the transition to a periodic phase leads to a tiling with the corresponding periodicity.

To illustrate the mechanism, we show in Fig. 1 a series of 6 patterns obtained by the standard dualization method [10, 11] for Schur angles between $\phi = 0^o$ ($\eta = 1$, octagonal case) and $\phi = 45^o$ ($\eta = 0$, periodic case with minimal period). The corresponding kinematic diffraction of δ-scatterers at vertex positions are presented in Fig. 2 (for details on the construction algorithm and the Fourier formulas used here, see [11, 12]). We have made an intensity cut at 1 % of the maximal intensity. Fig. 2a ($\phi = 0^o$) presents the eightfold symmetry of the octagonal pattern (Fig. 1a). In Fig. 2b ($\phi = 2^o$) the exact eightfold symmetry is broken leaving only fourfold symmetry behind, although many similarities to the eightfold case can still be seen. Fig. 2c ($\phi = 12^o$) and Fig. 2d ($\phi = 20^o$) show the transition within the fourfold symmetry. In Fig. 2e ($\phi = 30^o$) one can see an obvious step towards the periodic phase with the shortest lattice constant (Fig. 1f, Fig. 2f). The small spots which surround those with higher intensities will move towards the big spots and decrease simultaneously. Finally, this leads to the periodic case of Fig. 2f ($\phi = 45^o$).

Although we have not explicitly calculated the transition from the octagonal to the periodic phase by means of Landau theory, it is plausible that the Schur rotation with a single parameter provides a correct tool to do so and the dimensionless quantity η defined in Eq. 11 is the natural candidate for an order parameter which is directly accessible in experiment.

3. Four-, eight-, and twelvefold symmetric phases and the root lattice D_4

Another interesting application of the Schur rotation can be performed in the so-called root lattice D_4 to obtain a unified description of a transition between four-, eight-, and twelvefold symmetric phases [13]. This should prove useful for the description of alloys like $V_{15}Ni_{10}Si$ which can exist in eight- and twelvefold symmetric phases with the same stoichiometry [14, 15].

The root lattice D_4 is the checkerboard lattice in 4D. It can be constructed starting from the $\mathbb{Z}^4$ lattice and removing all points with an odd sum of indices, i.e.,

$$D_4 = \{\mathbf{x} \in \mathbb{Z}^4 \mid \quad \mathbf{x} = \Sigma_i n_i \mathbf{e}_i \quad \text{with} \quad \Sigma_i n_i \equiv 0 \bmod 2\}. \tag{13}$$

D_4 is generated by the integer linear combinations of the four vectors

$$\mathbf{e}_1 - \mathbf{e}_2, \quad \mathbf{e}_2 - \mathbf{e}_3, \quad \mathbf{e}_3 - \mathbf{e}_4, \quad \mathbf{e}_3 + \mathbf{e}_4, \tag{14}$$

where the $\mathbf{e}_i$ denote the standard Euclidean basis in $\mathbb{R}^4$. It provides the densest lattice packing in 4-D. The holohedry H of D_4 is larger than the hypercubic group $\Omega(4)$, namely

$$H = \Omega^+(4) \otimes_s S_3, \tag{15}$$

where $\Omega^+(4)$ is the subgroup of $\Omega(4)$ with even sign flips and S_3 is generated by the cyclic operation $(\mathbf{e}_1 - \mathbf{e}_2) \rightarrow (\mathbf{e}_3 - \mathbf{e}_4) \rightarrow (\mathbf{e}_3 + \mathbf{e}_4) \rightarrow (\mathbf{e}_1 - \mathbf{e}_2)$ and the reflection $\mathbf{e}_4 \rightarrow -\mathbf{e}_4$.

The canonical 4-D representation T of H is irreducible, i.e., it does not allow nontrivial invariant subspaces. However, several subgroups of H do, especially the groups $d_8 =$

$\langle\langle \mathbf{g}_8, \mathbf{s}\rangle\rangle$ and $d_{12} = \langle\langle \mathbf{g}_{12}, \mathbf{s}\rangle\rangle$ where $T(\mathbf{g}_8)$ and $T(\mathbf{s})$ are equal to Eq. 1 and

$$T(\mathbf{g}_{12}) := \frac{1}{2}\begin{pmatrix} -1 & 1 & -1 & -1 \\ 1 & 1 & 1 & -1 \\ 1 & 1 & -1 & 1 \\ -1 & 1 & 1 & 1 \end{pmatrix}. \tag{16}$$

The representation T, restricted to the subgroup d_8, decomposes into two inequivalent irreducible 2-D representations according to Eq. 2. Also the reduction matrix U_8 is the same as for $\mathbb{Z}^4$ (cf. Eq. 5).

Similarly, in the case of d_{12}, we find the decomposition

$$U_{12}\, T(\mathbf{g})\, U_{12}^{-1} = T^{red}(\mathbf{g}), \qquad \mathbf{g} \in d_{12}, \tag{17}$$

with

$$U_{12} := \sqrt{\frac{1}{1+\alpha^2}} \begin{pmatrix} 1 & \frac{\alpha}{\sqrt{2}} & 0 & -\frac{\alpha}{\sqrt{2}} \\ 0 & \frac{\alpha}{\sqrt{2}} & 1 & \frac{\alpha}{\sqrt{2}} \\ \alpha & -\frac{1}{\sqrt{2}} & 0 & \frac{1}{\sqrt{2}} \\ 0 & -\frac{1}{\sqrt{2}} & \alpha & -\frac{1}{\sqrt{2}} \end{pmatrix}, \qquad \alpha := 2\cos(\frac{\pi}{12}) = \frac{1+\sqrt{3}}{\sqrt{2}}, \tag{18}$$

and

$$T^{red}(\mathbf{g}_{12}) := \begin{pmatrix} c & -s & 0 & 0 \\ s & c & 0 & 0 \\ 0 & 0 & c' & -s' \\ 0 & 0 & s' & c' \end{pmatrix}, \qquad T^{red}(\mathbf{s}) := \mathrm{diag}(1,-1,1,-1), \tag{19}$$

with

$$\begin{aligned} c &:= \cos(\beta), & s &:= \sin(\beta), \\ c' &:= \cos(5\beta), & s' &:= \sin(5\beta), \end{aligned} \tag{20}$$

where $\beta := \frac{2\pi}{12}$. Again, the two irreps are inequivalent which is no longer true if we restrict T to the common subgroup $d_4 = \langle\langle \mathbf{g}_4, \mathbf{s}\rangle\rangle$ with $\mathbf{g}_4 = \mathbf{g}_8^2 = \mathbf{g}_{12}^3$.

Since $T^{red}(\mathbf{g}_4)$ is the matrix of Eq. 6, the irreps in $\mathbb{E}_{\|}$ and $\mathbb{E}_{\perp}$ are identical and we again have a phase freedom according to Schur's lemma (cf. Eq. 7). Therefore, we can derive

$$T^{red}(\mathbf{h}) = U(\phi) T(\mathbf{h}) U^{-1}(\phi), \qquad \mathbf{h} \in d_4, \tag{21}$$

and

$$U(\phi) = R(\phi) \cdot U_8, \tag{22}$$

with

$$U(0) = U_8, \quad U(\phi') = U_{12}, \quad \phi' = \arccos(\frac{\sqrt{3+\sqrt{3}} + \sqrt{3-\sqrt{3}}}{2\sqrt{3}}) \approx 17.6^0, \tag{23}$$

and $R(\phi)$ like in Eq. 7. The possibility of getting a twelvefold symmetric phase is due to the additional threefold element in H. There are other possible embeddings of twelvefold elements into H, cf. [16], but that is not our concern here.

As in the $\mathbb{Z}^4$ case, we find continuously many possible reduction points in between and beyond, all of which result in patterns with d_4 symmetry. Most of them will be truly quasiperiodic while a dense but countable subset results in rational reductions. We can also calculate the quantity η which is equal to that of Eq. 11 with the same physical interpretation.

The rational reductions are at $\tan(\phi) = (p - q\sqrt{2})/(p + q\sqrt{2})$ as in the $\mathbb{Z}^4$ lattice whereas the index of the translation subgroup compatible with $\mathbb{E}_{\|} \oplus \mathbb{E}_{\perp}$ is now twice as big due to the larger fundamental cell of D_4, i.e.,

$$I = [D^4 : \Gamma_{\|} \times \Gamma_{\perp}] = \left\{ \begin{array}{ll} 2(p^2 + 2q^2)^2, & \text{if } p \text{ odd} \\ \frac{1}{2}(p^2 + 2q^2)^2, & \text{if } p \text{ even} \end{array} \right\}. \tag{24}$$

Following the technique briefly described in Sec. 2, we illustrate the mechanism by an equivalent series of patterns and their Fourier transforms with δ-scatterers at vertex positions (Fig. 3 and Fig. 4). Here, only those patterns are presented which are obtained by the projections of the boundaries of the Delaunay cells. Of course, the same can be done for the patterns which are projections of the boundaries of the Voronoi cell. For details see [13], the group theoretic reduction scheme is identical and can be summarized in the following diagram:

$$\begin{array}{ccccc} & & H & & \\ & \swarrow & \downarrow & \searrow & \\ d_8 & & d_{12} & & d_4 \\ \phi = 0^o, \eta = 1 & & \phi \simeq 17.6^o, \eta = \alpha^{-1} & & \phi = 45^o, \eta = 0 \\ \text{8-fold} & & \text{12-fold} & & \text{4-fold, periodic} \\ & \searrow & \downarrow & \swarrow & \\ & & d_4 & & \end{array}$$

At this point we would like to mention another possibility for a Schur rotation that preserves sixfold symmetry. If we replace U_{12} by

$$V_{12} = \mathrm{diag}(1, 1, 1, -1) \cdot U_{12} \tag{25}$$

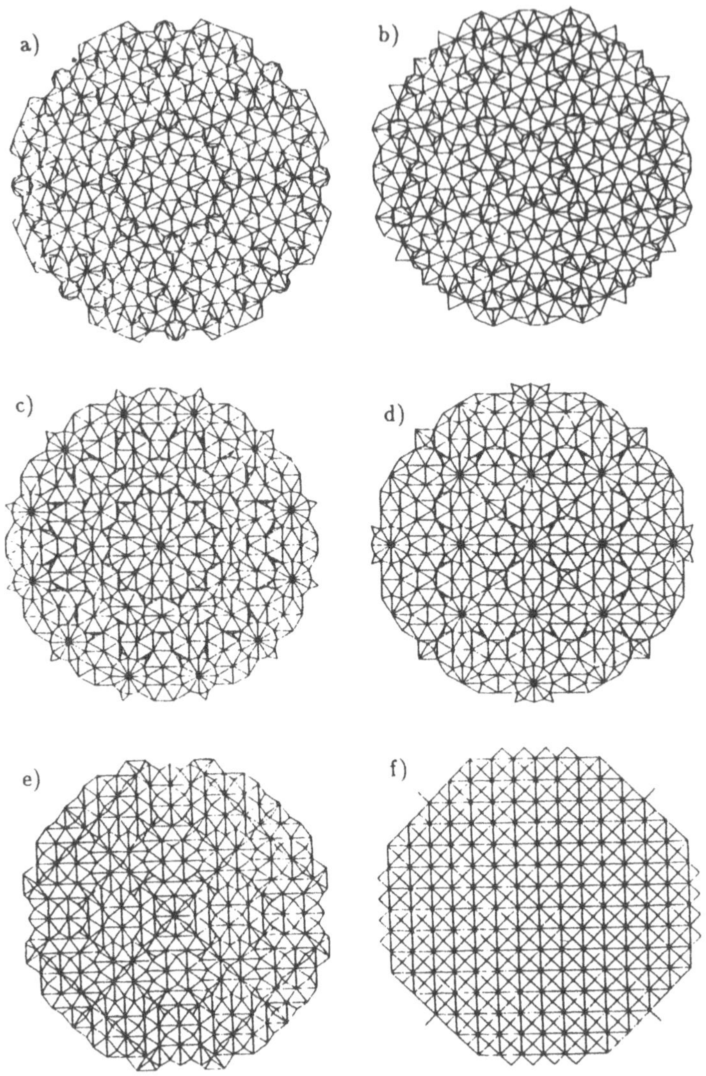

Fig. 3 a) The octagonal quasiperiodic pattern at $\phi = 0^o$, b) a quasiperiodic pattern with fourfold symmetry at $\phi = 2^o$, c) with twelvefold symmetry at $\phi = \arccos((\sqrt{3+\sqrt{3}} + \sqrt{3-\sqrt{3}})/(2\sqrt{3})) \approx 17.6^0$, d) with fourfold symmetry at $\phi = 20^o$, e) $\phi = 30^o$, f) a periodic pattern with fourfold symmetry at $\phi = 45^o$, all derived from D_4.

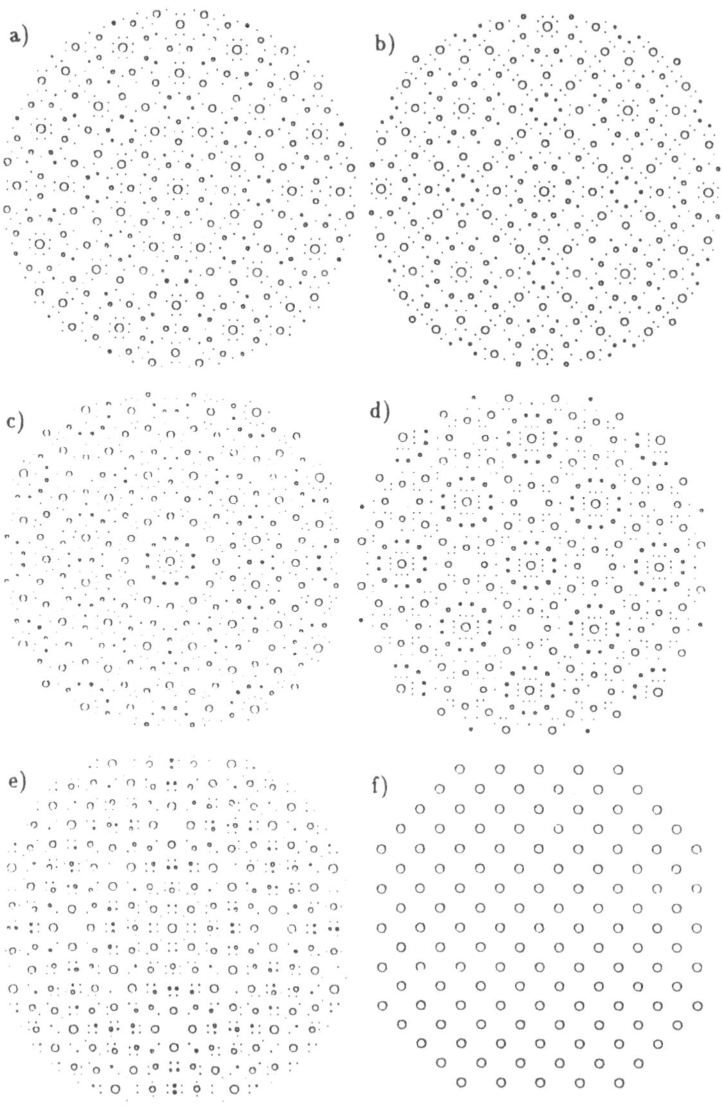

Fig. 4 a) The Fourier image of the octagonal quasiperiodic pattern at $\phi = 0^o$, b) of a quasiperiodic pattern with fourfold symmetry at $\phi = 2^o$, c) of the quasiperiodic pattern with twelvefold symmetry at $\phi = \arccos((\sqrt{3+\sqrt{3}} + \sqrt{3-\sqrt{3}})/(2\sqrt{3})) \approx 17.6^0$, d) of a quasiperiodic pattern with fourfold symmetry $\phi = 20^o$, e) $\phi = 30^o$, f) of a periodic pattern with fourfold symmetry at $\phi = 45^o$, derived from D_4.

we get $T(\mathbf{g}_{12})$ reduced to block form with rotation angle $\beta = 2\pi/12$ in the one 2D space and rotation angle 7β in the other. But $7 \equiv 1 \bmod 6$, so a restriction to $d_6 = \langle\langle \mathbf{g}_6, \mathbf{s}\rangle\rangle$ with

$$\mathbf{g}_6 = (\mathbf{g}_{12})^2, \qquad T(\mathbf{g}_6) := \frac{1}{2}\begin{pmatrix} 1 & -1 & 1 & -1 \\ 1 & 1 & -1 & -1 \\ -1 & 1 & 1 & -1 \\ 1 & 1 & 1 & 1 \end{pmatrix} \tag{26}$$

makes the blocks identical and thus again creates a phase freedom, i.e., we may equally well choose $V(\phi) = R(\phi)\cdot V_{12}$ with $R(\phi)$ from Eq. 7 as the reduction matrix. Periodic sections emerge if and only if the Schur angle has the form

$$\tan(\phi) = \pm\sqrt{2}\cdot\frac{p-(1+\sqrt{3})q}{(1+\sqrt{3})p+2q}, \quad (p,q)=1. \tag{27}$$

It turns out that we get hexagonal lattices in $\Gamma_{\|}$ and $\Gamma_{\perp}$ as one would expect but we will skip the details here.

4. Icosahedral and tetrahedral phases and the primitive hypercubic lattice $\mathbb{Z}^6$

Let us turn to the case of 3D quasicrystals and explain the situation with the icosahedral group in a little more detail, starting from the hypercubic lattice $\mathbb{Z}^6$ [6, 17]. Its point group is the group $\Omega(6)$ [18] which contains the icosahedral group Y_h in such a way that the canonical representation of $\Omega(6)$, when reduced to Y_h, splits into two inequivalent 3D irreps of Y_h [6]:

$$T^{red}(\mathbf{g}) = U\,T(\mathbf{g})\,U^{-1}, \quad \mathbf{g}\in Y_h, \tag{28}$$

with

$$U := \sqrt{\frac{1}{2}}\begin{pmatrix} 0 & c & s & 0 & c & -s \\ s & 0 & c & -s & 0 & c \\ c & s & 0 & c & -s & 0 \\ 0 & -s & c & 0 & -s & -c \\ c & 0 & -s & -c & 0 & -s \\ -s & c & 0 & -s & -c & 0 \end{pmatrix}, \tag{29}$$

and $c = \cos(\alpha), s = \sin(\alpha), \alpha = \frac{1}{2}\arctan(2)$.

On the other hand, when further restricted to the tetrahedral subgroup T, one obtains two identical 3D irreps and thus again the phase freedom for a Schur rotation. Let us take $U = U(\phi)$ as the corresponding reduction matrix where $\phi = \alpha$ gives back Eq. 29. Now, precisely for $\tan(\phi) = p/q$, we obtain a rational reduction and we find 3D sublattices $\Gamma_{\|}$ and $\Gamma_{\perp}$ of $\mathbb{Z}^6$ in $\mathbb{E}_{\|}$ and $\mathbb{E}_{\perp}$, respectively, with index

$$I = [\mathbb{Z}^6 : \Gamma_{\|} \times \Gamma_{\perp}] = \left\{ \begin{array}{ll} 2(p^2+q^2)^3, & \text{if } 0 \not\equiv p \equiv q \bmod 2 \\ 8(p^2+q^2)^3, & \text{if } p \not\equiv q \bmod 2 \end{array} \right\}, \tag{30}$$

where we have taken p and q coprime. We will now derive these findings explicitly.

If we interpret the first (the last) three lines of U columnwise as the projection of the six vectors $\mathbf{e}_1, \ldots, \mathbf{e}_6$ into $\mathbb{E}_{\|}$ ($\mathbb{E}_{\perp}$), respectively, we can explicitly determine the maximal translation lattice contained in $\mathbb{E}_{\|}$ ($\mathbb{E}_{\perp}$). The condition that a $\mathbb{Z}^6$ lattice vector lives in $\mathbb{E}_{\|}$ is expressed as

$$\pi_{\perp}(\Sigma_i n_i \mathbf{e}_i) = 0 \tag{31}$$

and results in

$$\begin{array}{rcl} (n_2+n_5)\cdot s &=& (n_3-n_6)\cdot c \\ (n_3+n_6)\cdot s &=& (n_1-n_4)\cdot c \\ (n_1+n_4)\cdot s &=& (n_2-n_5)\cdot c \end{array} \tag{32}$$

with $s = \sin(\phi), c = \cos(\phi)$ and integers $n_1, \ldots, n_6$. Consequently,

$$\tan(\phi) = \frac{n_1-n_4}{n_3+n_6} = \frac{n_2-n_5}{n_1+n_4} = \frac{n_3-n_6}{n_2+n_5} = \frac{p}{q}, \tag{33}$$

which proves the above statement on the appearance of rational sections. Without loss of generality, we may take p and q coprime, i.e., $(p,q) = 1$. Then, we can rewrite Eq. 33 as follows:

$$\begin{array}{l} n_1-n_4 = \lambda\cdot p, \quad n_2-n_5 = \mu\cdot p, \quad n_3-n_6 = \nu\cdot p \\ n_3+n_6 = \lambda\cdot q, \quad n_1+n_4 = \mu\cdot q, \quad n_2+n_5 = \nu\cdot q \end{array} \tag{34}$$

where λ, μ, ν must be integers because, e.g., $(p,q) = 1$ and $\lambda\cdot p, \lambda\cdot q \in \mathbb{Z}$ implies $\lambda \in \mathbb{Z}$.

Now, we get the central set of equations

$$\begin{array}{rcl} n_1 &=& \frac{1}{2}(\lambda\cdot p+\mu\cdot q) \\ n_2 &=& \frac{1}{2}(\mu\cdot p+\nu\cdot q) \\ n_3 &=& \frac{1}{2}(\nu\cdot p+\lambda\cdot q) \\ n_4 &=& \frac{1}{2}(-\lambda\cdot p+\mu\cdot q) \\ n_5 &=& \frac{1}{2}(-\mu\cdot p+\nu\cdot q) \\ n_6 &=& \frac{1}{2}(-\nu\cdot p+\lambda\cdot q). \end{array} \tag{35}$$

The sublattice $\Gamma_{\|}$ is now determined by all integer triples (λ, μ, ν) which yield integer numbers $n_1, \ldots, n_6$. Furthermore,

$$(\Sigma_i n_i \mathbf{e}_i)^2 = \Sigma_i n_i^2 = \frac{1}{2}(\lambda^2+\mu^2+\nu^2)(p^2+q^2). \tag{36}$$

Let us now, for given $(p,q)=1$, calculate the shortest lattice vectors of $\Gamma_{\|}$ (and hence of $\mathbb{Z}^6$) that lie in $\mathbb{E}_{\|}$. We cannot have $p \equiv q \equiv 0 \bmod 2$ because $(p,q)=1$ and stay thus with two cases. First, p and q odd results in $\lambda \equiv \mu \equiv \nu \bmod 2$ wherefore the shortest nonvanishing lattice vectors are obtained from $\lambda=\mu=\nu=1$, the next shell of lattice vectors belongs to $\lambda=2, \mu=\nu=0$ (and tetrahedral images). Together, they span a body-centered cubic lattice the fundamental cell of which has volume

$$\mathrm{vol}(FD)=\frac{1}{2}\{2(p^2+q^2)\}^{\frac{3}{2}}. \tag{37}$$

We have thus determined the maximal translation lattice *contained* in $\mathbb{E}_{\|}$ – being bcc in this case – rather than the fcc-structure of $\pi_{\|}(\mathbb{Z}^6)$, i.e., $\mathbb{Z}^6$ *projected* to $\mathbb{E}_{\|}$, as was done in ref. [6]. Second, p even and q odd (or vice versa) results in $\lambda \equiv \mu \equiv \nu \equiv 0 \bmod 2$ where the shortest nonzero vectors are obtained from $\lambda=2, \mu=\nu=0$ and tetrahedral images. They generate a primitive cubic lattice with lattice constant $\sqrt{2(p^2+q^2)}$, i.e., the fundamental cell has volume

$$\mathrm{vol}(FD)=\{2(p^2+q^2)\}^{\frac{3}{2}}. \tag{38}$$

The situation in $\mathbb{E}_{\perp}$ is very similar, for $\tan(\phi)=p/q$ with $(p,q)=1$ one derives the analogue of Eq. 35 to be

$$\begin{aligned} n_1 &= \tfrac{1}{2}(\mu\cdot p-\lambda\cdot q)\\ n_2 &= \tfrac{1}{2}(\nu\cdot p-\mu\cdot q)\\ n_3 &= \tfrac{1}{2}(\lambda\cdot p-\nu\cdot q)\\ n_4 &= \tfrac{1}{2}(-\mu\cdot p+\lambda\cdot q)\\ n_5 &= \tfrac{1}{2}(-\nu\cdot p+\mu\cdot q)\\ n_6 &= \tfrac{1}{2}(-\lambda\cdot p+\nu\cdot q) \end{aligned} \tag{39}$$

with the consequence, that — for given $(p,q)=1$ — $\mathbb{E}_{\perp}$ contains precisely the same type of lattice as $\mathbb{E}_{\|}$ and also with the same lattice constant! Because $\Gamma_{\|}$ and $\Gamma_{\perp}$ are thus equivalent as 3D lattices, one immediately obtains Eq. 30.

If we interpret the 6 columns of the matrix U as the 6 orthonormal basis vectors $\mathbf{e}_1,\ldots,\mathbf{e}_6$ of the hypercubic lattice $\mathbb{Z}^6$, we can write the transition to the 6D sublattice $\Gamma_{\|}\otimes\Gamma_{\perp}$ also in matrix form, namely

$$U\cdot Z=B, \tag{40}$$

where Z is a rational centering matrix with integer entries and where B can be interpreted as the basis matrix of the sublattice $\Gamma_{\|}\otimes\Gamma_{\perp}$. Explicitly, we find for $p \equiv q \equiv 1 \bmod 2$ the matrices

$$Z:=\begin{pmatrix} \mathbb{1}_3 & A\\ A & \mathbb{1}_3 \end{pmatrix}\cdot\begin{pmatrix} \frac{1}{2}(p+q)\cdot\mathbb{1}_3 & \frac{1}{2}(p-q)\cdot\mathbb{1}_3\\ \frac{1}{2}(-p+q)\cdot\mathbb{1}_3 & \frac{1}{2}(p+q)\cdot\mathbb{1}_3 \end{pmatrix}, \tag{41}$$

where

$$A = \begin{pmatrix} 0 & 1 & -1 \\ -1 & 0 & 1 \\ 1 & -1 & 0 \end{pmatrix},$$

and

$$B = \frac{a}{2} \cdot \begin{pmatrix} 1 & 0 \\ 0 & -1 \end{pmatrix} \otimes \begin{pmatrix} -1 & 1 & 1 \\ 1 & -1 & 1 \\ 1 & 1 & -1 \end{pmatrix}, \quad a = \sqrt{2(p^2 + q^2)},$$

wherefrom one can again recognize the bcc-structure of $\Gamma_{\|}$ and $\Gamma_{\perp}$. For $p \not\equiv q \bmod 2$, we find the matrices

$$Z := \begin{pmatrix} 0 & p & q & 0 & q & -p \\ q & 0 & p & -p & 0 & q \\ p & q & 0 & q & -p & 0 \\ 0 & -p & q & 0 & -q & -p \\ q & 0 & -p & -p & 0 & -q \\ -p & q & 0 & -q & -p & 0 \end{pmatrix}, \qquad B := a \cdot \mathbb{1}_6, \tag{42}$$

with $a = \sqrt{2(p^2 + q^2)}$. Here, the matrix Z is related to the reduction matrix U of Eq. 29 via $Z = a \cdot U^{-1}(\phi)$ for $\phi = \arctan(p/q)$.

The two situations correspond to a body centered cubic lattice ($p \equiv q \bmod 2$) or to a primitive cubic lattice ($p \not\equiv q \bmod 2$) in $\mathbb{E}_{\|}$. All 3D lattices $\Gamma_{\|}$ obtained from rational reduction possess cubic point symmetry which does, however, stem from a $\mathbb{Z}^6$ symmetry only for $\phi = k\pi/4, k \in 2\mathbb{Z} + 1$. Precisely in the latter cases also the complete structure obtained by the projection method will show the cubic symmetry. In all the other cases, the full cell structure has only tetrahedral point symmetry and the additional, non-tetrahedral point transformations cannot be lifted to a symmetry of $\mathbb{Z}^6$, if we keep the specific embedding of the tetrahedral group T. This embedding is required in order to get T simultaneously as a subgroup of Y_h and hence to get the link to the quasicrystalline icosahedral phase by Schur rotation.

Let us briefly describe the analogue of Eq. 11 where a candidate for an order parameter was given. Here, we have the vectors $\mathbf{e}_1 + \mathbf{e}_2 + \mathbf{e}_3$ and $\mathbf{e}_4 + \mathbf{e}_5 + \mathbf{e}_6$ as natural candidates, wherefore we define

$$\eta(\phi) = \frac{| \pi_{\|}(\mathbf{e}_1 + \mathbf{e}_2 + \mathbf{e}_3) |}{| \pi_{\|}(\mathbf{e}_4 + \mathbf{e}_5 + \mathbf{e}_6) |} = \frac{| \cos(\phi) - \sin(\phi) |}{| \cos(\phi) + \sin(\phi) |} = | \tan(\phi - \pi/4) | \,. \tag{43}$$

These findings are summarized in the following diagram, where — according to the previous remark — the reduction at $\phi = 45^o$ is singled out as the rational reduction via the full cubic group $\Omega(3) = O_h$ while the cubic symmetry of $\phi = 0^o$ is accidental and the reduction leads only to the tetrahedral group T.

$\Omega(6)$

↙ ↓ ↘

T	Y_h	O_h
$\phi = 0^o, \eta = 1$	$\phi \simeq 31.7^o, \eta = \tau^{-3}$	$\phi = 45^o, \eta = 0$
primitive cubic	icosahedral	bcc

↘ ↓ ↙

T

Here, $\tau = (1+\sqrt{5})/2$ is the golden mean. It is interesting to note that precisely the icosahedral and the bcc phase seem to be related experimentally [2].

5. F- and I-type icosahedral phases and the lattices D_6 and D_6^R

Let us now briefly describe the situation for the F-type icosahedral structure based on the root lattice D_6 [19] which is given as

$$D_6 = \{\mathbf{x} \in \mathbb{Z}^6 \mid \mathbf{x} = \Sigma_i n_i \mathbf{e}_i \text{ with } \Sigma_i n_i \equiv 0 \bmod 2\}. \tag{44}$$

Clearly, D_6 is a sublattice of $\mathbb{Z}^6$ of index 2 wherefore the general situation is almost the same as in Sec. 4, especially the reduction matrix remains unchanged and we do not get additional rational sections, the condition of which still reads $\tan(\phi) = p/q$, $(p,q) = 1$, where ϕ is the angle in the matrix U of Eq. 29.

From the general equations, we again obtain $n_1 = \frac{1}{2}(\lambda p + \mu q)$ etc. (cf. Eq. 35) with $\lambda, \mu, \nu \in \mathbb{Z}$, but additionally with the condition $\lambda + \mu + \nu \equiv 0 \bmod 2$ because $n_1 + \ldots + n_6 = (\lambda + \mu + \nu)q$, $n_1 + n_2 + n_3 - n_4 - n_5 - n_6 = (\lambda + \mu + \nu)p$. But we assume $(p,q) = 1$, so $p \equiv q \equiv 0 \bmod 2$ is impossible. Hence, $(\lambda + \mu + \nu)$ must be even in order to have both points in D_6. One can now determine the lattice vectors in the section and the corresponding translation group and the index of $\Gamma_{\|} \times \Gamma_{\perp}$ in D_6.

As a result, we find that all the rational sections have primitive, cubic translation groups in $\mathbb{E}_{\|}$ and $\mathbb{E}_{\perp}$ of equal lattice constant. We find

$$I = [D_6 : \Gamma_{\|} \times \Gamma_{\perp}] = \frac{\text{vol}(FD(\Gamma_{\|}))\text{vol}(FD(\Gamma_{\perp}))}{\text{vol}(FD(D_6))} = 4(p^2+q^2)^3. \tag{45}$$

The minimal index is $I = 4$ $(p = 0, q = 1)$. Again, the point group of the full periodic structure in $\mathbb{E}_{\|}$ and in $\mathbb{E}_{\perp}$ is, in general, not the cubic group $\Omega(3)$, but only the tetrahedral group T. The Al-Cu-Fe alloys [3] seem to provide an example where the transition from icosahedral to periodic phases can be observed without phase boundaries and without violation of tetrahedral symmetry.

Last but not least, let us briefly describe the situation for the I-type icosahedral structure based on the weight lattice D_6^R which is the reciprocal of D_6. One can define D_6^R as

$$\begin{aligned} D_6^R &= \mathbb{Z}^6 \cup (\mathbb{Z}^6 + \frac{1}{2}(\mathbf{e}_1 + \ldots + \mathbf{e}_6)) \\ &= \{\Sigma_i \mu_i \mathbf{e}_i \mid (\text{ all } \mu_i \in \mathbb{Z}) \text{ or } (\text{ all } \mu_i \in \mathbb{Z} + \frac{1}{2})\} \end{aligned} \tag{46}$$

wherefrom one can identify D_6^R with the body centered hypercubic lattice with $\mathrm{vol}(FD(D_6^R)) = 1/2$.

The geometric scenario remains very similar, reduction matrices do not change. We get again rational sections for $\tan(\phi) = p/q, (p,q) = 1$, and we have to solve the same general equations for the vectors in the sections, now also looking for the additional possibilities due to the body centers.

Let us summarize our findings here: The translation lattices for the rational sections are always of bcc-type (both in $\mathbb{E}_{\|}$ and in $\mathbb{E}_{\perp}$, and with identical lattice constant), and the index is

$$I = [D_6^R : \Gamma_{\|} \times \Gamma_{\perp}] = \frac{\mathrm{vol}(FD(\Gamma_{\|}))\mathrm{vol}(FD(\Gamma_{\perp}))}{\mathrm{vol}(FD(D_6^R))} = 4(p^2 + q^2)^3. \tag{47}$$

It is remarkable that none of the three examples presented allow a one-parameter transition to a cubic fcc phase. The relation of the I-type icosahedral phase to bcc phases only might be useful for the search for experimental candidates of I-type icosahedral quasicrystals which have not been found so far.

6. Concluding Remarks

Motivated by experimental observations we have outlined the simplest possible scheme of transitions between different phases that maintain the maximal common symmetry. This was achieved by the so-called Schur rotation. Although an operation in higher-dimensional space, the rotation angle gives rise to a measurable quantity in physical space $\mathbb{E}_{\|}$. Therefore we think that it could be useful for an explicit structure model as well as for understanding structural similarities between 'neighbouring' phases, such as fourfold and eightfold symmetric ones or 4-, 8-, and 12-fold symmetric ones in 2D quasicrystals and primitive cubic, bcc, and icosahedral phases in 3D quasicrystals.

7. Acknowledgement

The authors are grateful to M. Schlottmann and D. Zeidler for helpful discussions. This work was supported by Deutsche Forschungsgemeinschaft and Alfried Krupp von Bohlen und Halbach Stiftung.

References

[1] K. H. Kuo, in "Proceedings of the Anniversary Adriatico Research Conference on Quasicrystals, ICTP, Trieste, Italy, 1989", eds. M. V. Jaric and S. Lundqvist, World Scientific, Singapore, 1990

[2] M. Audier and P. Guyot, *Phil. Mag.* **B 52** (1985) L15
V. Elser and C. L. Henley, *Phys. Rev. Lett.* **55** (1985) 2883

[3] W. Liu and U. Köster, in: *Phase Trans.*, to be published

[4] J. L. Birman, *Phys. Rev. Lett.* **17** (1966) 1216

[5] M. Baake, D. Joseph, P. Kramer, and M. Schlottmann, *J. Phys.* **A** (1990) L961

[6] P. Kramer, *Acta Cryst.* **A 43**, (1987) 486

[7] J.-C. Toledano and P. Toledano, "The Landau Theory of Phase Transitions", World Scientific, Singapore, 1989

[8] R. Ammann, in B. Grünbaum and G. C. Shephard, "Tilings and Patterns", W. H. Freeman & Co., New York, 1987, p. 556f, R. Ammann, B. Grünbaum, and G. C. Shepard, *Discr. Comput. Geom.* **8**, (1992) 1

[9] F. P. M. Beenker, Eindhoven, TH-Report 82-WSK-04 (1982)
Z. M. Wang and K. H. Kuo, *Acta Cryst.* **A 44**, (1988) 857

[10] P. Kramer, *Mod. Phys. Lett.* **B 1** (1987) 7, *Int. J. Mod. Phys.* **B 1** (1987) 145, *J. Math. Phys.* **29** (1988) 516

[11] P. Kramer and M. Schlottmann, *J. Phys.* **A 22**, (1989) L1097

[12] M. Baake, P. Kramer, M. Schlottmann, and D. Zeidler, *Mod. Phys. Lett.* **B 4** (1990) 249 and *Int. J. Mod. Phys.* **B 4** (1990) 2217

[13] M. Baake, D. Joseph, and M. Schlottmann, *Int. J. Mod. Phys.* **B5**, (1991) 1927

[14] N. Wang, H. Chen, and K. H. Kuo, *Phys. Rev. Lett.* **59** (1987) 1010

[15] H. Chen, D. X. Li, and K. H. Kuo, *Phys. Rev. Lett.* **60** (1988) 1645

[16] L.-C. Chen and J. L. Birman, *J. Math. Phys.* **12**, (1971) 2454

[17] A. Knupfer, Diplom-thesis, Tübingen (1990)

[18] M. Baake, *J. Math. Phys.* **25** (1984) 3171

[19] P. Kramer, Z. Papadopolos, and D. Zeidler, in: "Proc. Symp. Group Theory in Physics", eds. A. Frank et al., AIP Conference Proc., to appear, *J. Non-Cryst. Solids*, in press

D. Zeidler, Phd-thesis, Tübingen (1992)

Crystal-Quasicrystal Transitions
M.J. Yacamán and M. Torres (Editors)

Quasicrystals and their approximants. A unified view

D. Romeu and J. L. Aragón.

Instituto de Física U.N.A.M., P. O. Box 20-364, México D. F. 01000

Abstract

The Decahedral Recursive model is used to explain at atomic level structures of quasicrystals and approximant structures for decagonal and icosahedral phases. Possible mechanisms for the quasicrystal-crystal transformation by means of phason hops which change both the set of fundamental tiles and the symmetry of the tiling are discussed. Electron microscope diffraction and high resolution images from simulated structures are presented and compared with experimental results.

1. INTRODUCTION.

Real quasicrystals are described by a set of fundamentals units or *tiles*, decorated with atoms, filling space without gaps, plus rules to pack them so that they force non periodic order. Most models, however, prefer to consider atomic clusters as the important entities and to describe the quasicrystalline structure as a set of *nodes*, connected by *linkages*, where identical clusters are placed. The tiling description can be recovered since the set nodes plus linkages can be related with vertices and edges of a tiling model. In this case, vertices are decorated with clusters and matching rules are replaced by recipes to pack neighbouring clusters.

In this work we shall be dealing with a cluster description of quasicrystalline structures and their associated approximants for both icosahedral and decagonal systems, that requires: i) realistic clusters (hereafter referred as *basic clusters*) and ii) rules to pack them to cover the space propagating either periodic or non-periodic long range order. As we shall see, the model, called Decahedral-Recursive (DR) provides recipes to realize these two necessities and construct quasicrystalline structures.

The DR model is a general atomistic model that simulates crystalline and quasicrystalline atomic structures of any symmetry [1,2]. Following a small set of general (symmetry independent) energy minimizing rules, it describes how atoms aggregate to form low energy clusters (decahedral stage) which are good candidates for basic clusters, and how these coalesce (recursive stage) to form macroscopic structures by minimizing the cluster-cluster interfacial energy. This model has been successfully applied to construct atomic models of quasicrystals. In particular, models of several decagonal systems with different periodicities together with their associated approximants, namely AlCoCuSi [3], AlCuCo [4], AlPdMn [5], reproducing electron microscopy observations. Also, icosahedral systems (with approximants) such as AlCuLi, have also been modeled using the same basic ideas.

This work is organized as follows. Section 2 is intended as an introduction to essential ideas of DR model, namely, the decahedral and recursive stages, and a higher-dimensional approach which connect the model with the general theory of quasicrystals. In section 3 the model is used to build several decagonal and parent crystalline phases with different periodicities and 2D structures. Additionally a mechanism of phase transformation via phason flips is proposed. Finally, section 4 describes the structure of the icosahedral AlCuLi phase by following DR recipes.

2. THE DECAHEDRAL-RECURSIVE MODEL.

In this section we shall describe how the two fundamental ingredients of a cluster-based quasicrystal description, namely, basic cluster and packing rules, can be obtained from the DR model. The growth of a basic cluster is a matter of the decahedral stage, and the answer to how clusters coalesce to form a macroscopic structure is given by the recursive stage. It will also be explained how the recursive procedure can be viewed in the framework of the cut and projection method, with a peculiar window function.

2.1 Decahedral stage.

Free jet expansion experiments [6] and quasi-static cooling simulations [7] from small rare gas clusters have shown that particles with 13, 19, 23, 26, 29,... atoms, known as magic numbers, are more stable than others. The observation that magic number particles are composed of a complete number of irregular decahedra (pentagonal bipyramids) [8], constitutes the basis of the decahedral stage of the DR model.

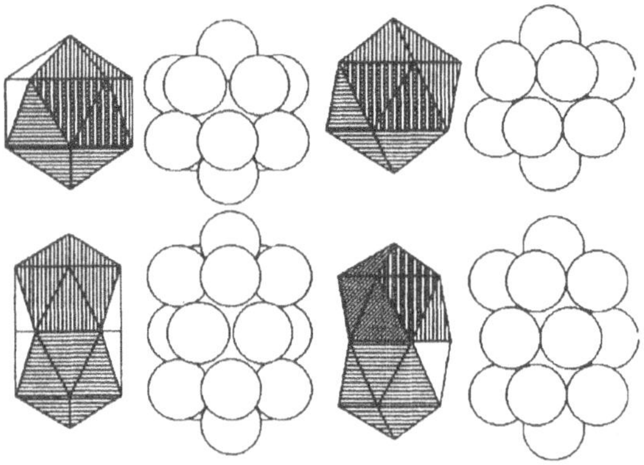

Figure 1. Magic numbers of 13 and 19 atoms. Composing interpenetrated pentagonal bipyramids are shaded.

Under the hypothesis that magic number results can be extrapolated to metals, the DR model assumes that growth completing decahedra, or decahedral growth, follows a minimum energy path and is therefore preferred [9]. This means that atoms in a given structure are shared by several tetrahedra simultaneously, and given the efficiency of tetrahedral packing, there is an energy gain

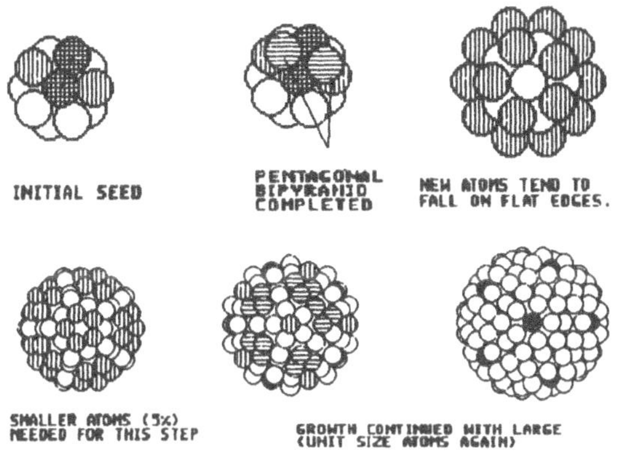

Figure 2. Decahedral growth of the icosahedral phase by completing pentagonal bipyramids at each step.

which provides the driving force for the process. When growth starts from the liquid phase, small clusters are assumed to develop from magic number seeds present in the super cooled liquid as heterophase transitions. The first two magic numbers with 13 and 19 atoms shown in Figure 1 give rise to structures with icosahedral (I) and decagonal (D) symmetries respectively [1,9,10], other magic numbers give rise to different phases [1]. Under decahedral growth, relatively large clusters can be generated, its symmetry depending only on that of the selected seed. These clusters will turn out to be the basic clusters composing quasicrystals and their approximants. Figure 2 exemplifies atom by atom, the decahedral growth of a 13 atom seed (icosahedron) leading to a large cluster with icosahedral symmetry. If the 19 atom magic number seed (M19) is selected, clusters with decagonal symmetry are generated by the same mechanism [9]. The growth process is carried out using atoms of the same size until the build up of geometric frustration forces the use of smaller (about 5%) atoms, so that a minimum of 2 atomic species are required to maintain stability in quasicrystals [9].

2.2 Recursive stage.

During decahedral growth, special sites develop in the cluster surface which become good nucleation centres: growth can re-start from these sites as if they where new origins. These special sites are called O-points following Bollman [11] and in both the icosahedral and decagonal cases, they are located at the centre of (normally distorted) icosahedra [1]. Since O-point sites appear in the solid-liquid interface that at the centres of incomplete icosahedra, there is an energy gain when atoms in the liquid complete these icosahedra, recovering the initial seed and thus restarting the growth process again, using the O-point site as a new origin. Thus, decahedral growth proceeds by completing decahedra, constantly yielding new possible centres (O-points) which the system may or may not use as new origins, depending on external conditions.

O-points have the extra property that when the cluster is shifted into them, a relatively large number of coincidences arise between shifted and unshifted atoms. If after shifting the cluster into these points, atoms in the intersecting volume are removed, the result is a larger composite cluster with low interfacial energy. This last assertion is based on the Coincidence Site Lattice Model [12] which predicts a low energy (good atomic fit) boundary between crystals (clusters) shifted (and/or rotated) into positions such that the coordinates of some shifted/unshifted atoms coincide. The recursive stage enlarges a cluster by displacing it into successive O-point generations until a sufficiently large structure results.

Some O-points are special in that they give rise to a larger percentage of coincidences than others, and have the same surroundings up to a larger volume. Atoms in this volume define a repetitive pattern that composes the structure. In the spirit of a cluster-based model, this pattern constitutes the basic cluster of the structure, and the special O-points (the centres of the basic cluster), are the nodes of the model. Note that all nodes are O-points but the converse is not necessarily true. As a matter of fact, a node can "migrate" into an adjacent O-point, transforming it into a node through small atomic displacements (see section 3). This migration of nodes along the set of O-point sites (coincidence quasilattice) by diffusive atomic displacements produces the well known phason flips or hops, and they lie at the heart of crystal-quasicrystal phase transforms. In section 3.5 a phase transformation mechanism between decagonal and associated (parent) crystalline phases is proposed based on the above phason hops.

Summarizing, the DR model provides the ingredients of a cluster-based quasicrystal model, namely, a basic cluster and packing rules, by combining decahedral and recursive stages, in the following way:

1) Starting from a magic number seed, a larger cluster is grown by completing decahedra, until the first set of O-points appears. These O-points are used to grow the next generation cluster by shifting the cluster of the previous generation into its O-point sites, according to the rules of the recursive stage, until the first nodes appear. The cluster so generated is the basic cluster of the structure.

2) Once the basic cluster is defined, further growth proceeds by shifting it into those of its nodes that maximize the percentage of coinciding atoms.

2.3 The DR model in higher dimensions.

The cut and projection method [13-15] has been the basic tool to generate quasiperiodic structures and study its long-range order and diffraction properties. In this approach, a quasicrystalline structure is viewed as a projection onto three-dimensional (3D) physical space of the subset of points of a hyperlattice (including a well defined decoration) that fall within a strip. Models of quasicrystal formation can be phrased in this language with distinct strip selection. Perfect quasicrystals [16] can be obtained with a plane strip parallel to physical space $E^{\parallel}$ (see Figure 3a). Random tiling quasicrystals [17-19] are generated with a rippled but unbroken strip, as is illustrated in Figure 3b; the symmetry of the projected structure is related to the average orientation

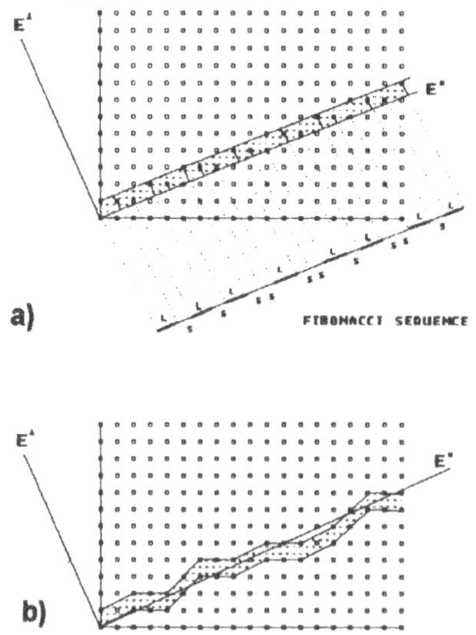

Figure 3. 2 dimensional cut and projection scheme for a) perfect quasicrystal and b) random tiling quasicrystal.

of the strip [19]. Diffraction properties, such as diffuse scattering, are governed by the fluctuations around the average orientation of the strip, along $E^{\perp}$: the orthogonal space to $E^{\parallel}$. Finally, an icosahedral glass model [20] corresponds to a rippled and broken strip [19].

In order to inscribe the DR model within this framework, lets start with a DR quasicrystal structure. Now, let $\mathcal{L}\subset\mathbf{R}^N$ be an N dimensional lattice (usually, N=6 for the icosahedral case and 5 for the decagonal, octagonal and dodecagonal cases [21]), whose vertices project onto the O-points of the structure. The decoration of the lattice projects onto the small cluster contained within the initial set of O-points which will be referred as 'initial cluster' (see inset in figure 4). In the icosahedral case, for example, this cluster consists of a unit icosahedron and a dodecahedron [1,2] referred to as ICO1 and DODE in section 4.2. Since the decoration of each lattice point is the initial cluster lifted to N dimensional space, we can define a portion S_0 of the strip called 'Basic Slab' (see inset in Figure 4) as the set:

$$S_0 = \{\ \mathbf{x}=(\mathbf{x}^{\parallel},\mathbf{x}^{\perp})\ \mid\ \mathbf{x}^{\perp}\in\Omega,\ \|\mathbf{x}^{\parallel}\|\le r_0\ \}$$

where Ω is the projection of the unitary cube onto $E^{\perp}$, called acceptance domain, and r_0 is the outer radius of the initial 3D cluster. According to its definition, S_0 contains only the lifted initial cluster as depicted in Figure 4 in the trivial two dimensional case.

Since O-points in 3D space ($\mathbf{o}^{\parallel}$) correspond to lattice points in hyperspace ($\mathbf{o}=\mathbf{o}^{\parallel}+\mathbf{o}^{\perp}$), the N dimensional counterpart of the recursive stage consists in shifting the Basic Slab S_0 onto lattice points $\mathbf{o}_i\in\mathcal{L}$. Discarding atoms in 3D is equivalent to a non intersecting requirement in hyperspace of the projection, onto $E^{\parallel}$, of neighbouring slabs. The jth-step of this procedure gives:

$$S_j = S_{j-1} \cup (S_0 + \mathbf{o}_j) - \{\mathbf{x}' \in\ (S_0 + \mathbf{o}_j)\ \mid\ \mathbf{P}^{\parallel}(\mathbf{x}') \in \mathbf{P}^{\parallel}(S_{j-1})\ \}$$

where $\mathbf{o}_i\in\mathcal{L}$, and $\mathbf{P}^{\parallel}$ denotes the orthogonal projector onto $E^{\parallel}$. The subtracted set is the portion of the slab containing points that would project onto "occupied volume" in 3D. Following this procedure we can generate an infinite staircase-like strip where different slabs S_j (with equal $E^{\perp}$ but different $E^{\parallel}$ widths) are related by a shift through the perpendicular component of a lattice point. Figure 4 shows the strip in the 2D case for a particular selection of O-points.

The average orientation of the strip depends on the selection of O-points during growth, and is given by the *phason strain tensor* **E**, defined by $\mathbf{x}^{\perp}\approx\mathbf{E}\mathbf{x}^{\parallel}+\mathbf{h}$, where **h** is the displacement of the average strip from the origin [19]. Fluctuations of the strip around the average orientation are defined by $<\mathbf{x}^{\perp 2}>=<|\,\mathbf{x}^{\perp}(\mathbf{x}^{\parallel})-\mathbf{E}\mathbf{x}^{\parallel}-\mathbf{h}\,|^2>$ [19].

Some generalities can be mentioned about the diffraction properties of the structures generated by this procedure. Let ρ_0 be the density of the decoration associated with each lattice point $\mathbf{r}\in\mathcal{L}$. The Fourier transform of this density can be written as [2]:

$$\hat{\rho}_0(\boldsymbol{k})=\sum_{i=1}^{s}\hat{\rho}_i^3(\boldsymbol{k}^{\parallel})\exp\{-2\pi i\boldsymbol{k}\boldsymbol{r}_i\}$$

where $\mathbf{r}_1, \mathbf{r}_2, .., \mathbf{r}_s$ are the positions of the (hyper-)atoms of the decoration with respect to the unit cell of $\mathcal{L}$, and $\rho^3_i(\mathbf{r}^{\parallel})$ is the density of an ordinary 3-dimensional atom. It can be proved [2] that the general expression for the Fourier transform of the projected structure is:

$$\hat{\rho}_4(\boldsymbol{k}^{\parallel})=V^*\sum_{l\in\mathcal{L}^*}\sum_{i=1}^{s}\hat{\rho}_i^3(\boldsymbol{l}^{*\parallel})\exp\{-2\pi i\boldsymbol{l}^{*\parallel}\boldsymbol{r}_i\}\ \hat{W}(\boldsymbol{k}^{\parallel}-\boldsymbol{l}^{*\parallel},-\boldsymbol{l}^{*\perp})$$

where * is used to identify reciprocal vectors, V^* is the reciprocal space unit cell volume, and $\hat{W}$ is the Fourier transform of the strip function.

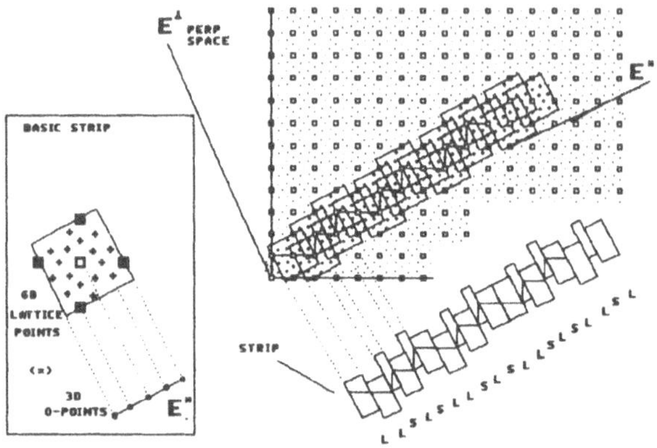

Figure 4. Two dimensional example of the strip construction according to the DR model. (see text).

Strict delta functions are obtained from the previous equation if W does not depend on $\mathbf{r}^\parallel$, or if the up-and-down jumps of the band are themselves quasiperiodic with wave vectors that are also wave vectors of the corresponding perfect quasicrystal, but in general, the diffraction pattern consists of peaks centred on Γ^* whose shapes are given by $\hat{W}$. When W is chosen according to the needs of the DR model to produce quasicrystals (see Figure 4), $\hat{W}$ will display short wavelength components (due to the jumps of the band to reach the O-points) but it will not display long wavelength components (since over distances much longer than the distances between O-points, the band is rather flat and follows the manifold $E^\parallel$ closely). Consequently the diffraction pattern of the DR structures will consist of Bragg peaks plus diffuse scattering.

3. THE DECAGONAL PHASE

3.1 The basic cluster

The decagonal phase seed is the 19 atom magic number M19 shown in figure 1. M19 is the smallest particle having the same Laue symmetry of the decagonal phase. M19 can be regarded as two icosahedra (or three decahedra) interpenetrated along the c axis sharing 6 atoms. However, the icosahedra in M19 are not regular; the distance between atoms along the c axis has been shrunk about 12% to approximately equal the circumradius $2b$ of the pentagons that define them ($b = 5^{-1/2}$ is approximately equivalent to 0.1 nm); these values

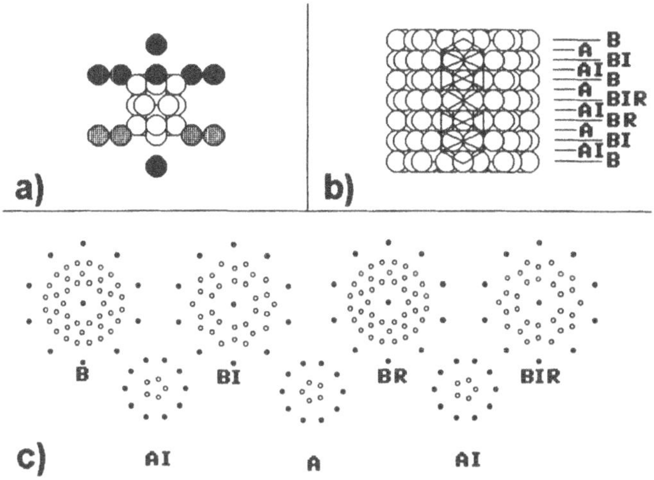

Figure 5. a) Decagonal O-point set with M19 as reference. b) 1.2 nm period cluster D_1 c) 1.2 nm basic cluster planes

were determined by relaxing a chain of M19 clusters using the conjugate gradient method and a Mie 4/7 potential as described in ref. [9]. As we shall see, the D phase consists of a 2D network of such chains parallel to c. For brevity, the shrunk icosahedra will be called pseudo-icosahedra. In the above relaxation, all atoms were assumed to be Al, so actual values in real alloys may be slightly different. In fact, exact atomic coincidence is obtained in larger decagonal clusters by scaling M19 in the radial direction by 1.187 after relaxation.

Large decagonal structures are generated using as O-points, two pseudo-icosahedra of opposite left-right parity (rotated 36°) scaled by 3, called O-pseudo-icosahedra (Figure 5a). Growth proceeds by finding successive O-point generations by displacing both O-pseudo-icosahedra into previous generations, and then displacing M19 into them. The resulting structures are periodic along the c axis but can be periodic or quasiperiodic in the radial direction. Thanks to this periodicity, it is not necessary to continue growth indefinitely along the c axis. As for the radial direction, the second O-point generation, already contains cluster centre sites (nodes) that provide the means to generate macroscopically sized structures. The size of decagonal basic clusters is limited radially by the distance to the origin of the first appearing set of cluster centres and longitudinally by the period of the particular system being considered.

Figure 5b shows a decagonal cluster called D_1, together with the planes of the basic cluster (section 3.2) used to build it. D_1 was obtained through the DR model in ref. [10] and has a 1.2 nm period, although other periods are possible (see below). As we shall see in the next section, there can only be 6 different planes in any decagonal structure which shall be called A, B, AI, BI, BR and BIR in accordance to the names of the planes of the basic cluster that compose them (Figure 5c). Letters I and R in plane names, stand for Inverse parity and Rotation respectively, meaning that pentagons in planes B and BI have reversed parities and plane BR are planes B rotated 36° (note A = AIR and AR = AI). Nodal O-points (possible cluster centres) are located at the vertices of a regular decagon with circumradius $12b$ (1.21

nm) on planes type B shown with full circles in Figure 5c, as well as regular O-point sites on planes A and AI. D_1 has the basic decagonal features: interpenetrated M19 clusters forming columns of pseudo-icosahedra, and it has five glide and five mirror planes parallel to c in addition to a 5_2 axis. Note that atomic sites in the basic cluster agree exactly with those determined from scanning tunnelling observations [22,23], and that it has no mirror plane normal to c in contrast to common belief. The existence of such a mirror would increase the elastic energy of the system by breaking the pseudo-icosahedral stacking sequence and as we shall see when we discuss diffraction results, this mirror is not needed to explain experimental diffraction data.

3.2 The origin of different periods

In order to understand the origin of the various periodicities in multiples of 0.4 nm observed in the Decagonal phase, let us consider the decagonal prism resulting from the stacking of D_1 clusters along c. Such decagonal column consists of a set of pseudo-icosahedral columns parallel to c, with an O-point set given by decagons on planes type A and decagons of nodal O-points plus a single O-point the centre on planes type B as shown in Figure 6. The symmetry of this semi-infinite network is 10/mmm, but the symmetry of the complete structure is limited by its decoration (M19) which is only 5/mmm. The symmetry of the 2D O-point skeleton (coincidence quasilattice) depends on how the columns are packed (periodically or quasiperiodically), the symmetry of each O-point column remaining 10/mmm. It is important to note that altering the decoration of each plane does not change the coincidence quasilattice, which is assumed to be the same for all decagonal phases, so diffraction patterns from different systems must have the same symmetry, as observed.

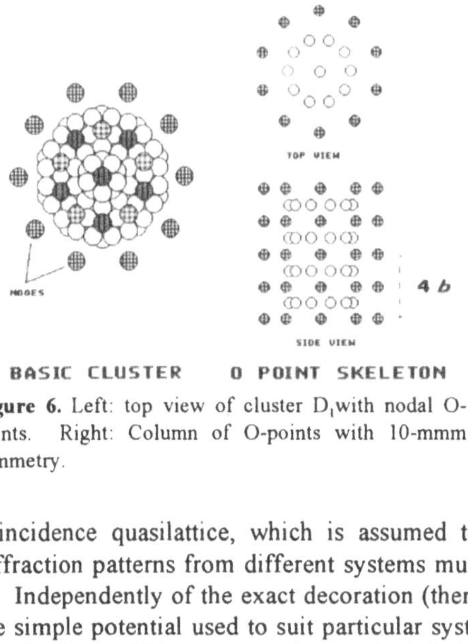

Figure 6. Left: top view of cluster $D_1$with nodal O-points. Right: Column of O-points with 10-mmm symmetry.

Independently of the exact decoration (there may be decoration details not predictable by the simple potential used to suit particular systems, for example, the centres of some pseudo icosahedra could be empty as in icosahedral AlCuLi), the number of different atomic planes (normal to c) that can exist in any decagonal structure is limited to six, whatever the 2D structure is. To understand this, note that atomic planes are transversal cuts through pseudo-icosahedral columns. As a consequence, and assuming stacking faults are forbidden, it follows that when the first pseudo-icosahedron develops through radial growth in a given column, the structure of that column is entirely fixed with period $4b$. This is so because when atoms are arranged on a plane at the vertices of a pentagon centred at a given (x,y) point, adjacent planes should contain isolated atoms on that point, the next pair of planes two pentagons of opposite parity and so on. This limits the number of possible planes to six, (A, B, AI, BI, BR, BIR) as shown in figure 5c. Note planes A force planes B and BIR, located immediately above and below, to have single atoms where A has pentagons and vice versa, and the following planes, AI, to have pentagons of opposite parity. Once a plane, say BI is known, the others can be deduced by placing single atoms on pentagons and viceversa.

No matter what the detailed composition of each plane is, the full structure still consists of a 2D network of pseudo icosahedral columns, so if the structure of any plane, say BI, is changed by arbitrarily changing the parity of any of its pentagons, the parity of all decahedra in the corresponding column must also be changed if energy costly staking faults are to be avoided, thus limiting the number of different possible planes. Since the structure along each column repeats every 4*b*, the matching rule between planes actually groups them in 8 possible units 0.4 nm thick, called U1-U8, each containing five planes as shown in table 1. Note units U5 through U8 are actually units U1 through U4 reflected through a plane so there are only four fundamental units. As we shall see, the different ways in which these units are stacked along the c axis determine the different periods observed in decagonal phases.

Table 1
Fundamental units defined by decagonal planes

UNIT	SEQUENCE OF PLANES				
U1	B	A	BI	AI	B
U2	BR	A	BIR	AI	BR
U3	B	A	BIR	AI	BR
U4	BR	A	BI	AI	B
U5	BI	AI	B	A	BI
U6	BIR	AI	BR	A	BIR
U7	BI	AI	B	A	BIR
U8	BIR	AI	BR	A	B

Once more, in order to avoid stacking faults, fundamental units must in turn be stacked following a simple packing rule: the head of one unit must coincide with the tail of the neighbour. For example, the sequences {U5} (...U5-U5-U5...) and {U6}, both yielding a 0.4 nm. period structure are both allowed since U5 and U6 heads and tails are equal. However, the 0.8 nm sequence {U5-U6} (...U5-U6-U5-U6...) is forbidden. By stacking fundamental units following this simple matching rule we can generate all observed periods of the decagonal phase (Figure 7). For example, the sequence {U6-U7} has a 0.8 nm period, {U7-U6-U8} or {U8-U5-U7} 1.2 nm, and {U5-U7-U6-U8} has a 1.6 nm period.

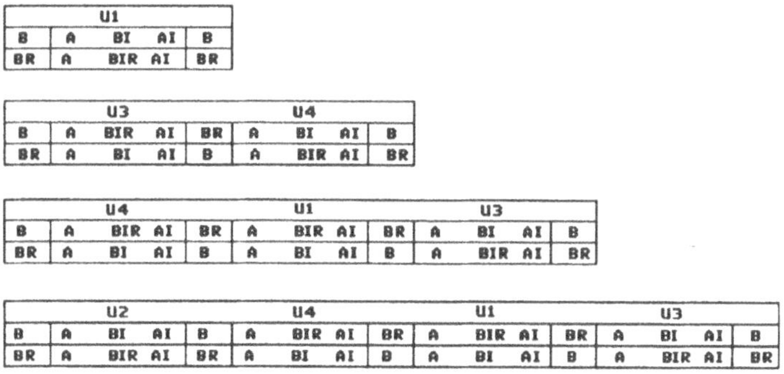

Figure 7. Stacking of fundamental units that produce the periods 0.4, 0.8, 1.2 and 1.6 nm.

There are 2 combinations of the units that produce 0.4 nm and 1.2 nm periods, but only one for the 0.8 nm and 1.6 nm periods. Of course many kinds of mistakes are in principle possible contributing to disorder in the periodic direction such as stacking faults or antiphase boundaries. For example, there could exist an "extra" U5 unit in the 1.6 nm sequence: ...(U5-U7-U6-U8)-(U5)-(U5-U7-U6-U8)... The presence of such extra units could account for the observation of very faint intensity sheets in the diffraction patterns normal to the c axis which tend to disappear in some samples after some heat treatment [10]. These sheets would be produced by the insertion of extra units causing the alloy to have a local 0.8 nm. period at places.

3.3 Two-dimensional structure

The 2D structure can be generated by laterally shifting the basic cluster of a given period into the nearest available nodal O-points on planes type B (discarding overlapping atoms) so that the number of atomic coincidences is maximized [1]. Lateral displacements of the decagonal basic cluster into any of the 10 decagonal nodal O-points give rise to 100% coincidence between atoms in all planes, and therefore to a coherent, minimum energy, cluster-cluster interface [10]. Although all nodal O-points are equally good candidates to become cluster centres, having chosen one, the selection of others is no longer arbitrary. This restriction defines a simple set of packing rules for clusters (not for tiles) that produce decorated space covering cluster networks with maximum atomic coincidence. Since in DR structures every atom belongs to a basic cluster, the maximum density condition is satisfied and a (random) tiling is defined [19].

Packing rules between clusters, are best understood by following the growth of a maximum coincidence structure from a single cluster (Figure 8). For brevity we will restrict ourselves to the 0.4 nm period case here, but the same reasoning applies to the 0.8, 1.2 and 1.6 nm period structures since coincidence maximizing rules apply to all planes. Minimum elastic energy (100% coincidence in all planes) is achieved when node linkages (tile edges), make an angle of either 108° or 144° as shown in Figure 8 (left), other choices result in energy costly non coincident atomic overlap and should be considered as structural defects. This is illustrated at the centre-right of Figure 8, showing a section of planes BI and AI with some defects. Note that the 72° and 36° rhombic tiles have non coincident overlapping atoms on planes AI and BI respectively, effectively ruling them out of

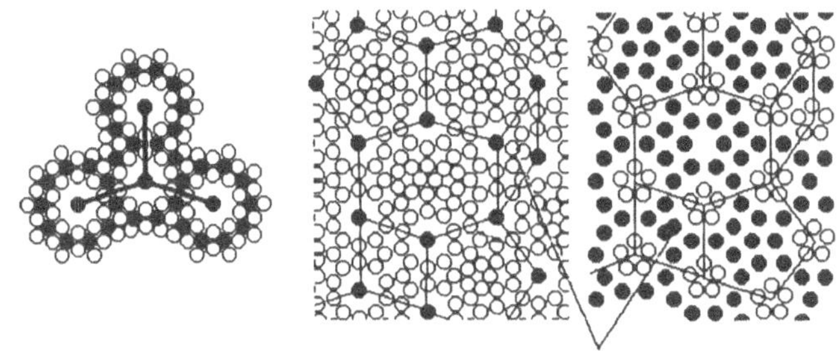

Figure 8. Left: plane BI of basic cluster in 100% coincidence configuration. Centre-right: Planes BI and AI with arrowed defects inside 36° and 72° rhombi.

the ground state tiling. The energy cost of the 36° rhombus is less severe since a relatively low strain structure (with lower density) can still grow by removing the two overlapping atoms. Nevertheless, experimentally, better quasicrystals are normally obtained from AlCoCuSi alloys which do not contain the 36° rhombus (see below) than from AlCoCu alloys, where it is observed. Choosing opposite linkages (180°) forces the eventual appearance of the above rhombi. In a defect free structure, only 3 out of the 10 possible adjacent O-points must be chosen as cluster centres, so that the resulting cluster linkages define 108° or 144° angles. Within this limitation, the selection of centres from available O-points is entirely random since there is no evident physical reason to prefer one coincidence maximizing configuration from another. We shall call the resulting low-energy network a random coincidence network, or tiling to differentiate it from an entirely random tiling, with no energy minimizing restrictions. Later we shall see that O-point sites that are not chosen as nodes, may eventually become so through phason flips, whose net result is to shift a cluster centre into an adjacent O-point through small diffusive displacements of a few atoms.

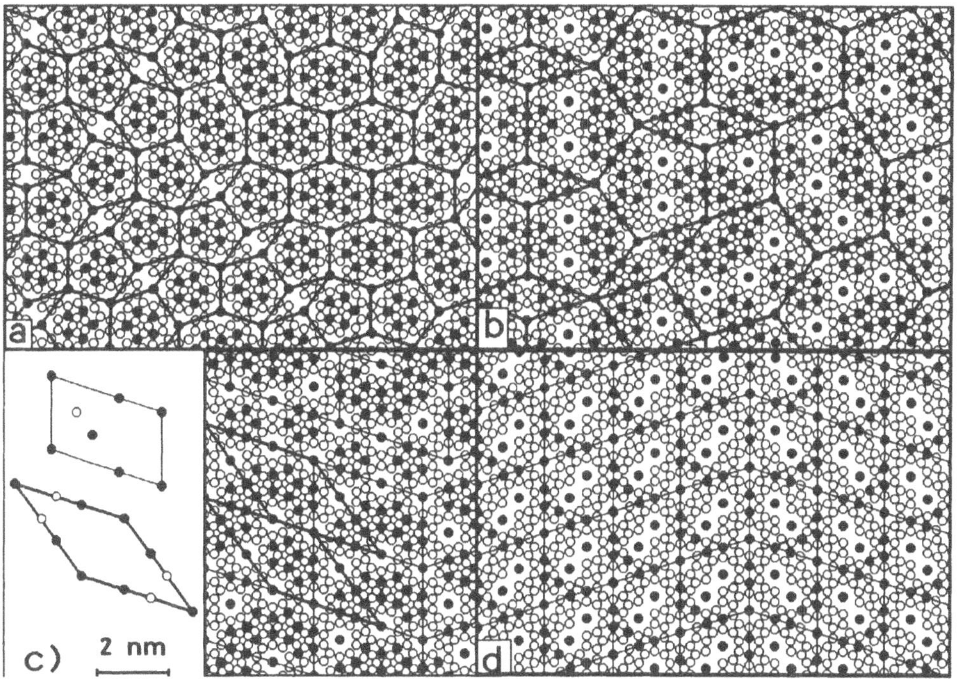

Figure 9. a) Orthorhombic: O_1(left) and O_2 (right) of $Al_{65}Cu_{20}Fe_{10}Cr_5$. b) Orthorhombic O_1' (left) and O_2' (right) phases of $Al_{65}Cu_{20}Co_{15}Si_2$. c) Monoclinic phase of $Al_{70}Pd_{13}Mn_{17}$, inset shows different cell decorations d) Monoclinic phase of $Al_{65}Cu_{15}Co_{20}$.

The above coincidence maximization packing rules for clusters (not tilings), determine the properties of the resulting network. In principle, both crystalline and quasicrystalline structures are allowed. For example, small ordered zones or microcrystalline domains might be present in a randomly grown structure, if these domains are large enough, microcrystalline coincidence networks with well defined orientation relationships and coherent boundaries will form as shown in Figure 9, with nodal O-point sites shaded. For clarity, only BI planes are shown since they permit to see the tiling more clearly and the whole 3D structure can be deduced from them as explained above. These microcrystalline networks have been observed and identified as the orthorhombic O_1 (Figure 9a left),O_2 (Figure 9a right) ,O_1' (Figure 9b left), O_2'(Figure 9b right) phases of $Al_{65}Cu_{20}Fe_{10}Cr_5$ and $Al_{65}Cu_{20}Co_{15}Si_2$ [3], and the monoclinic phases of $Al_{70}Pd_{13}Mn_{17}$ (Figure 9c) and $Al_{65}Cu_{15}Cu_{20}$ (Figure 9d) [4,5]. It is interesting to note that the O_2 phase, which contains the defective 36° rhombic tile of edge length 1.24 nm, occurs in a much smaller proportion to the O_1 phase [3] where it is not observed.

Maximally disordered networks, in the sense of having no discernible microdomains such as that shown in Figure 10a, are also possible. This highly disordered network shows Fibonacci spaced lines defining a quasiperiodic pentagrid when viewed at a glancing angle, line segments along vertical Fibonacci lines are drawn at the bottom of the figure as a visual aid. These lines have been observed in $Al_{65}Cu_{20}Co_{15}Si_2$ HREM images [25] and are also evident in high resolution image simulations of the network (see Figure 16). Note that weak line contrast is observed where closer than normal line segments at the bottom of the figure are found, making it difficult to tell where to place the line as experimentally observed [25].

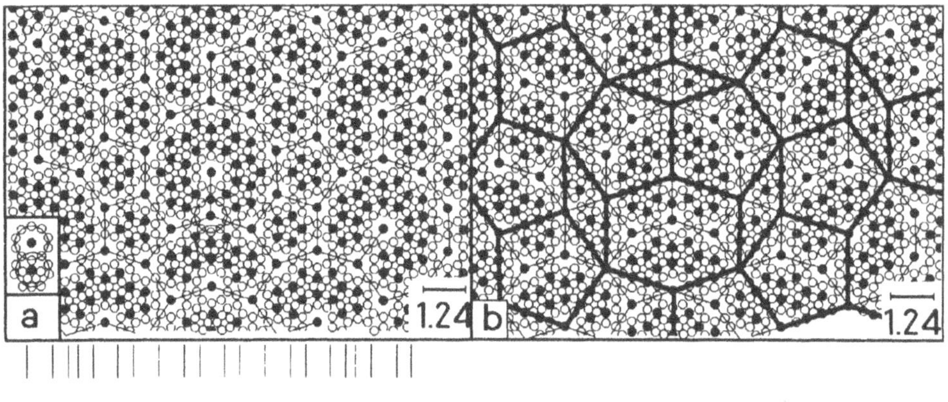

Figure 10. a) Plane BI of a random coincidence network with no crystalline domains. Inset at bottom left details atomic configuration around O-points circled at the centre before and after a phason flip shifting the cluster centre from one O-point to the other. b) Quasiperiodic superstructure in thick lines. Scale in nm.

Although the random coincidence network of Figure 10a looks somewhat chaotic in the sense of being built out of a large number of tiles, it has in fact a much simpler quasiperiodic superstructure associated with it, defined by pentagonal, hexagonal and 36° rhombic tiles, shown in Figure 10b with thicker lines. This quasiperiodic superstructure can be generated by the following recursive rule: 1. Take the nodal O-point sites of the basic cluster on a BI plane (Figure 5c) as starting O-point set. 2. Scale the current O-point set by τ^2. 3.- Decorate the scaled set with the starting O-point set. 4. Repeat from step 2. A portion of the resulting quasilattice is shown in Figure 11a. Shaded points correspond to cluster centres which inflate with τ^2, the complete set inflating with τ. One could decorate again the O-points appearing in step 3 before going back to step 2 to produce the set of all O-point sites, accessible via a cluster linkage from another. Normally, this extra step cannot be repeated further since the 100% coincidence condition would be violated. Figure 11b shows a decagonal section from the above quasilattice that we call the basic decagon since the whole quasilattice can be deduced from it by deflating the pentagonal, hexagonal and rhombic tiles as shown by thin lines or by overlapping shifted and rotated copies of the basic decagon on itself so that the rhombic tiles coincide.

A quasiperiodic tiling obeying a τ^2 inflation is thus associated to a highly disordered (no crystalline microdomains) coincidence network. The random coincidence network of Figure 10a can be obtained by randomly selecting cluster centres from the above quasilattice, preserving the angular relationships between tile edges. Different choices are equivalent, in a cut and projection scheme, to different shifts of the strip along $E^{\perp}$, preserving the local isomorphism. Note the superstructure edge lengths are τ times larger ($\simeq$ 2 nm) and that at this scale, the 36° and 72° rhombic tiles do not break the 100% coincidence rule, making the approximant of Figure 9d possible. There has been a report of a Decagonal phase in AlPdMn [4] where the 36° rhombus is absent from the 2 nm tiling, however, closer examination of the micrographs indicates that the tiling may be similar to that of Figure 10b, with some poorly defined decagonal wheels. Note that care must be taken to insure that all tiles are decorated identically, selecting nodes as centres of decagonal wheels from poorly detailed electron micrographs may lead to errors in the determination of the nature of the tiling.

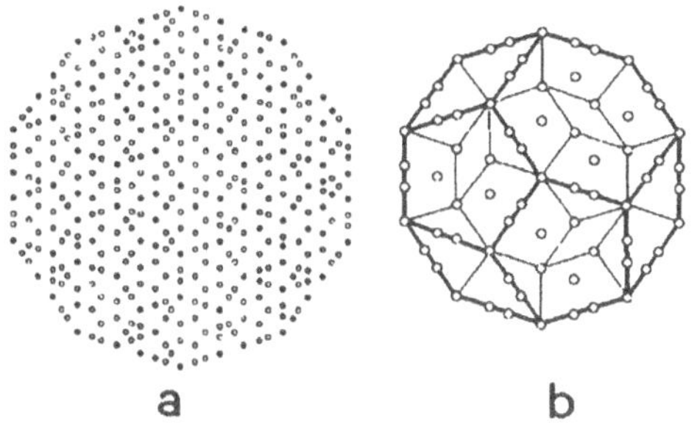

Figure 11. a) Section of quasilattice obtained recursively (see text). b) Basic decagon (central portion of a).

3.4 Electron microscopy and diffraction

Quasicrystal models are normally assessed by their ability to reproduce electron and/or X ray diffraction patterns, however this is an insufficient test since phase information is absent from them. As a consequence, there are different structures having the same diffraction intensities. A model can only be considered complete if besides describing diffraction data accurately, it is also able to reproduce high resolution electron microscope images.

In order to asses the similarity of DR decagonal structures with actual alloys, both electron diffraction and high resolution images have been simulated from a small section of a quasicrystalline network of 0.4 nm period with planes ...BI-AI-B-A-BI--- separated approximately 0.1 nm. Figure 12 shows this section plane by plane, with the exception of plane A which is identical to AI except for the pentagons of atoms above and below nodes

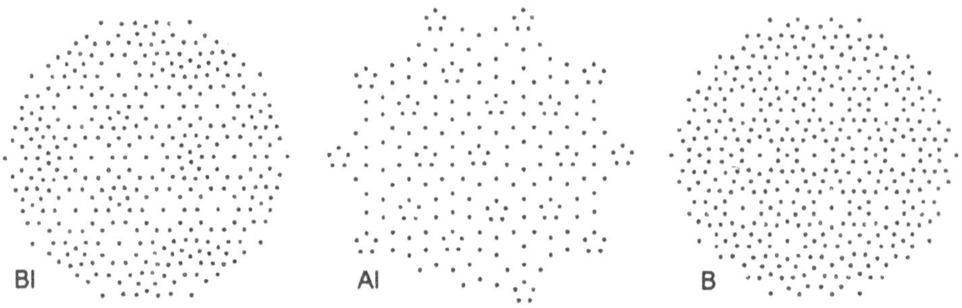

Figure 12. Planes BI, AI and B of the 0.4 nm cluster used for diffraction and high resolution image calculations.

on planes BI and B, which are inverted to preserve the stacking of pseudo-icosahedra. Note again that all planes have O-points over pentagons and viceversa, defining columns of distorted icosahedra. Figure 13 shows kinematical electron diffraction calculations from a columnar cluster composed of 20 periods of the above cluster in the 10-fold and the 2-fold D and P orientations. A smaller section of the cluster was used for the 10 fold calculations to obtain wider spots, so that they could be clearly seen. The pattern was calculated with the formula

$$I_g = \sum_{cluster} f_i(g)\ \exp(2\pi i\ g.r_i)$$

assuming identical atoms. This assumption and the small radial size of the cluster makes intensities hard to compare with experimental values. Nevertheless the patterns reproduce the positions of the most intense experimental spots completely accurately and calculated intensities qualitatively match X ray and Electron Backscattering (EBS) experiments performed on AlCuCoSi [26]. This is to be expected since EBS patterns are not subject to a strong multiple scattering effect, so intensities approach those of the kinematical calculation.

The D and P zone axes are the most important (and also the hardest to reproduce) since reciprocal planes normal to c every $n^{\cdot}(4\ A)^{-1}$, with integer n, are extinct. The observed 10/mmm Laue symmetry of the decagonal phase together with these extinctions have led to the conclusion [27] that the only possible space groups for this phase are $P10_5mc$ and $P10_5mmc$, both having a mirror plane normal to c. Note that the stacking sequence ...BI-AI-B-A-BI... of the cluster does not have such mirror plane, so some discussion is in order.

First, although there is an alternative basic cluster compatible with the DR model by which the mirror plane could be obtained by the stacking ...BI-AI-B-AI-B.... this opposes the direct experimental evidence of Kortan et al [22,23] which observe the ...BI-AI-B-A-BI... stacking predicted by our model and breaks the stacking sequence of pseudo-icosahedra which explains the origin of the fundamental units and the different periods in multiples of 0.4 nm. The answer to this dilemma is that a 10/mmm Laue symmetry with the observed extinctions is compatible with structures having no mirror plane, ie, diffraction alone does not suffice to guarantee the existence of this mirror as we shall see next.

The Laue symmetry of the atomic conglomerate of Figure 12 is 10/mmm (see Figure 13) and has the extinctions observed experimentally. In fact there is a slight intensity difference between some of the weakest spots in the D pattern, on either side of the pattern centre in Figure 13c. These intensities are interchanged when the columnar cluster defined by the sequence ...BI-A-B-AI-BI... (former sequence mirrored) is diffracted. When both stacking sequences are included in the calculation, (...BI-AI-B-A-BI... and ...BI-A-B-AI-BI...), the intensities on either side of the pattern centre are the same, and the patterns show an exact 10/mmm symmetry with extinctions. Now, the difference between planes A and AI is only the orientation of the pentagons above and below nodes, so the two sequences above correspond to a of 0.2 nm shift of the columns of pseudo-icosahedra running along nodes. Since the elastic energy of the system is not altered by such a shift, there should be in principle the same number of (shifted and unshifted) columns ...BI-A-B-AI-BI... and ...BI-AI-B-A-BI... in any decagonal quasicrystal, yielding the observed 10/mmm Laue symmetry. Even if the shifted and unshifted pseudo-icosahedral columns in an experimental sample were not exactly balanced, the intensity differences would be too small to be detected (less than 5% for the brightest spots for the unbalanced cluster of figure 12), since the number of atoms that contribute to the asymmetry is a small fraction of the atoms in a basic cluster. Differences larger than this are normally observed in 2 fold patterns simply as a consequence of small misalignments of the Ewald sphere.

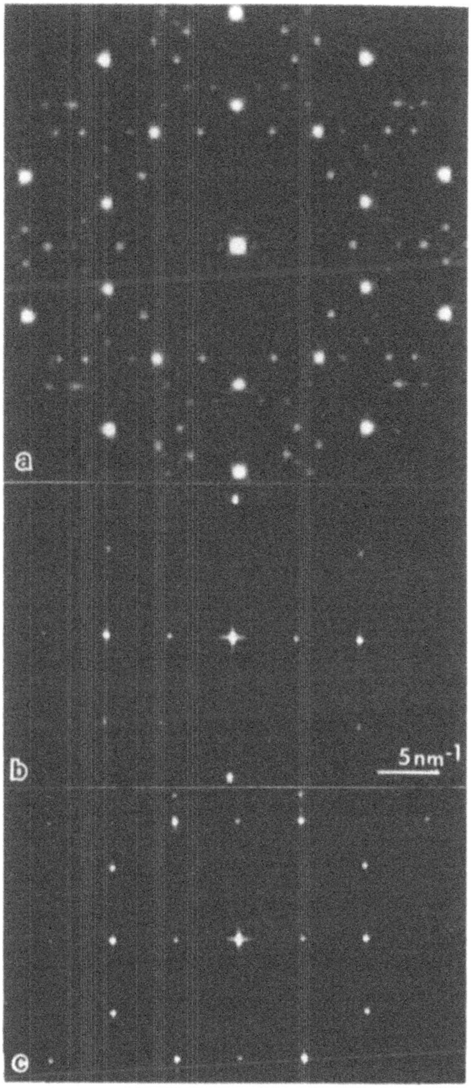

Figure 13. Calculated diffraction patterns from cluster in Fig. 12 in the 10 fold, P and D orientations.

As we mentioned at the start of the section, image simulations must accompany diffraction calculations from atomistic models as a means of checking the phase information loss in diffraction experiments. Figure 14 shows calculated HREM images from the cluster in figure 12, using the multislice method [24] for two different diffraction conditions. Figure 14a taken at the Scerzer condition shows that columns of

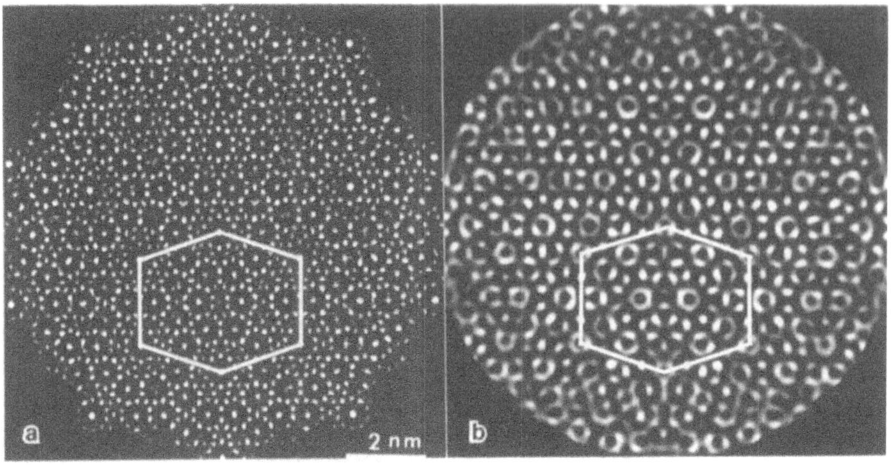

Figure 14. Simulated HREM images from cluster in Figure 12. a) Scherzer condition. b) 10 nm defocus below Scherzer. Hexagonal tile shown for reference.

distorted icosahedra are responsible for the observed wheels of bright spots. Note the similarity with real alloy images [4]. Since most decagonal systems show the same decagonal wheel contrast, it appears that they have a similar basic cluster structure, differing only in chemical composition which in turn determines the nature of the underlying network. Fig 14b shows an image from the same cluster with a defocus of 10 nm below Scherzer, note Fibonacci spaced dark lines can be seen in the picture as experimentally observed [25].

It is important to note that the tiles one would infer from photographs 14a and 14b are different. For example, it would have appeared odd to chose the centre of Figure 14b as a tile vertex, since at this defocus, it does not appear to have a 10 fold symmetry. From this, we can conclude that infering tiles from electron micrographs alone can lead to serious errors in the determination of the proper tiling.

3.5 Phason hops and crystal-quasicrystal phase transformations

We have seen that under DR coincidence maximizing rules, the atomic structures of the decagonal phase and its associated approximants can be reproduced. In this section we shall discuss the mechanisms that can produce a transformation from one to the other, and the conditions under such transformations may occur.

It has been pointed out [25] that decagonal crystal-quasicrystal transformations may occur through phasons. Under the DR model framework, a phason hop or flip is defined as the shift of a network node (cluster centre) into an adjacent O-point as illustrated in figure 15 showing planes BI, AI and B of a section of the cluster in figure 12, before (left) and after (right) such a hop. Phason hops occur between O-points since in this way maximum coincidence between clusters is maintained. Note the atomic configuration around nodes before and after the hop is quite similar. This makes phason flips possible with only small diffusive displacements of each atom provided atoms move cooperatively, ie they occupying the positions others have left vacant.

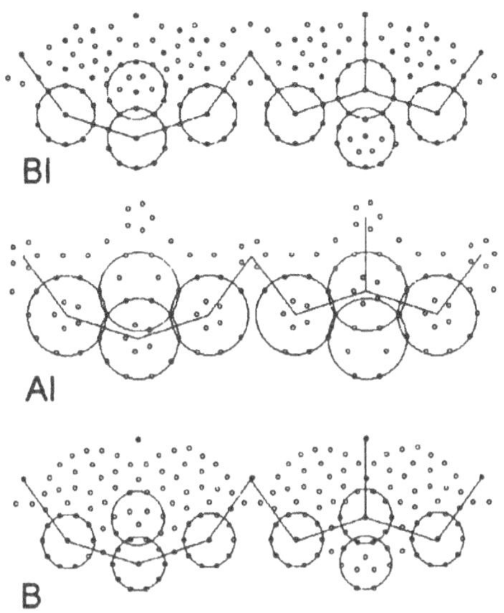

Figure 15. Planes BI, AI and B of a section of the cluster in figure 12, before (left) and after (right) a phason hop.

In general, phason hops between O-points should occur randomly at high temperature, this would have the effect of reducing the size of microcrystalline domains, thereby increasing the configurational entropy of the system. A reduction of the size of microcrystalline domains increases the Shannon entropy of the system as originally suggested in refs. [28-31]. This entropy can be calculated for arbitrary structures in either physical or perp space and measures the degree of disorder of the system, reaching a maximum in the quasicrystalline state where there are no microdomains. It is therefore likely that the microcrystalline to decagonal transformation, is produced by a randomization of the structure at high temperature trough phason flips like these.

Given that the structure across tile edges is the same in all planes, all intertile energies must be nearly equal. It is therefore hard to envisage a mechanism that should force the reverse quasicrystal-crystal transformation on decreasing the temperature, unless the energy density is different for different tiles. In this case one could think of a transformation that would increase the relative proportion of low energy tiles at the expense of others at low temperatures, provided enough thermal energy is available for diffusion. This might explain the quasicrystal-crystal transformation observed in AlCoCu by Hiraga et al. [4], where pentagons, hexagons and 36° rhombi, give way to 72° rhombi. Note the phason hops defined above, can actually produce both a rearrangement of the tiles and the change in their geometry required to produce the above transformation.

4. ICOSAHEDRAL PHASES

4.1. The crystalline R-AlCuLi phase

The production of large grains of the highly stable AlCuLi icosahedral phase [33] raised hopes of understanding the structure of quasicrystalline phases, since single-grain X-ray and neutron diffraction data became applicable. Other properties of this phase stimulated new attempts to model this quasicrystalline alloy: besides the icosahedral phase, a crystalline bcc phase, with almost the same composition and density is formed simultaneously (2.47 g/cm^3 in icosahedral $Al_{0.570}Cu_{0.108}Li_{0.322}$ vs. 2.46 g/cm^3 in crystalline $Al_{0.564}Cu_{0.116}Li_{0.32}$). X-ray and neutron studies have shown similarities between the short and medium range order of

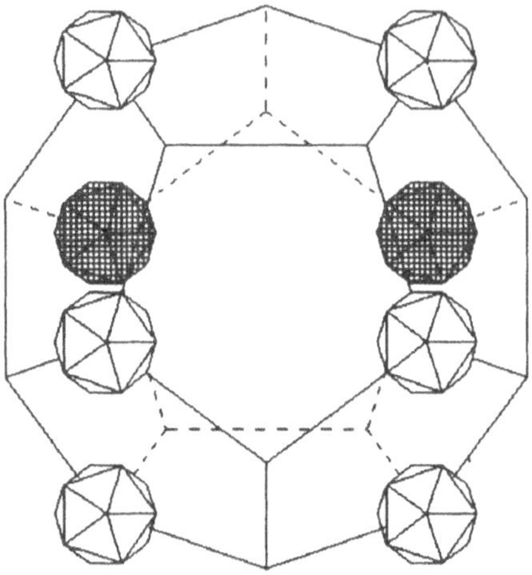

Figure 16. D_{14} dodecahedral shell with icosahedra sitting on bcc positions.

icosahedral (T2) and bcc (R) phases [34]. The close structural relationship between these structures had been previously noted by Henley and Elser [35] who conjectured that the icosahedral phase is formed from the same atomic units of the R-phase, and that the R-phase can be approximated as a <1/1> rational approximant of the icosahedral phase in the cut and projection formalism. This structural affinity between both phases arises naturally in the DR model for the T2-phase.

The structure of the crystalline R-AlCuLi phase was solved many years ago [36,37] and has been recently refined by X-ray and neutron diffraction [38]. It consists of large interpenetrated clusters with pseudo-icosahedral symmetry arranged in a bcc structure with lattice parameter **a**=1.39056 nm. The Al, Cu and Li atoms are distributed over shells around the origin. Starting from the origin (presumably vacant), 14 coordination polyhedra are necessary to reach the body centre position [38]; the last shell is a Li dodecahedron of circumradius 1.20426 nm, hereafter referred to as D_{14}. Eight of the 20 vertices of this dodecahedron are in the body centre positions (Figure 16 shows the D_{14} shell and small icosahedra sitting on the aforementioned bcc positions). The key difference between icosahedral T2 and crystalline R phases can be found in the seventh shell of this large cluster. This shell is a cube truncated with {111} and {110} planes and is the only one with cubic symmetry [39]. As we shall see in what follows, in the icosahedral case, this shell is completed with 6 extra atoms and transforms the truncated cube into an icosidodecahedron.

4.2 Modelling AlCuLi T2 and R phases

A large icosahedral cluster can be grown starting from the 13 atom magic number seed (icosahedron) by completing decahedra as described (see Figure 2). After three iterations, the resulting cluster consists of three shells: an icosahedron (the seed), referred to as ICO1, a dodecahedron (DODE) with atoms sitting on the faces of ICO1 (together defining the initial cluster), and a second icosahedron (ICO2) with circumradius twice that of ICO1. Note that this cluster is the Bergman unit that describes AlCuLi type alloys whereas AlCuFe alloys require a Mackay icosahedron. Both stars are, however, obtainaible from the DR model since both are composed of a complete number of interpenetrated decahedra. The Mackay icosahedron is produced by shifting the star ICO1 into its own surface atoms [1].

The vertices of ICO2 are centres of partially completed and slightly distorted icosahedra, so this shell constitutes the first set of O-points. The recursive stage is applied to this set and in subsequent steps, several families of O-points arise. In order to preserve quasiperiodic long-range order we discard some of the families by applying a minimum phason path criterion. This is carried out as follows: each shell of the cluster can be lifted to 6D space. ICO2 atoms lift into the vertices of a unit 6D cube; DODE lifts into the body centre position and ICO1 into the mid-edges of the cube [2]. Therefore, in 6D the vertices of ICO1, DODE and ICO2 have coordinates of the form {100000}, {110000} and {200000}, respectively. Each point **r** in 6D can be decomposed into parallel and perpendicular components ($\mathbf{r}^{\parallel}, \mathbf{r}^{\perp}$). Only those O-point families with minimal perpendicular component in 6D, ie minimal phason component $\mathbf{r}^{\perp}$, are selected to continue recursive growth.

By following this minimum phason path, we generate a large cluster composed of shells which are identical to those experimentally determined for the R-phase (except the seventh one), moreover, the vertices of D_{14} are special O-points (nodes) ie, they are maximum coincidence positions and the structure of the original cluster is recovered around them. Therefore, we shall consider as basic cluster for the T2 phase the complete 498 atom cluster, up to the D_{14} shell of nodes. Once the basic cluster is defined, the process is greatly simplified since we have only to shift the basic cluster into the positions of D_{14}. Subsequent nodal families appear at ensuing stages of the process and we select at each

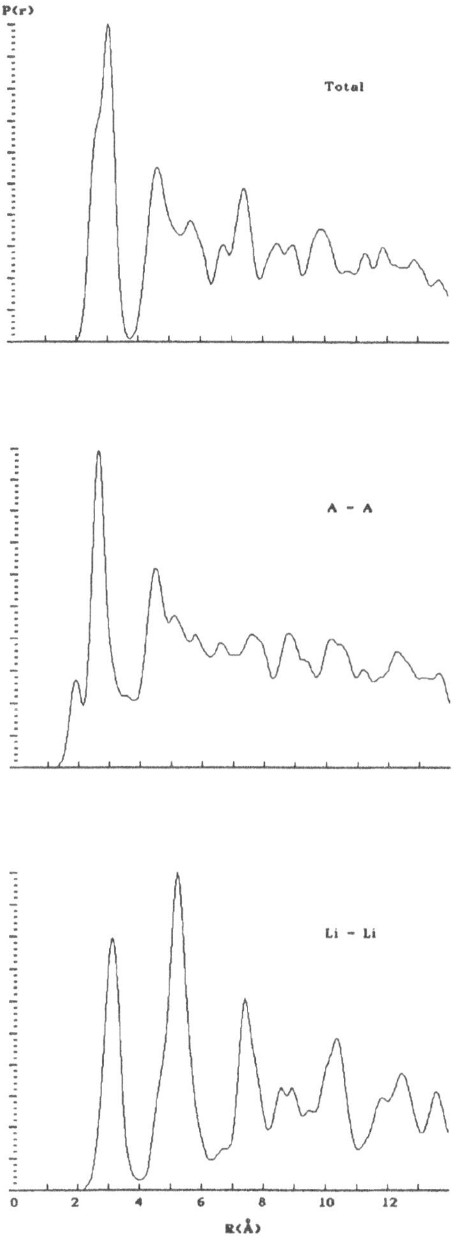

Figure 17. Pair distribution functions of the AlCuLi quasicrystal model described in text. A stands for Al and Cu atoms.

stage those families that yield the maximum coincidence ratio. Once maximum coincidence positions are determined in each iteration, we can select them all or some of them randomly. In the first case, we obtain a nearly perfect icosahedral structure, whilst the random selection yields a disordered structure exhibiting the properties of a random tiling model. In what follows we will concentrate on the first case.

A cluster containing 10330 atoms was generated with the described procedure and relaxed with a Mie 4/7 potential. Figure 17 shows the calculated pair distribution functions of the model, note that in the figure A stands for Al and Cu atoms. All atomic distances and weights are reproduced without unphysically short distances, and the results match almost exactly the reported experimental data [40] for short distances. Reproduction of data for large interatomic distances requires a much larger cluster. Figure 18 shows a HREM multislice simulation of a 3500 atom slice taken from the cluster of the model that matches exactly experimental high resolution images (see for example [41]). The simulation are: 400Kev and 40 nm overfocus. Finally, simulated diffraction pattern of this model can be indexed, according Cahn et al. scheme [42], with a quasilattice constant of $\mathbf{d}_0$=1.91 nm (compare Reference [43]), as explained in detail elsewhere [44].

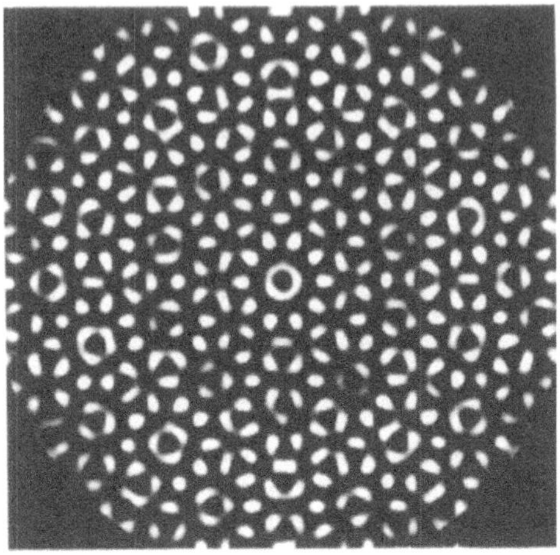

Figure 18. HREM image of the model, with 40 nm overfocus.

The set of centre clusters were lifted to 6D space and then the equation, $\mathbf{x}^{\perp}=\mathbf{E}\mathbf{x}^{\parallel}+\mathbf{h}$, was fitted by a least squares procedure, giving:

$$E = \begin{bmatrix} 0.000015 & 0.000039 & 0.000000 \\ 0.000000 & 0.000024 & 0.000000 \\ 0.000001 & -0.000021 & 0.000015 \end{bmatrix} .$$

This value for **E** conforms to an almost perfect icosahedral symmetry of the projected structure, and correspond to a zero phason strain. Fluctuations of the $\mathbf{x}^{\perp}$ around this $\mathbf{E}^{\parallel}$ average orientation, defined by $\langle \mathbf{x}^{\perp 2} \rangle = \langle \mid \mathbf{x}^{\perp}(\mathbf{x}^{\parallel})-\mathbf{E}\mathbf{x}^{\parallel}-\mathbf{h} \mid^2 \rangle$ [19], show a fast convergence to a constant value close to zero (fig.19).

Some remarks on the relationship between R and T2 phase can be done. As it was pointed out, the basic cluster for the T2 phase coincides exactly with the R-phase cluster up to D_{14}, except for the seventh shell. In the former case this shell is an icosidodecahedron, and in the latter a truncated cube. Analyzing the truncated cube, we can notice that its 24 vertices can be combined with the six vertices of a, suitably oriented, octahedron of the same circumradius to get an icosidodecahedron as depicted in Figure 20, where the dashed faces of the icosidodecahedron correspond to the triangular faces of the truncated cube. Since T2 and R phases always coexist [34], it is reasonable to think that a slight composition ratio during the cooling process can lead to form either a truncated cube or an icosidodecahedron, giving place to the formation of R and T2 phases respectively. Note also that the extra atoms in the basic cluster explains the slight difference in composition between quasicrystalline and crystalline phases.

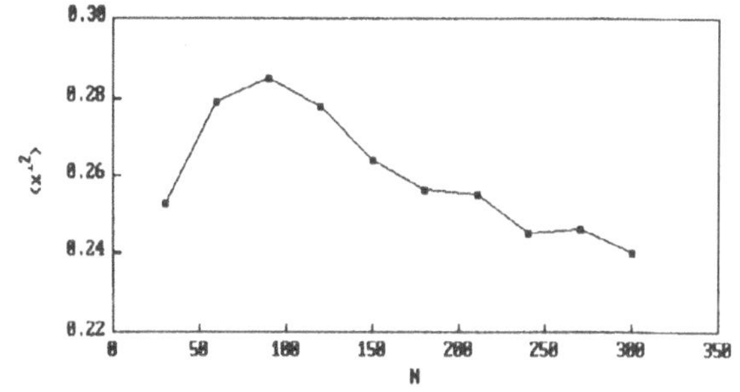

Figure 19. Phason fluctuations of the nodes as function of system size.

If the R-phase cluster (with a truncated cube in the seventh shell) is used as basic cluster, the recursive process leads directly to the structure of the R-phase, ie coincidence maximization leads to the crystalline structure. In this case, not all nodes of D_{14} have the same coincidence ratio; the eight nodes at bcc positions (see Figure 16) have, actually, a 100% of coincidences as a consequence of the translational symmetry of the phase. In other words, these sites are now the only nodes of the structure yielding the maximum possible 100% coincidence, and using them leads naturally to a b.c.c. structure composed of big interpenetrating clusters located at the lattice sites, that is, the R-phase. In this way, both T2 and R phase can be growth with the same basic rules of DR model.

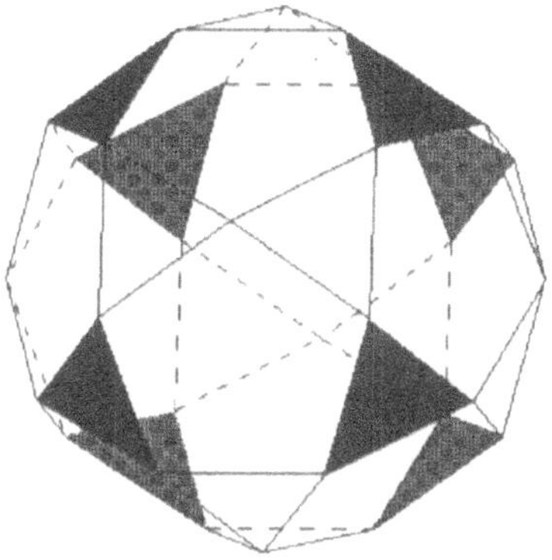

Figure 20. Dashed faces of the icosidodecahedron are triangular faces of a inscribed truncated cube.

5. CONCLUSIONS

The DR model is not specifically designed to describe specific alloys, in contrast with all other atomistic models. DR structures are derived from general principles where the emphasis is placed in the energetics of their formation rather than in their crystallographic description. As a consequence, it offers an explanation for the origin of the basic clusters and their packing rules leading to the formation of quasicrystals and approximant phases as well as the mechanisms (phason hops) involved in the transformation of one into the other. This ability to describe general properties is the most important feature of the DR model. The fact that icosahedral and decagonal quasicrystals can be generated by the same set of ideas, means that regardless of its detailed structure, quasicrystals can be understood, in principle, in terms of very simple and basic physical ideas that also hold for crystals.

Due to this generality alloy-specific refinements are needed to describe in detail particular systems. However, as it stands, the model has reproduced accurately measurable properties of the icosahedral AlCuLi system as well as decagonal AlCoCu scanning tunnelling observations [22,23], and is in basic agreement with the AlCoCu model of [45], (also composed of columns of distorted icosahedra) based on direct experimental observations.

6. ACKNOWLEDGEMENTS

The authors are grateful to Manuel Torres and his group, in collaboration with which a large portion of the ideas portrayed in this work were developed. We thank Mr. A. Sánchez for photographical support. Financial assistance from DGAPA-UNAM trough grant IN-104989 is acknowledged.

7. REFERENCES.

1 D. Romeu. Acta. Met. et Mater. **38**, (1990), 113.
2 J. L. Aragón, D. Romeu and A. Gómez. Phys. Rev. B, **44**, (1991), 584.
3 C. Dong, J.M. Dubois, S.S. Kang and M. Audier, Phil. Mag. B, 65 (1992) 107.
4 K. Hiraga, W. Sun and J. Lincoln, Jap. J. Appl. Phys., 30 (1991) L302.
5 K. Hiraga, W. Sun, J. Lincoln, M. Kaneko and Y. Matsuo, Jap. J. Appl. Phys., 30 (1991) 2028.
6 I.A. Harris, R.S. Kidwell and J.A. Northby, Phys. Rev. Lett., 53 (1984) 2390.
7 J.J. Sáenz, J.M. Soler and N. García, Chem. Phys. Lett., 114 (1985) 15.
8 J. Farges, M.F. De Feraudi, B. Raoult and G. Torchet, Surf. Sci., 156 (1985) 370.
9 D. Romeu. Mat. Sci. Forum., 22-24 (1987) 257; Int. J. Mod. Phys., B2 (1988) 265; Int. Jou. Mod. Phys., B2 (1988) 77.
10 D. Romeu. Phil. Mag. B, in press.
11 W. Bollmann, Crystal Defects and Crystalline Interfaces, Springer-Verlag, N.Y. 1970.
12 S. Ranganathan, Acta Cryst., 21 (1966) 197.
13 P.A. Kalugin, A.Y. Kitaev and L.S. Levitov, JETP Lett., 41 (1985) 145.
14 M. Duneau and A. Katz, Phys. Rev. Lett., 54 (1985) 36.
15 V. Elser, Acta Cryst., A42 (1985) 36.
16 D. Levine and P.J. Steinhardt, Phys. Rev. Lett., 53 (1984) 2477.
17 V. Elser, Phys. Rev. Lett., 54 (1985) 1730.
18 Di Vicenzo D. P. and P. J. Steinhardt (eds.), Quasicrystals, the state of the art. Directions in Cond. Mat. Phys., World Scientific, Singapore 1991.
19 C.L. Henley, p. 429 of ref. 18.
20 P.W. Stephens and A.I. Goldman, Phys. Rev. Lett., 56 (1986) 1168; 57 (1986) 2331.

21 T. Janssen, Acta Cryst., A42 (1986) 261.
22 A. R. Kortan, R. S. Becker, F. A. Thiel and H. S. Chen. Phys. Rev. Lett. 64 (1990) 200.
23 Kortan in ref 18
24 J. M. Cowley, Diffraction Physics, North Holland 1975.
25 M. Audier and B. Robertson, Phil. Mag. Lett. 64 (1991) 401.
26 D. Romeu and D. J. Dingley, To be published.
27 W. Steurer W. and K. H. Kuo K., Acta Cryst. B46 (1990) 703.
28 M. Torres, G. Pastor, I. Jiménez, J. L. Aragón and M. José-Yacamán, Phil. Mag. Lett., 62 (1990) 349.
29 M. Torres, J. L. Aragón and G. Pastor, Scripta Met. et Mater., 25 (1991) 1123.
30 M. Torres p.620 in ref 32.
31 J. L. Aragón and M. Torres M, Europhys. Lett. 15 (1991) 203.
32 J. M. Perez-Mato, F. J. Zuñiga and G. Madariaga (eds). Methods of structural Analysis of modulated structures and quasicrystals. Lekeitio, Spain, World Scientific 1990.
33 B. Dubost, J.M. Lang, M. Tanaka, P. Sainfort and M. Audier, Nature, 324 (1986) 48.
34 M. Audier, C. Janot, M. de Boissieu and B. Dubost, Phil. Mag., B60 (1989) 437.
35 C.L. Henley and V. Elser, Phil. Mag., B53 (1986) L59.
36 G. Bergman, J.L.T. Waugh and L. Pauling, Acta Cryst., 10 (1957) 254.
37 E.E. Cherkashin, P.I. Kripyakevich and G.I. Oleksiv, Sov. Phys. Cryst., 8 (1964) 681.
38 M. Audier, J. Pannetier, M. Leblanc, C. Janot, J.M. Lang and B. Dubost, Physica, B153 (1988) 136.
39 L.A. Aslanov, Acta Cryst., A47 (1991) 63.
40 M. de Boissieu, C. Janot, J.M. Dubois, M. Audier and B. Dubost, J. Phys. France, 50 (1989) 1689.
41 T. Fujiwara and T. Ogawa (eds). Quasicrystals. Springer, Berlin 1990. p.68.
42 J.W. Cahn, D. Shechtman and D. Gratias, J. Mater. Res., 1 (1986) 13.
43 F. Dènoyer, G. Heger, M. Lambert, J.M. Lang and P. Sainfort, J. Physique, 48 (1987) 1357.
44 J.L. Aragón, D. Romeu, M. Torres and J. Fayos, J. Non-Cryst. Solids, in press.
45 T.L. Daulton and K.F. Kelton., Phil. Mag. B, 66 (1992) 37.

Crystal-Quasicrystal Transitions
M.J. Yacamán and M. Torres (Editors)

ACOUSTIC QUASICRYSTALS TO ANALOGICALLY STUDY QUASICRYSTAL-CRYSTAL TRANSITIONS

M. Torres[a], F. Montero de Espinosa[b], G. Pastor[a], E. Velázquez[a], I. Jiménez[a] and Alan L. Mackay[c]

[a] Instituto de Electrónica de Comunicaciones
[b] Instituto de Acústica
[a,b] Consejo Superior de Investigaciones Científicas
Serrano 144, 28006 Madrid, Spain.
[c] Department of Crystallography, Birkbeck College, (University of London), Malet Street, London WC1E 7HX, UK

Abstract

The recently found acoustic quasicrystals and corresponding acousto-optic diffraction patterns are theoretical and experimentally studied and used here to analogically generate quasicrystal-crystal transitions. An acousto-optic device is developed to show the transition from an octagonal acoustic quasicrystal to its rational approximant acoustic crystals. The configurational entropy changes along this transition are also studied.

1. INTRODUCTION

The density wave of a quasicrystal or its parent crystalline structure is characterized by the Fourier expansion

$$\rho(r) = \sum_{q(\theta)\in L_R} A_{q\parallel}\exp[iq_\parallel(\theta)\cdot r]$$

where L_R is the reciprocal hyperlattice (usually cubic) in the associated N-dimensional virtual hyperspace, $A_{q\parallel}$ are complex amplitudes for each $q_\parallel$ (q vector restriction into the parallel physical space), and θ parametrizes the adequate rotation [1-3] (and references therein) or more general affine mapping [4-6] (and references therein) in the hyperspace which connects quasiperiodic to periodic structures in the parallel space. The projection of the canonical basis of L_R into the parallel space is denominated the vector star. For a certain value $\theta = \theta_Q$, the star symmetry coincides with those of the quasicrystal and the corresponding point group is non-crystallographic. In this work, the parallel physical space will be 2-dimensional (2D). For $\theta = \theta_Q$, N-2 vectors of the star are expressed as irrational combinations of the other two which are considered as a basis, being ϕ the related irrational number ($\phi=1+\surd 2$ for octagonal quasicrystals, $\phi=\tau=(1+\surd 5)/2$, the golden number, for icosahedral and decagonal ones and $\phi=2+\surd 3$ for dodecagonal ones, [7]).

For certain values of θ, ϕ is replaced by one of its rational approximants, the star becomes commensurate and crystalline approximants of the quasicrystal are reached. Uncountable many intermediate states are generated along the transitions from the quasicrystal to its approximant crystals. To study these transitions from a qualitative point of view, it is sufficient to consider a Fourier expansion only extended to the N star vectors [8].

In these conditions, the theoretical [9] and experimentally [10,11] recently found acoustic quasicrystals are a good tool to analogically study the quasicrystal-crystal transitions.

2. ACOUSTIC QUASICRYSTALS

We have recently produced an acoustic quasi-crystalline standing wave pattern with five-fold symmetry [10,11]. It is a standing wave field of five compression waves in the body of a liquid where the nodes and anti-nodes form a two-dimensional pattern in which, as in the Penrose tiling, there are two incommensurable periods (in the ratio of the golden number) in the same direction. It is a nodal pattern formed in the body of a liquid for which the slow acoustic scanning of the standing structure, the optical diffraction pattern and the corresponding low contrast image of the instantaneous structure have been performed.

A PVC cylindrical cell, internally coated with silicone rubber loaded with alumina to reach good anechoic levels, has been developed . The thickness of this skin is 10 mm, being its specific acoustic impedance very close to that of the water and the resultant cell internal diameter is 60 mm. The cylinder is filled with de-gasified water at room temperature, with five piezo-electric ceramic transducers of diameter 20 mm backed with air and placed according to a regular pentagon in the meridian section. These transducers work at resonant conditions at 5.33 MHz and its overtones.

The cavity is closed with two methacrylate lids that can be removed. By removing one of the lids and by using a needle piezo-ceramic hydrophone that registers pressure maxima with an active tip of 0.2 mm in diameter , we slowly scan the standing pressure field in the cavity meridian section with acquisition steps of 0.02 mm. The XY scanning is performed by an automatized mechanical device with two stepped motors having a precision of 0.01 mm. The reflected echoes are 26 dB down the direct signal when only one transducer is excited and placing the hydrophone at the cavity center. This result indicates good anechoicity conditions. When the five transducers are simultaneously working at 5.33 MHz, with a wave length λ_s about 0.28 mm, a reproducible pressure standing wave pattern, with nodes a wavelength apart, is registered. Two regions of this pattern in near-field conditions exhibiting quasicrystalline fivefold symmetry are shown in figures 1 and 3. The figure 1 shows decagonal clusters of antinodes and an scheme of its topology is shown in the figure 2. In the figure 3, four regular pentagons of antinodes respectively scaled up τ, τ^2 and τ^3 which indicate local selfsimilarity are shown. In both figures 1 and 3 a region about 4×2.5 mm^2 is shown.

A transversal acoustic scanning showing a channels structure has also been

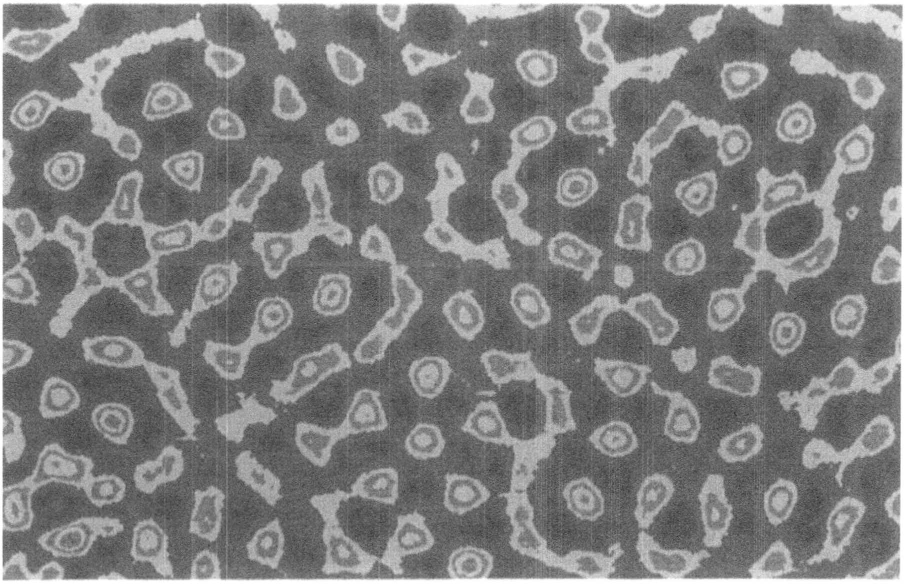

Fig.1 A region of an acoustic quasicrystal showing decagonal clusters.

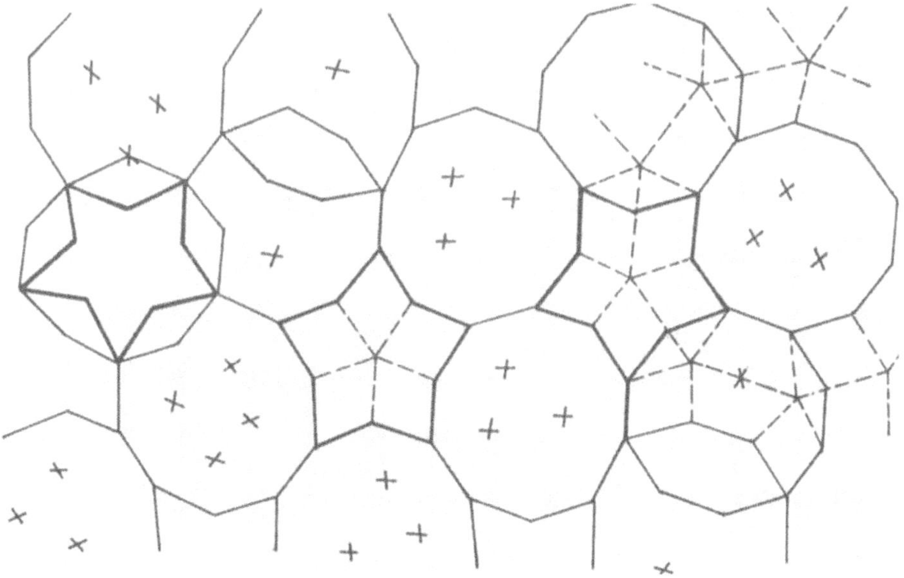

Fig.2 The geometry of the figure 1 is schematically shown.

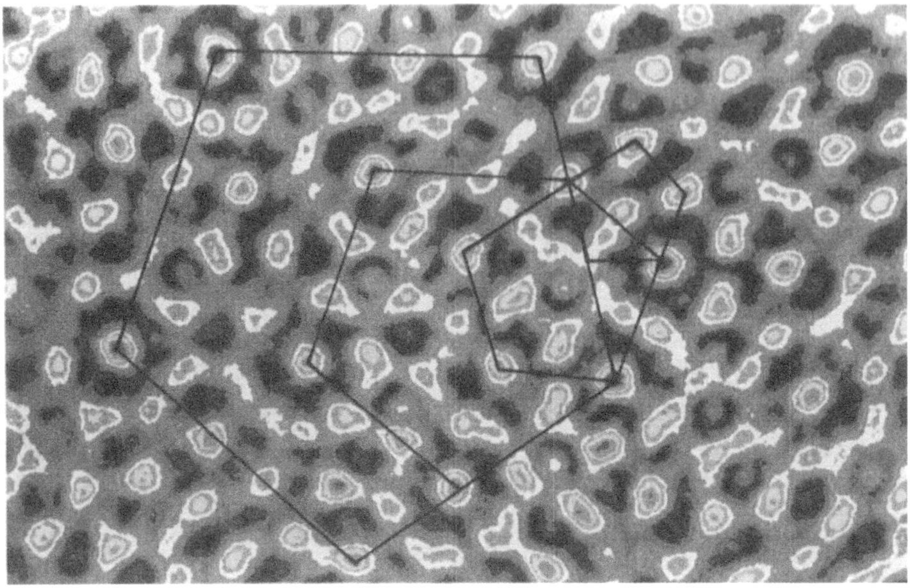

Fig.3 Pressure pattern of an acoustic quasicrystal showing selfsimilarity.

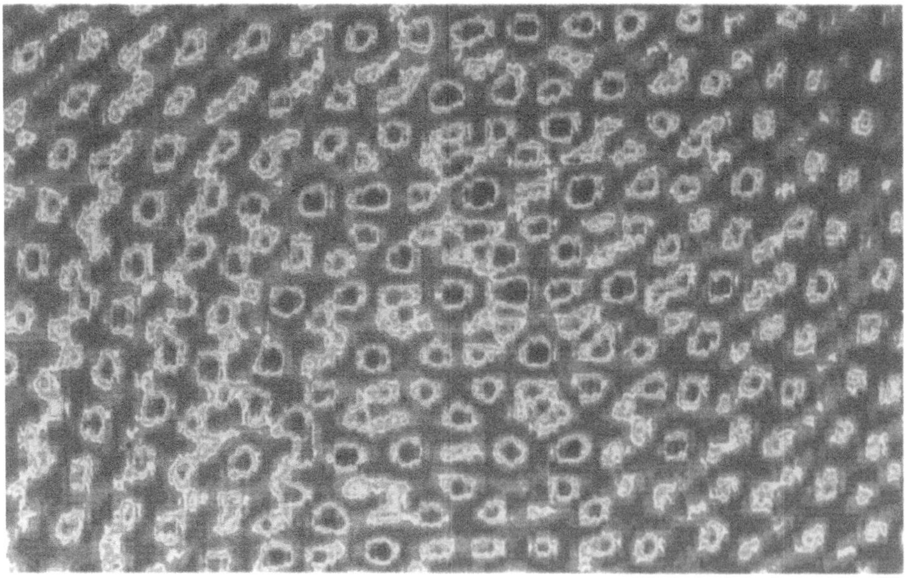

Fig.4 Pressure pattern obtained under far-field conditions.

performed. The structure is periodic along this transversal direction, hence, this acoustic quasicrystal is a decagonal one.

We repeated the experiment placing a diaphragm at the surface of each one of the five transducers to work in far-field conditions. We show the result of this experiment in the figure 4. The pressure pattern of this figure shows a region about 5.3×3.3 mm^2.

3. THE ACOUSTO-OPTIC DIFFRACTION PATTERN

We have performed diffraction experiments, when the cell is closed, by means of an He-Ne (λ_l=633 nm) laser beam with focusing capabilities. The laser beam is directed along the cell cylinder axis and focused onto a translucent screen. The diffraction pattern corresponding to an ultrasonic frequency of 5.33 MHz is shown in the figure 5. In this figure, the center is masked to suppress overexposure and the laser beam embraces a sample circle with a diameter about 5 mm. The measured scattering angle $\hat{\underset{\sim}{\theta}}$ corresponding to the lighter decagonal ring fits with high accuracy the foreseen one by the quantum phonon-photon Brillouin scattering theory in a one-phonon process for the anti-Stokes component (phonon absorbed) and for the Stokes one (phonon emitted) of the scattered radiation [12]. The frequency shift corresponding to the scattered radiation is negligible and the elastic approximation can be used. It is straightforward to deduce that the scattering angle corresponding to each spot of the principal decagonal diffraction ring is given by $\hat{\underset{\sim}{\theta}} \approx f_s\lambda_l/v_s$, where f_s is the ultrasonic frequency, λ_l is the laser wave length in the air and v_s is the sound velocity in the water. Starting from the observed $\hat{\underset{\sim}{\theta}}$, we measure v_s = 1,49×10^3 m/sec.

The weaker rings of the diffraction pattern can be easily interpreted as resulting of the Brillouin scattering in two-phonons processes for all the possible combinations of two vectors of the principal ring. Intensities corresponding to three-phonons and multiphonons scattering decay and are not observed.

The laser beam intensity I is split as $I = I_0 + I_1 + I_2 + \ldots$, where I_0 is the not scattered beam intensity, I_1 is the scattered intensity corresponding to one-phonon processes, I_2 is the corresponding to two-phonon ones, and so on.

I_1 is uniformly distributed on each one of the ten spots of the diffraction pattern principal lighter ring (see the figure 5). These ten spots define the fundamental vectors according to the clockwise orientation: $\mathbf{v}_1$, $-\mathbf{v}_4$, $\mathbf{v}_2$, $-\mathbf{v}_5$, $\mathbf{v}_3$, $-\mathbf{v}_1$, $\mathbf{v}_4$, $-\mathbf{v}_2$, $\mathbf{v}_5$ and $-\mathbf{v}_3$. I_2 is uniformly distributed on each one of the 10^2 binary additions $\pm\mathbf{v}_i\pm\mathbf{v}_j$, i,j = 1,...,5, that in his turn fall on the weak rings of the diffraction pattern shown in the figure 5, except for null additions that fall onto the central spot. The intensity corresponding to three-phonon processes is not detected in our experiment, but its behaviour would go on the same described rules.

The diffraction pattern can be indexed with five integers, and it is theoretically dense due to the incommensurate character of the vector set $\{\mathbf{v}_i\}_{i=1,\ldots,5}$, although it is experimentally observed as discrete [9]. The two-indices undulatory theoretical study (based on standard Fourier optics and under the scalar approximation [13]) of the reference [9] (see also the references therein) can be generalized to our five-indices experimental case. Now, the liquid refractive index field is $\mu(\mathbf{r}) = \mu_0 + \mu_1 \Sigma\cos(\mathbf{v}_i.\mathbf{r})$, i = 1,...,5 and the diffraction pattern amplitudes are

$$\psi(q) = \frac{1}{2\pi}\exp(ikD(\mu_0-1))\int_{-\infty}^{\infty}\int_{-\infty}^{\infty}\exp(i[t[\cos(v_1.r)+...+\cos(v_5.r)]-q.r])dxdy$$

being $k=2\pi/\lambda_l$ and D the light path length through the ultrasonic field. And, by using the development

$$\exp(it[\cos(v_1.r)+...+\cos(v_5.r)]) =$$
$$\sum_{l_1,...,l_5} i^{(l_1+...+l_5)}J_{l_1}(t)...J_{l_5}(t)\exp(i(l_1v_1+...+l_5v_5).r)$$

where the $l_1, ..., l_5$ are integers, the J_{li} are the l_i^{th} order Bessel functions of the first kind and $t=kD\mu_1$, we obtain

$$\psi(q) = \exp(ikD(\mu_0-1))\sum_{l_1,...,l_5} i^{(l_1+...+l_5)}J_{l_1}(t)...J_{l_5}(t)\delta(q-l_1v_1-...-l_5v_5) \quad , \forall q\in R^2$$

and the corresponding intensities

$$\psi^*_{l_1...l_5}\psi_{l_1...l_5} = J^2_{l_1}(t)...J^2_{l_5}(t)$$

We have measured the intensities of the laser beam without applying the ultrasonic field, the undiffracted central beam and the corresponding ones to the lighter decagonal ring by using an optical-electrical converter. Two experiments with different ultrasound amplitudes have been made. The mathematical model fits well our experiments for $t=0.3$ and for $t=0.5$.

By blocking the not diffracted central beam, and taking an image at sufficiently long distance from the focal plane (under Fraunhofer conditions) the inverse Fourier transform of the diffraction pattern is optically generated and the direct optical image showed in the figure 6 is obtained. This pattern, showing fivefold symmetry, agrees with the figure 4.

The above mentioned theoretical results about the optical diffraction pattern admit of a direct generalization for other quasicrystalline beams arrangements (icosahedral, octagonal and dodecagonal ones) and even for the case of ultrasounds with different amplitudes and frequencies. In such a case, the intensity corresponding to the spot $q=l_1v_1+...+l_Nv_N$ is $J_{l1}^2(t_1)...J_{lN}^2(t_N)$.

A parallel quantum study of our acousto-optic diffraction patterns can be made following the methods pointed out in the reference [14] based on the elegant Feynman diagrams and reaching identical results.

Obviously, for only one ultrasound beam, the intensity corresponding to the spot with index l is $J_l^2(t)$, and, taking into account the well known identity $\Sigma J_l^2(t)=1$, where the addition is extended over all the integers, the intensity can be considered as a discrete

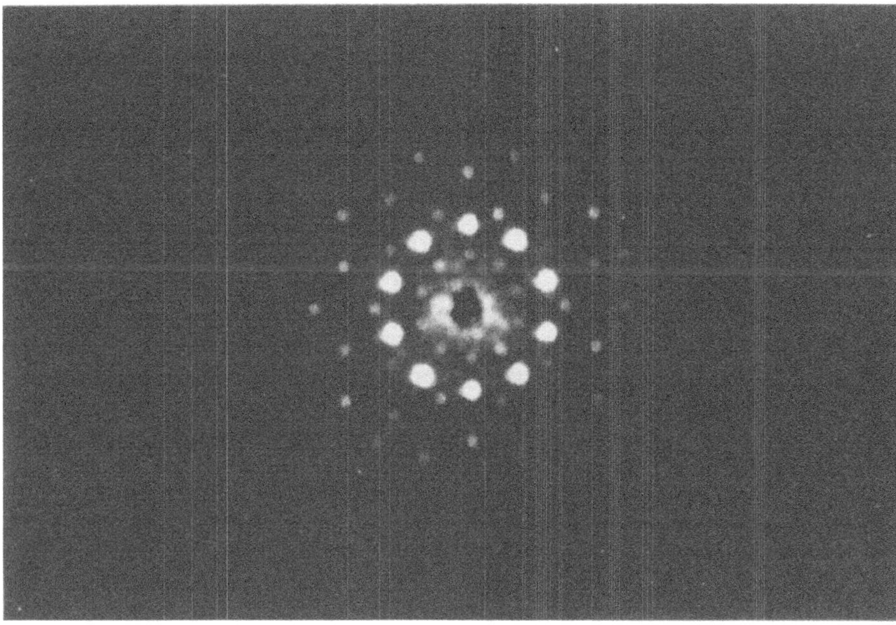

Fig.5 Optical diffraction pattern of a fivefold acoustic quasicrystal.

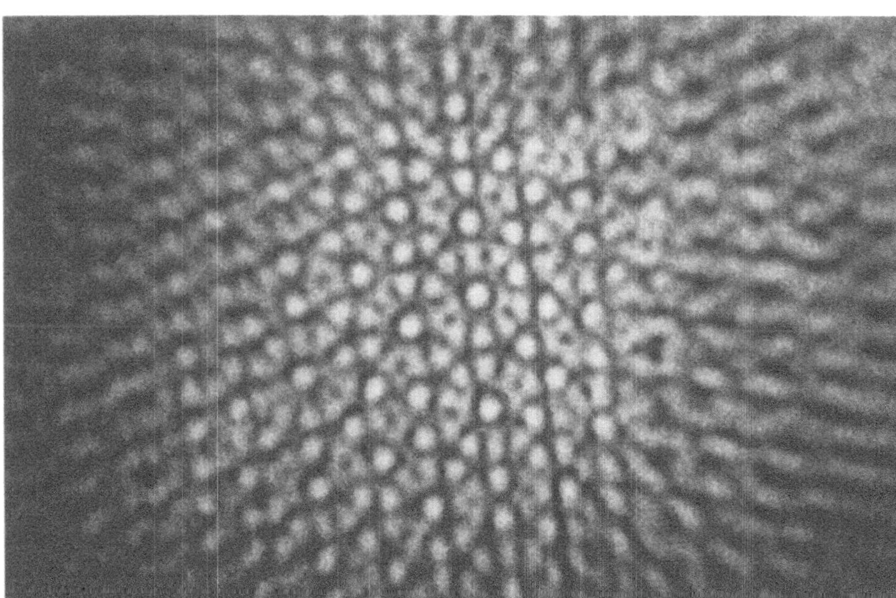

Fig.6 Optical image of a fivefold acoustic quasicrystal.

probability distribution. Hence, the above mentioned results for several ultrasound beams actually are the expression of the joint probability for independent events.

4. FROM AN OCTAGONAL ACOUSTIC QUASICRYSTAL TO ITS RATIONAL APPROXIMANT ACOUSTIC CRYSTALS.

We have developed other anechoical acoustic cell with four transducers respectively perpendicular to the vectors: $\mathbf{u}_1=(1,0)$, $\mathbf{u}_2=(\sqrt{2}/2,\sqrt{2}/2)$, $\mathbf{u}_3=(0,1)$ and $\mathbf{u}_4=(-\sqrt{2}/2,\sqrt{2}/2)$, i. e., according to an hemioctagonal arrangement, working at the frequencies: $f_1=f_3=2ncf_0$ and $f_2=f_4=naf_0\sqrt{2}$, where $f_0=5.33$MHz, $c=\cos\theta$, $a=\cos\theta+\sin\theta$, $n=(2\cos 22.5°)^{-1}$ is a normalization factor and $0\leq\theta\leq 45°$. Now, we define the vectors $\pm\mathbf{F}_i=\pm f_i\mathbf{u}_i$, $i=1,...,4$. This one-parameter frequency transformation can be interpreted as the shadow of a rotation [3] in the 4D frequency virtual hyperspace over the 2D frequency space. This scheme is obviously based on the fact that the length of the vectors of the acousto-optic diffraction pattern is directly proportional to the ultrasound frequency according to the above mentioned quantum phonon-photon Brillouin scattering theory. In the above mentioned definition of the $\mathbf{F}_i$, the sign + corresponds to the phonon absorbed case and the sign - to the phonon emitted one.

When the ultrasound frequencies vary along the resonance bands, the ultrasound amplitudes are forced to be constant during the transformation by changing the applied voltage to the transducers. In these conditions, the amplitudes of the four acoustic waves remain constant during the experiment.

For $\theta=22.5°$, $\mathbf{F}_2=(\sqrt{2}/2)(\mathbf{F}_1+\mathbf{F}_3)$, $\mathbf{F}_4=(\sqrt{2}/2)(-\mathbf{F}_1+\mathbf{F}_3)$, $f_1=f_2=f_3=f_4=f_0$ and the octagonal acoustic quasicrystal is generated. The Fraunhofer optical image of this quasicrystal is obtained as above mentioned and shown in the figure 7. To reach the rational approximants of this quasicrystalline structure, we need to express $\mathbf{F}_2$ and $\mathbf{F}_4$ as rational combinations of $\mathbf{F}_1$ and $\mathbf{F}_3$ by replacing $\sqrt{2}$ by its rational approximants in the above mentioned expressions except for a trivial normalization factor. Two square acoustic crystals 45° rotated are reached at $\theta=0$ and 45°. One of them is shown in the figure 8. The figures 9 and 10 show two rational approximant acoustic crystals that lie between the structures of the figures 7 and 8. Many others rational approximants and uncountable many intermediate structures are generated by changing the frequency set according to the above mentioned transformation. The experimental results agree with the theoretically foreseen ones of the reference [3].

5. CONFIGURATIONAL ENTROPY

The structures generated by modulation of waves with the *forbidden* incommensurate arrangements that exhibit quasicrystals are the boundaries between the order and chaos [15]. Even the distributions of the singular points (hyperbolic and elliptic ones) of these surfaces are not well known from a mathematical point of view [16]. For a quasicrystalline wave modulation, the heights of the maxima are distributed along a continuum and the organization of the modulated wave surface is highly disordered. Intuitively, we can say that this surface has a high entropy. However, when a rational approximant of the above mentioned structure is reached, the spectrum of maxima collapses into a discrete one and the configurational entropy diminishes.

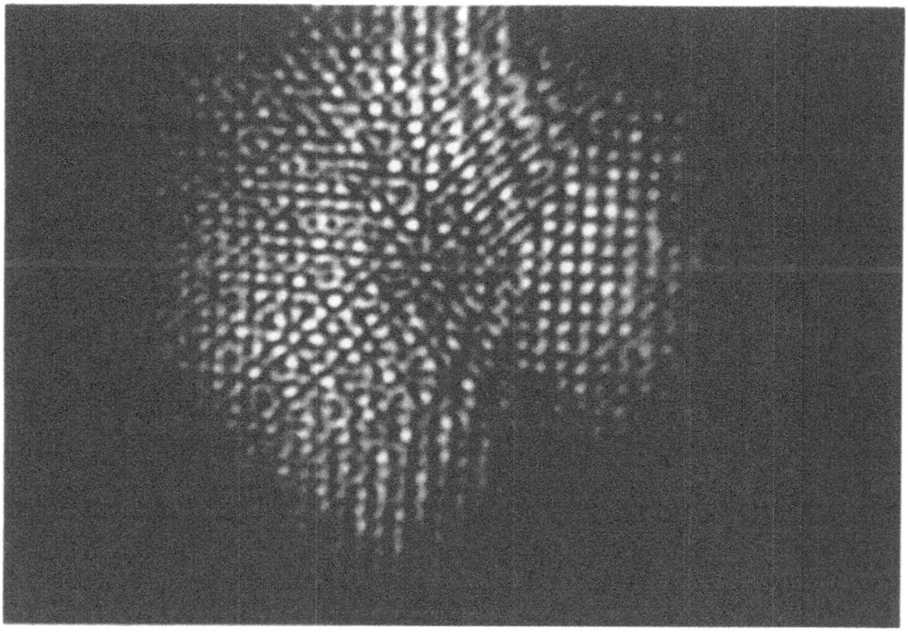

Fig.7 Optical image of an octagonal acoustic quasicrystal.

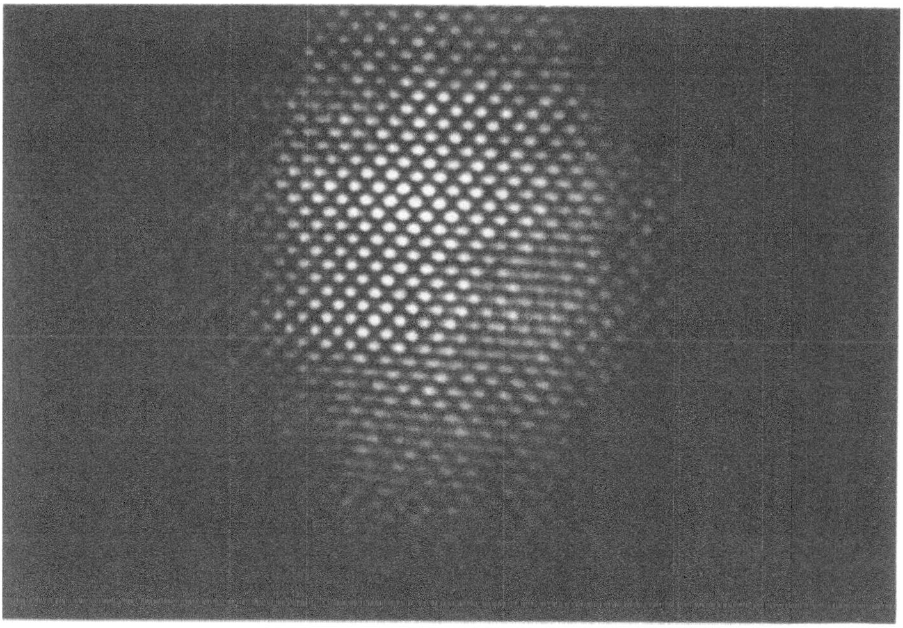

Fig.8 Optical image of the square crystalline approximant.

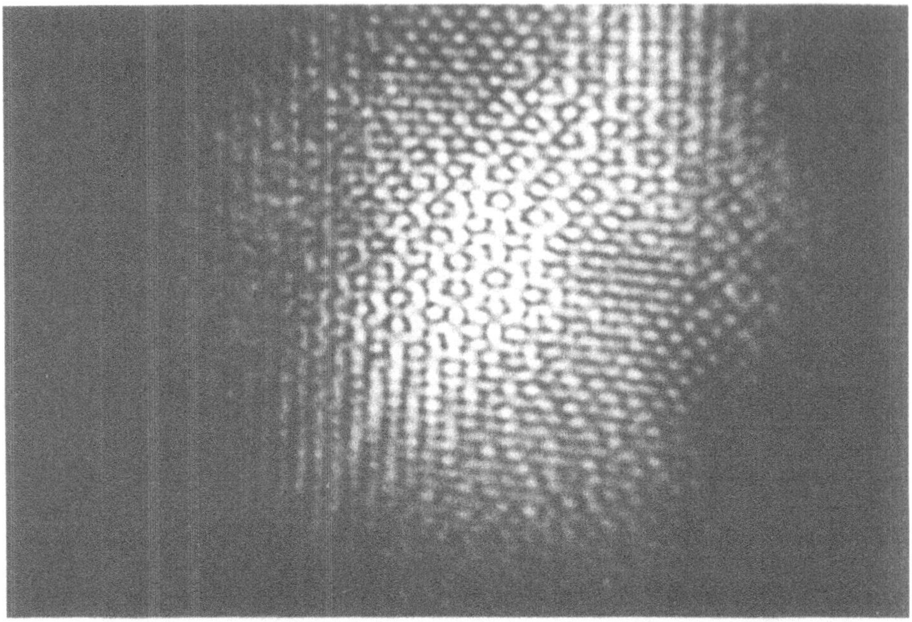

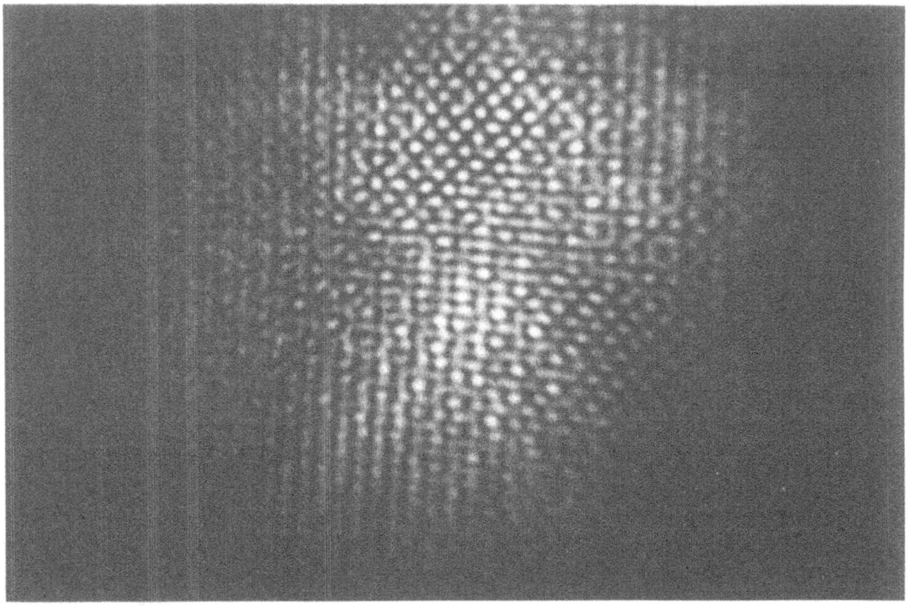

Figs.9 and 10 Optical images of two acoustic crystalline approximants.

The Shannon configurational entropy is a generalized entropy measure [17] and can be directly related to the thermodynamic entropy [8,18,19]. For acoustic quasicrystals, we can define the Shannon entropy as $S(\theta) = -\Sigma\varphi_i ln\varphi_i$, $i = 1,2,...,M$, with $\Sigma\varphi_i = 1$, where φ_i is the relative frequency of the antinode corresponding to a pressure maximum P_i and $M = M(\theta)$. The entropy can also be expressed as a function of the rational approximant a_n of the irrational number that characterizes to the quasicrystal and can be experimentally measured by means of the above mentioned hydrophone probe. The more the approximant order increases the more the size of the periodic cell of the approximant structure increases. For a sufficiently high order of a_n, the periodic cell size corresponding to a_{n+1} scales up ϕ^2 ($\phi = 1 + \surd 2$ for octagonal quasicrystals) in relation with the a_n cell. On the other hand, the more the approximant order increases the more M and S grow. For an infinite quasiperiodic sample, M and $S \rightarrow \infty$. The entropy S is essentially discontinuous along the quasicrystal-crystal transitions due to the discontinuous character of the Dedekind cuts [6].

The figure 11 is a non-hyperspatial computer simulation that shows the location of maxima corresponding to the acoustic octagonal quasicrystal. The size of this figure is about $46\pi \times 46\pi$, where 2π corresponds to an ultrasound wavelength of the beams 1 and 3. This figure coincides with the standard octagonal Penrose tiling usually generated starting from an hyperspatial projection. Although the didactic character of this hyperdimensional representation, the above mentioned coincidence reinforces the well known criticism of David Mermin about the non necessity of using the hyperspace to study quasicrystals. In the figure 12 the continuous spectrum of the heights of maxima of the acoustic octagonal quasicrystal is shown. The φ relative frequency scale is arbitrary and the P pressure scale is [0,4] in shortened units, i. e., for unit pressure amplitudes. The figures 13-16 show the locations of maxima and the figures 17-20 their discrete height spectra for different acoustic approximant crystals of the octagonal quasicrystal. These approximant structures derive from the series 1/1, 3/2, 7/5, 17/12, ... of rational approximants of $\surd 2$. There is other alternative series of rational approximants of $\surd 2$: 2/1, 4/3, 10/7, 24/17, ... Starting from this alternative series, the same above mentioned structures can be generated but these structures appear 45° rotated.

The size of the figures 13-16 is $100\pi \times 100\pi$ in shortened units. These figures respectively correspond to the rational approximants 1/1, 3/2, 7/5 and 17/12, and the sides of the shown periodic cells are respectively scaled up 2/1, 5/2 and 12/5.

The discrete spectra of the figures 17-20 respectively correspond to the same above mentioned rational approximants but now the sample size is $214\pi \times 214\pi$. The scales of the axes are taken as in the figure 12.

For the sample whose size is $214\pi \times 214\pi$, we obtain the entropies $S(1/1) = 0.6931$, $S(3/2) = 0.9651$, $S(7/5) = 2.0905$, $S(17/12) = 3.6643$, ..., $S(\surd 2) = 7.6252$. For samples with very large sizes, we observe the following approximate asymptotic law: $S(a_{n+1}) = S(a_n) + 2ln\phi$, $n \rightarrow \infty$, where a_{n+1} and a_n are two contiguous rational approximants and $\phi = 1 + \surd 2$. The entropy S measures the statistical homogeneity of the system, reaching a maximum in the quasicrystalline state. The higher homogeneity is, the higher disorder and the lower information are.

Acknowledgement

Financial support from DGICYT, project PB 91/0103, is gratefully acknowledged.

Fig.11 Location of maxima of an octagonal quasicrystal.

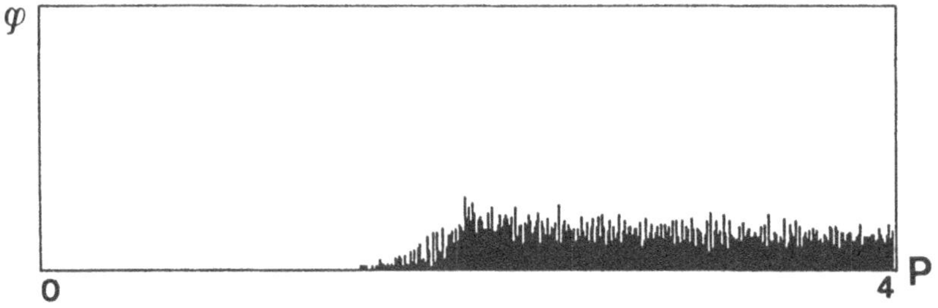

Fig.12 Continuous spectrum of maxima of an octagonal quasicrystal.

Fig.13 Location of maxima of the 1/1 approximant. In the center of every periodic cell, there is an smooth maximum not detected by the program.

Fig.14 Location of maxima of the 3/2 approximant.

Fig.15 Location of maxima of the 7/5 approximant.

Fig.16 Location of maxima of the 17/12 approximant.

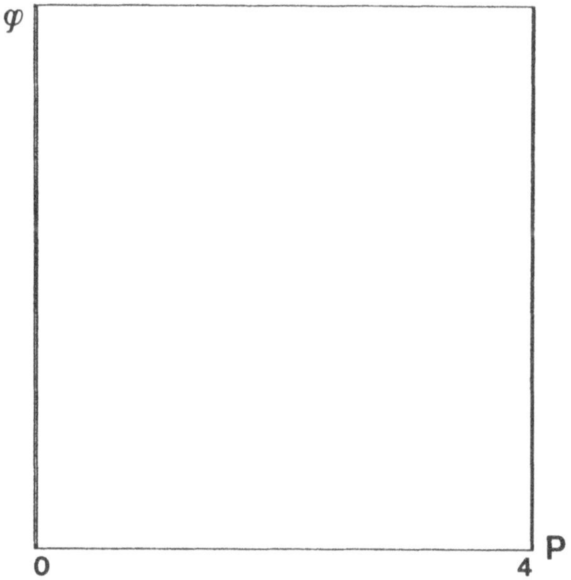

Fig.17 Spectrum of the 1/1 approximant.

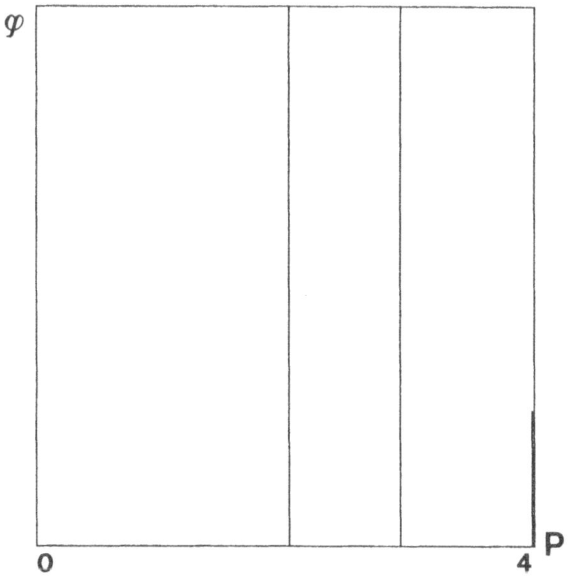

Fig.18 Spectrum of the 3/2 approximant.

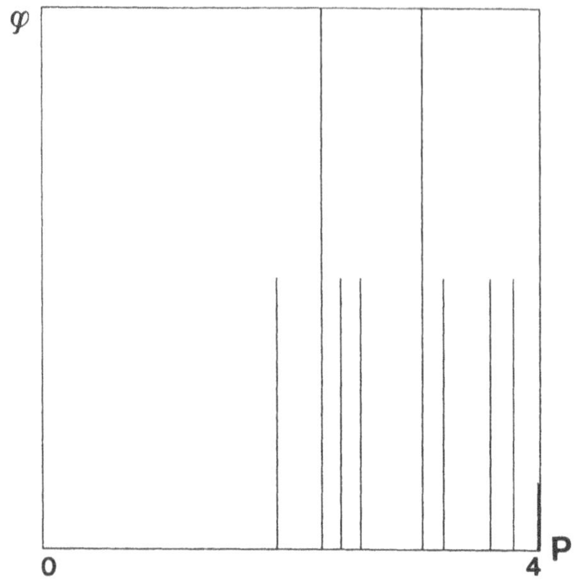

Fig.19 Spectrum of the 7/5 approximant.

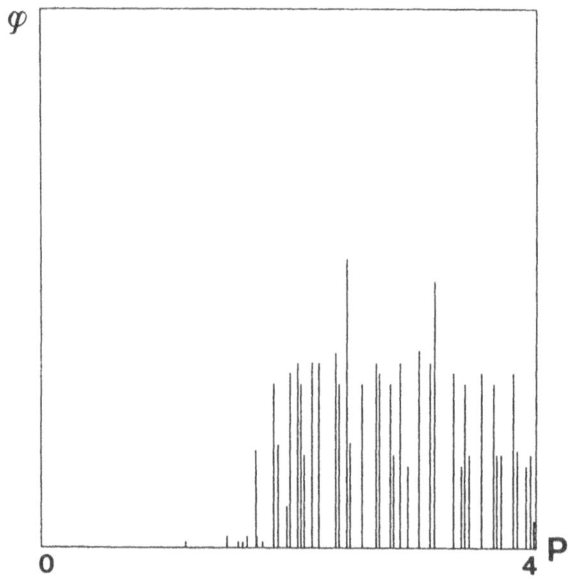

Fig.20 Spectrum of the 17/12 approximant.

REFERENCES

1 M. Torres, G. Pastor, I. Jiménez and J. Fayos, *Phil. Mag. Lett.* **59** (1989) 181.
2 A.L. Mackay, *Nature* **344** (1990) 21.
3 M. Torres, G. Pastor, I. Jiménez, J.L. Aragón and M. José-Yacamán, *Phil. Mag. Lett.* **62** (1990) 349.
4 M. Torres, G. Pastor and I. Jiménez, *Acta Cryst.* **A46** (Suppl.) (1990) C-390.
5 M. Torres, J.L. Aragón, I. Jiménez, E. Velázquez and G. Pastor, *Europhys. Lett.* **18** (1992) 601.
6 M. Torres, G. Pastor, I. Jiménez, J.L. Aragón and D. Romeu, *Scripta Metall.* **27** (1992) 83.
7 M. Torres, J.L. Aragón and G. Pastor, *Scripta Metall.* **25** (1991) 1123.
8 M. Torres, J.L. Aragón, D. Romeu, I. Jiménez, E. Velázquez and G. Pastor, *J. Non-Cryst. Solids* **153&154** (1993), in press.
9 R. Mosseri and F. Bailly, *J. Phys. I France* **2** (1992) 1715.
10 F. Montero de Espinosa, M. Torres, G. Pastor, M.A. Muriel and A.L. Mackay, *Chaos, Solitons & Fractals* **3** (1993) in press.
11 F. Montero de Espinosa, M. Torres, G. Pastor, M.A. Muriel and A.L. Mackay, submitted to *Europhys. Lett.*
12 N.W. Ashcroft and N.D. Mermin, *Solid State Physics* (Saunders College, Philadelphia, 1976).
13 J. W. Goodman, *Introduction to Fourier Optics* (McGraw-Hill, New York, 1968).
14 J. Xu and R. Stroud, *Acousto-Optic Devices* (John Wiley & Sons, New York, 1992).
15 G.M. Zaslavsky, R.Z. Sagdeev, D.A. Usikov and A.A. Chernikov, *Weak Chaos and Quasi-Regular Patterns* (Cambridge University Press, Cambridge, 1991).
16 V.I. Arnol'd, *Physica D* **33** (1988) 21.
17 A. Wehrl, *Rev. Mod. Phys.* **50** (1978) 221.
18 L. Brillouin, *La Science et la Theorie de l'Information* (Masson et Cie., Paris, 1959).
19 M. Torres, *Methods of Structural Analysis of Modulated Structures and Quasicrystals*, edited by J.M. Pérez-Mato, F.J. Zúñiga, and G. Madariaga (World Scientific, Singapore, 1991).

AUTHOR INDEX

Made in United States
Cleveland, OH
09 September 2026

44089029R00138